Up and Running with AutoCAD® 2010

Elliot Gindis

AMSTERDAM • BOSTON • HEIDELBERG • LONDON
NEW YORK • OXFORD • PARIS • SAN DIEGO
SAN FRANCISCO • SINGAPORE • SYDNEY • TOKYO
Academic Press is an imprint of Elsevier

Academic Press is an imprint of Elsevier
30 Corporate Drive, Suite 400, Burlington, MA 01803, USA
525 B Street, Suite 1900, San Diego, California 92101-4495, USA
84 Theobald's Road, London WC1X 8RR, UK

Notices

Knowledge and best practice in this field are constantly changing. As new research and experience broaden our understanding, changes in research methods, professional practices, or medical treatment may become necessary. Practitioners and researchers must always rely on their own experience and knowledge in evaluating and using any information, methods, compounds, or experiments described herein. In using such information or methods they should be mindful of their own safety and the safety of others, including parties for whom they have a professional responsibility. To the fullest extent of the law, neither the Publisher nor the authors, contributors, or editors, assume any liability for any injury and/or damage to persons or property as a matter of products liability, negligence or otherwise, or from any use or operation of any methods, products, instructions, or ideas contained in the material herein.

Library of Congress Cataloging-in-Publication Data
Application submitted

British Library Cataloguing-in-Publication Data
A catalogue record for this book is available from the British Library.

ISBN: 978-0-12-375719-7

For information on all Academic Press publications
Visit our Web site at www.elsevierdirect.com

Printed in the United States of America
09   10   11   9   8   7   6   5   4   3   2   1

Up and Running with AutoCAD® 2010

CONTENTS

Contents

Contents

Contents

Contents

Contents

Acknowledgements

A textbook of this magnitude is rarely a product of only one person's effort. The author wishes to thank early reviewers of this text, including efforts of Jeremy Lucas and Nick Guarriello to improve the original editions, and Chris Ramirez for extensive research and ideas when most needed, as well as using the text in his classroom. Additional thanks to Pratt Institute of Design, New York Institute of Technology, Russell and Titu Sarder of Netcom Information Systems, RoboTECH CAD solutions, RM Architect, and other companies and schools for ongoing support.

Extensive gratitude also to Joseph P. Hayton, Anne McGee, Becky Pease and the rest of the team at Elsevier for believing in the project and for their invaluable support in getting the book out to market. Finally I would like to thank my friends and family, especially my parents and Colleen Sullivan for the patience and encouragement as well as standing by me as months of work turned into years. We're not out of the woods yet, but we're halfway there!

This book is dedicated to the hundreds of students that have passed through my classrooms and made teaching the enjoyable adventure it has become.

About the Author

Elliot Gindis started out using AutoCAD in a New York City area civil engineering company in September of 1996, moving on to consulting work shortly afterwards. He has since drafted in a wide variety of fields ranging from all aspects of architecture and building design to electrical, mechanical, civil, structural, and rail design. These assignments, including lengthy stays with IBM and Siemens Transportation Systems, totaled over 50+ companies to date.

In 1999 Elliot began teaching part-time at Pratt Institute of Design, followed by positions at Netcom Information Systems, RoboTECH CAD solutions and more recently at New York Institute of Technology. In 2003 Elliot formed Vertical Technologies Consulting and Design *(www.VTCDesign.com)*, an AutoCAD training firm that has trained corporate clients in using and optimizing AutoCAD.

Elliot holds a bachelor's degree in aerospace engineering from Embry Riddle Aeronautical University. He currently resides in the Atlanta area and continues to be involved with AutoCAD education and CAD consulting. *"Up and Running with AutoCAD 2010,"* which carefully incorporates lessons learned from 14 years of teaching and industry work, is his first textbook on the subject. He can be reached at: *Elliot.Gindis@gmail.com.*

Preface

➤ What Is AutoCAD?

AutoCAD is a PC-based drafting and design software package developed and marketed by Autodesk®, Inc. As of 2010, it has been around for approximately 28 years – several lifetimes in the software industry. It has grown from modest beginnings to an industry standard, often imitated, sometimes exceeded, but never equaled. The basic premise is simple and is one of the reasons for AutoCAD's success. Anything you can think of, you can draw quickly and easily. For many years it helped if this "anything" was 2D in nature as AutoCAD is essentially a superb electronic 2D drafting board. In recent releases, however, its 3D capabilities finally matured, and AutoCAD is now also considered an excellent 3D visualization tool.

The software has a rather steep learning curve to become an expert, but a surprisingly easy one to just get started. Most importantly, it's well worth learning. This is truly global software that has been adopted by millions of architects, designers and engineers worldwide. Over the years Autodesk has expanded that reach by introducing add-on packages that customize AutoCAD for industry-specific tasks such as electrical, civil, mechanical and many others. However, underneath all of these add-ons is still plain AutoCAD. This software remains hugely popular. Learn it well as it is still one of the best things you can add to your resume and skill set!

➤ About This Book

This book is not like most on the market. While many authors certainly fancy their particular text as being unique and novel in its approach, I have rarely reviewed one that was clear to a beginner student and distilled AutoCAD concepts down to basic, easy to understand, explanations. The problem may be that many of the available books are written by either industry technical experts or teachers, but rarely by someone who is actively both. One really needs to interact with the industry and the students, in equal measure, to bridge the gap between reality and the classroom.

After years of AutoCAD design work in the daytime and teaching nights and weekends, I set out to create a set of classroom notes that outlined, in an easy to understand manner, exactly how AutoCAD was used and applied, not theoretical musings or clinical descriptions of the commands. These notes eventually were expanded into the book that you're holding now. The rationale was simple: I need this person to be up and running as soon as possible to do a job. How do we make this happen?

➤ Teaching Methods

This book has its roots in a certain philosophy developed while attending engineering school many years ago. While there I had sometimes been frustrated with the complex presentation of what in retrospect amounted to rather simple topics. My favorite quote was: "Most ideas in engineering are not that hard to understand, but often become so upon explanation." The moral of that quote was that concepts can usually be distilled to their essence, and explained in an easy and straightforward manner. That is the job of a teacher: not to blow away their students with technical expertise, but to use their experience and top-level knowledge to sort out what is important and what is secondary, and explain the essentials in plain language.

Such was the approach to this AutoCAD book. I wanted everything here to be highly *practical* and easy to understand. There will be few descriptions of procedures or commands that are rarely used in practice. If we talk about it, you will likely need it. The first thing you must learn is how to draw a line. You will see this command on the first few pages of Chapter 1. It is essential to present the "core" of AutoCAD, or essential knowledge common to just about *any* drafting situation, all of it meant to get you up and running quickly. This stripped down approach proved effective in the classroom and was carefully incorporated into this text.

Preface

> ## ➤ Text Organization

This book comes in two parts: **Level 1 – Beginner to Intermediate** and **Level 2 – Intermediate to Advanced.**

Level 1 (Chapters 1 - 10) is meant to give you a wide breadth of knowledge on many topics; a sort of "mile wide" approach. These ten chapters comprise, in my experience, the complete essential knowledge set of an intermediate user. You will then be able to work on, if not necessarily set up and manage, moderate to complex drawings. If your CAD requirements are modest, or if you will not be required to draft full-time, then this is where you stop.

Level 2 (Chapters 11 - 20) is meant for advanced users who are CAD managers, full-time AutoCAD draftspersons, architects, or self-employed and must do everything themselves. The goal here is depth, as many features not deemed critically important in Level 1 will be revisited to explore additional advanced options. Also introduced will be advanced topics necessary to set up and manage complex drawings.

Throughout the book, the following methods are used to present material:

- Explain what the new concept or command is and why it is important.
- Cover the command step-by-step (if needed), with your input and AutoCAD responses shown so you can follow and learn them.
- Give you a chance to apply just-learned knowledge to a real-life exercise, drawing or model.
- Test yourself with end-of-chapter quizzes and drawing exercises that ask questions about the essential knowledge.

You will not see an extensive array of distracting "learning aides" in this text. You will, however, see some common features throughout such as:

Commands – These will be presented in almost all cases in the form of a command matrix such as the one shown here for a line. You can choose any of the methods for entering the command.	*Keyboard:* Type in **line** and press Enter *Cascading menus:* **Draw➔Line** *Toolbar icon:* *Ribbon:* **Home tab➔Line** Line
Tips and Tricks – These will mostly be seen in the first few chapters and one is shown here. They are very specific and deliberate suggestions to smooth out the learning experience. Do take note!	*TIP #1: The Esc (escape) key in the upper left hand corner of your keyboard is your new best friend while learning AutoCAD. It will get you out of just about any trouble you get yourself into. If something doesn't look right, just press the Esc key and repeat the command. Mine was worn out learning AutoCAD, so expect to use it often.*

Step by Step Instructions – These will be featured whenever practical and show you exactly how to execute the command, such as the example with **line** here. What AutoCAD says will be in the default font: `Courier New`. The rest of the steps will be in the standard print Times New Roman font.	**Step 1** Begin the **line** command via any of the above methods • AutoCAD will say: `Specify first point:` **Step 2** Using the mouse, left-click anywhere on the screen • AutoCAD will say: `Specify next point or [Undo]:` **Step 3** Move the mouse elsewhere on the screen and left-click again. You can keep repeating Step 2 as many times as you wish. If you're done, click Enter or Esc.
Learning Objectives and Time for Completion – Each chapter will begin with this, which builds a "roadmap" for you to follow while progressing through the chapter, as well as sets expectations of what you will learn if you put in the time to go through the chapter. Time for completion is based on classroom teaching experience, but is only an estimate. If you're learning AutoCAD in school, your instructor may choose to cover part of a chapter or more than one at a time.	In this chapter we will introduce AutoCAD and discuss the following: • Introduction and the basic commands • The *Create Objects* commands • The *Edit/Modify Objects* commands • The *View Objects* commands • Etc… By the end of the chapter you will have….. **Estimated time for completion of chapter: 3 hours**
Summary, Review Questions, Exercises – Each chapter will conclude with these. Be sure to not skip these pages and review everything you learned.	**Summary** **Review Questions** **Exercises**

> **What Your Goal Should Be**

Just learning commands is not enough; you need to see the big picture and truly understand AutoCAD and how it functions for it to become effortless and transparent. The focus after all is on your design. AutoCAD is just one of the tools to realize it.

A good analogy is ice hockey. A professional player doesn't think about skating – to him its second nature. He's focused on strategy, scoring a goal and getting by the defenders. This mentality should be yours as well. You must become proficient through study and practice, to the point where you are working with AutoCAD, not struggling against it. It will then become "transparent" and you will focus only on the design to truly perform the best architecture or engineering work you are capable of.

If you are in an instructor-led class, take good notes. If you are self-studying from this text, pay very close attention to every topic; there is nothing that is unimportant. Don't skip or cut corners, and complete every drawing assignment. And most importantly, you have to practice, daily if possible, as there is no substitute for sitting down and using the software! Not everyone these days has the opportunity to learn while working and getting paid; companies want ready-made experts and don't want to wait. If that's the case you have to practice on your own in the evening or weekends. Just taking a class or reading this book alone is not enough!

It may seem like a big mountain to climb right now, but it is completely doable. Once on top you will find that AutoCAD is not the frustrating program it may have seemed like in the early days, but an intuitive software package that, with proficiency of use, will become a natural extension of your mind when working on a new design. That in the end is the mark of successful software; it helps you do your job easier and faster. Good luck!

Elliot Gindis - Fall 2009

Contact the author at:
Elliot.Gindis@gmail.com

Level 1

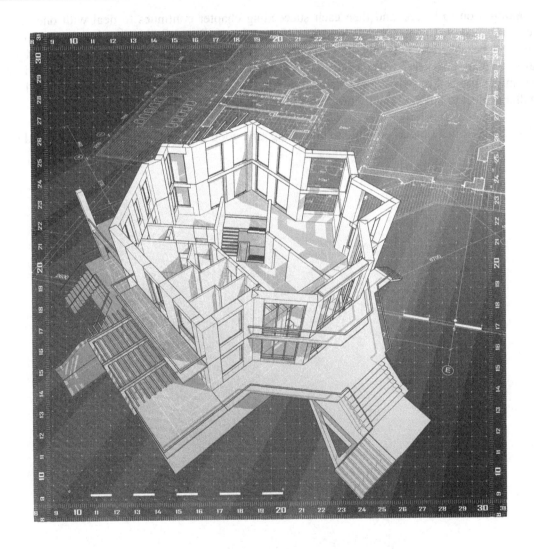

Chapters 1 - 10

Introduction to Level 1

Level 1 will be the very beginning of your studies. No prior knowledge of AutoCAD is assumed, only basic familiarity with computers and some technical aptitude. You are also at an advantage if you have hand-drafted before, as many AutoCAD techniques flow from the old paper and pencil days, a fact that will be alluded to later in the chapters.

We will begin Chapter 1 by outlining the basic commands under "Create objects" and "Modify objects" followed by an introduction to the AutoCAD environment. We will then introduce basic accuracy tools of Ortho and OSNAP. Chapter 2 will continue the basics by adding units and various data entry tools. These first two chapters are the most important, as success here will ensure you will understand the rest, and will be able to function in the AutoCAD environment.

Chapter 3 continues on to layers, and then each succeeding chapter continues to deal with one or more major topics per chapter: text in Chapter 4, hatching in Chapter 5 and dimensioning in Chapter 6. During these six chapters you will be asked to not only practice what you learned, but also apply the knowledge to a basic architectural floor plan. Chapter 7 will introduce blocks, and Chapter 8 arrays. At this point you will be asked to draw another major project, this time a mechanical device. Level 1 concludes with isometric in Chapter 9 (with another, smaller project to do), and finally printing and output in Chapter 10.

Be sure to dedicate as much time as possible to practicing, and good luck in your studies of AutoCAD!

CHAPTER 1

AutoCAD Fundamentals
Part I

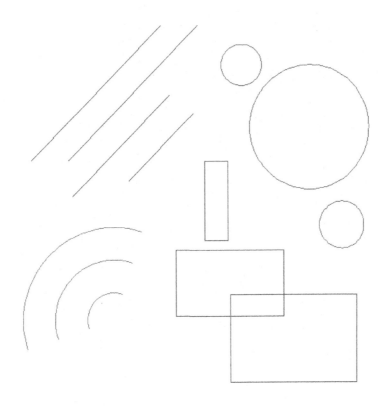

Chapter 1

Learning Objectives

In this chapter we will introduce AutoCAD and discuss the following:

- Introduction and the basic commands
- The *Create Objects* commands
- The *Edit/Modify Objects* commands
- The *View Objects* commands
- The AutoCAD environment
- Interacting with AutoCAD
- Practicing the *Create Object* commands
- Practicing the *Edit/Modify* commands
- Selection Methods – Window and Crossing
- Accuracy in drafting – Ortho
- Accuracy in drafting – OSNAPs

By the end of the chapter you will have learned the essential basics of creating, modifying and viewing objects, the AutoCAD environment and accuracy in the form of straight lines and precise alignment of geometric objects via OSNAP points.

Estimated time for completion of chapter: 3 hours

Sec 1.1 - Introduction and Basic Commands

AutoCAD 2010 is a very complex program. If you are taking a class or reading this textbook this is something you probably already know. The commands available to you, along with their submenus and various options, number in the thousands. So how do you get a handle on them, and begin using the software? Well, you have to realize two important facts.

First, you must understand that on a typical workday, *90% of your AutoCAD drafting time is spent using only 10% of the available commands, over and over again*. So getting started is easy; you only need to learn a handful of key commands and as you progress and build confidence, you can add depth to your knowledge by learning new ones.

Second, you must understand that even the most complex drawing is essentially *made up of only a few basic fundamental objects that appear over and over again* in various combinations on the screen. Once you learn how to create them and edit them, you'll be able to draw surprisingly quickly. Understanding the above is the key to learning the software. We are going to strip away the perceived complexities of AutoCAD and reduce it to its essential core. Let's go ahead now and develop the list of the basic commands.

For a moment, view AutoCAD as a fancy electronic hand-drafting board. In the old days of pencil, eraser and T-square, what was the simplest thing that you could draft on a blank sheet of paper? That of course would be a line. In your notes write the following header: **CREATE OBJECTS**, and below it add **line**.

So what other geometric object can we draw? Think of basic building blocks – those that cannot be broken down any further. A **circle** will qualify and so will an **arc**. Because it's so common and useful, let's throw in a **rectangle** as well (even though you should note that it's a compound object, made up of four lines). So here's the final list of fundamental objects the way you should have them written down in your notes:

CREATE OBJECTS:

- **Line**
- **Arc**
- **Circle**
- **Rectangle**

As surprising as it may sound, these four objects, in large quantities, make up the vast majority of a typical design, so already you have the basic tools. We will create these on the AutoCAD screen in a bit. For now, let's keep going and get the rest of the list down on paper.

So now that you have the objects, what can you do with them? You can **erase** them, which is probably the most obvious. You can also **move** them around your screen, and in a similar manner **copy** them. The objects can **rotate**, and you can also **scale** them up or down in size. With lines, if they are too long, you can **trim** them, and if they are too short, **extend** them. **Offset** is a sort of precise copy, and is one of the most useful commands in AutoCAD. Then there is **mirror**, used as the name implies to make a mirror-image copy of an object. Finally there is **fillet**, used to put a radius on two intersecting lines, among other things. There are a few more useful commands which we will learn a bit later. But for now, under the header **EDIT/MODIFY OBJECTS**, list the commands just mentioned.

EDIT/MODIFY OBJECTS:

- **Erase**
- **Move**
- **Copy**
- **Rotate**
- **Scale**
- **Trim**
- **Extend**
- **Offset**
- **Mirror**
- **Fillet**

Once again, as surprising as it may sound, this short list represents almost the entire set of basic edit/modify commands that you will need once you begin to draft. Start memorizing them!

To finish up, let's add several **VIEW OBJECTS** commands. With AutoCAD, unlike paper hand drafting, you don't always see your whole design in front of you. You may need to **zoom** in for a close-up or out to see the big picture. You may also need to **pan** around to view other parts of the drawing. With a wheeled mouse, so common on computers these days, it's very easy to do both as we will soon see. To this list we will add the **regen** command. It stands for *regenerate*, and it simply refreshes your screen, something you may find useful later. So here's the list for **VIEW OBJECTS**.

VIEW OBJECTS:

- **Zoom**
- **Pan**
- **Regen**

So this is it for now, just 17 commands making up the basic set. Here's what you need to do:

1) As mentioned before, memorize them so you know what you have available.
2) Understand the basic idea, if not the details behind each command. This should be easy to do because (except for maybe *offset* and *fillet*) the commands are intuitive and not cryptic in any way; "erase" means erase, whether it's AutoCAD, a marker on a whiteboard or a pencil line!

We're ready now to start up AutoCAD, discuss how to interact with the program and try them all out.

Sec 1.2 - The AutoCAD Environment

It is assumed that your computer, whether at home, school or training class, is loaded with AutoCAD 2010. It is also assumed that AutoCAD starts up just fine (via the AutoCAD icon or start menu) and everything is configured right. If not, ask your instructor, as there are just too many things that can go wrong on a particular PC or laptop, and it's beyond the scope of this book to cover these situations. If all is well, fire up AutoCAD and you should see the screen depicted in Figure 1.1 (your particular screen may vary slightly, which we will discuss soon).

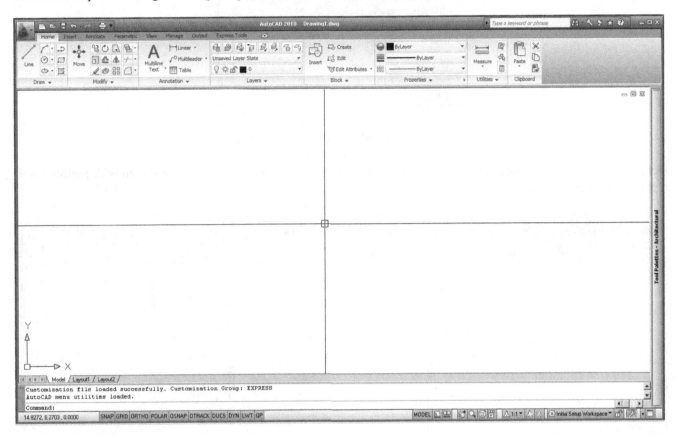

Figure 1.1 – AutoCAD 2010

This is your basic "out of the box" AutoCAD screen for the Initial Setup Workspace. From both learning and teaching points of view, the new screen layout is a big improvement over the previous versions of AutoCAD and Autodesk has done an admirable job in keeping things clean and simple.

AutoCAD has also gone through a major facelift that began last year with Release 2009. If you have caught a glimpse of earlier versions, you may have noticed toolbars present. They are still around, but what we now have, dominating the upper part of the screen, is called the "Ribbon". It's a new way of interacting with AutoCAD, and we will discuss it in detail soon. There are other screen layouts or "Workspaces" available to you, including one with toolbars and they can be accessed through this menu at the bottom right (Figure 1.2).

Figure 1.2 – Workspace Switching

If you click on that menu, you will see the following Workspace Switching Menu (Figure 1.3).

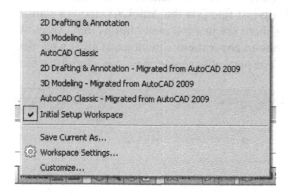

Figure 1.3 – Workspace Switching Menu

Here you can switch to "AutoCAD Classic," which will remove the Ribbon and load the screen with toolbars and a palette, as seen in Figure 1.4.

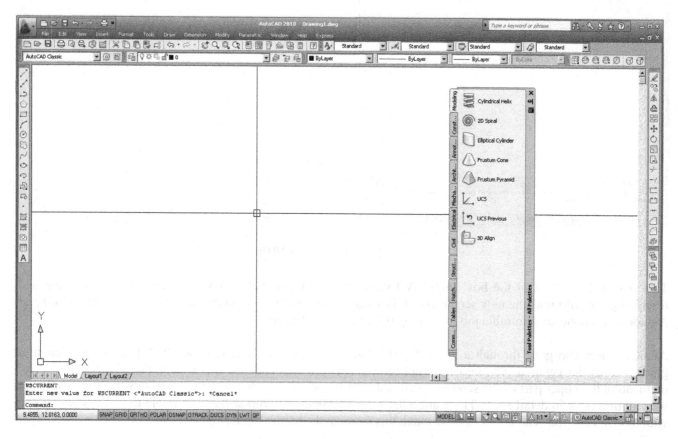

Figure 1.4 – AutoCAD 2010 (Classic)

So which workspace to use? Well, for now switch back to "Initial Setup Workspace" with the Ribbon. We will shortly discuss how to interact with AutoCAD, and give you some choices. Just before we do that though, let's go over the other features of the screen as seen in Figure 1.5.

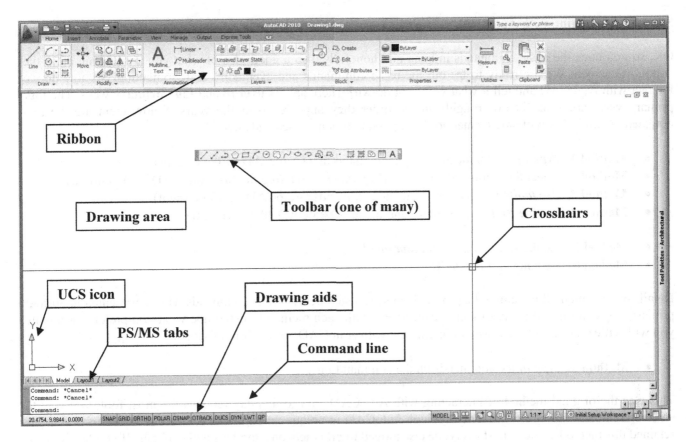

Figure 1.5 – AutoCAD 2010 screen elements

Here is a brief description of what is labeled above.

- **Drawing area -** Takes up most of the screen and is usually colored black (white in this text). It's where you will work and your design will appear. If you wish to change it to black to ease eye strain, you will need to right-click into "Options…", then choose "Display" and "Colors". We will cover this in advanced chapters, so you should ask your instructor for assistance in the meantime, if necessary.
- **Command line -** Right below the drawing area, usually colored white. It's where the commands may be entered and also where AutoCAD tells you what it needs to continue. You need to always keep an eye on what appears here as this is how AutoCAD communicates with you.
- **UCS icon -** A basic X-Y-Z (Z isn't visible) grid symbol. Not important for now, but will be later in advanced studies and 3D. It can be turned off as will be shown later.
- **MS/PS tabs -** Model Space/Paper Space tabs. Not important for now, but will be in advanced studies.
- **Toolbar -** Toolbars contain icons that can be pressed to activate commands. They are an alternative to typing and the Ribbon, and most commands can be accessed this way. AutoCAD has dozens of them.
- **Crosshairs -** The mouse cursor. It can be full-size and span the entire screen, or a small (flyspeck) size. You can change the size of the crosshairs if you wish, but full-screen is recommended.
- **Drawing aids –** These various settings will assist you in drafting and modeling. We will introduce them as necessary. An additional set of advanced drawing aids are just to the right of these basic ones.
- **Ribbon –** A new way of interacting with AutoCAD's commands, to be discussed soon.

Sec 1.3 - Interacting with AutoCAD

OK, so you have the basic commands in hand and hopefully also a good idea of what you are looking at on the AutoCAD screen. We are ready to try out the basic commands and eventually draft something. So how do we interact with AutoCAD and tell it what you want drawn? There are **six overall** and **four primary** ways. The four primary ways are listed below, roughly in the order they appeared over the years. Below them are the two outdated methods that will only be mentioned in passing as a historical side note.

- **Method 1:** *Type in* the commands on the command line (AutoCAD v1.0 – current)
- **Method 2:** Select the commands from the ***drop-down cascading menus*** (AutoCAD v1.0 - current)
- **Method 3:** Use *toolbar* icons to activate the commands (AutoCAD 12/13 - current)
- **Method 4:** Use the ***Ribbon*** tabs, icons and menus (AutoCAD 2009 – current)

- Method 5: Use the screen side-menu (*outdated*)
- Method 6: Use a tablet (*outdated*)

Details of each method including the pros and cons are listed below. Most commands will be presented in all four primary ways and it will be up to you to experiment with each method to determine what you prefer. Eventually you will settle on one particular way of interacting with AutoCAD, or a hybrid of several.

- **Method 1:** *Type in* the commands on the command line.

This was the original method of interacting with AutoCAD, and to this day remains the most foolproof way to enter a command, good old fashioned typing! AutoCAD is unique among leading CAD software in that it has retained this method while almost everyone else moved to graphic icons, toolbars and Ribbons. If you hate typing, this will probably not be your preferred choice.

However, don't discount keyboard entry entirely; AutoCAD has kept it for a reason! When the commands are abbreviated to one or two letters (Line = "L", Arc = "A", etc.) input can be incredibly fast. Just watch a professional typist for proof of the speed with which one can enter data via a keyboard. Other advantages to typing are that you no longer have toolbars or a Ribbon cluttering up precious screen space (there is never enough of it), and that you no longer have to take your eyes off the design to find an icon, instead, the command is literally at your fingertips. Finally, this method is the only way to enter a few of the commands (mostly in 3D). The disadvantage is of course that you have to type!

To use this method, simply type in the desired command (spelling counts) at the command line, as seen in Figure 1.6 and press Enter. The sequence will initiate and you can proceed. There are a number of shortcuts built into AutoCAD (try using just the first letter or two of a command), and we will learn how to make our own shortcuts in advanced chapters. This method is still preferred by many "legacy" users (a kind way to say they've been using AutoCAD forever).

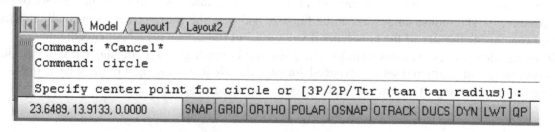

Figure 1.6 – AutoCAD 2010 command line ("circle" typed in)

- **Method 2:** Select the commands from the ***drop-down cascading menus***

This method has also been around since the beginning. It presents a way to access virtually every AutoCAD command, and indeed many students start out by checking out every one of them as a quick crash course on what's available (a fun but not very effective way of learning AutoCAD). Go into the AutoCAD Classic workspace and examine the cascading menus at the top of the screen (so named because they drop out like a waterfall). These are similar in basic arrangement to other software and you should be able to navigate through them easily. We will refer to them on occasion in the following format: **Menu→Command**. So for the sequence shown in Figure 1.7, you would read **Draw→Circle→3 Points**. If for any reason you cannot see the cascading menu bar, you can bring it up via the down arrow at the very top left of the screen (next to the printer symbol). Select "Show Menu Bar" and the menu will appear.

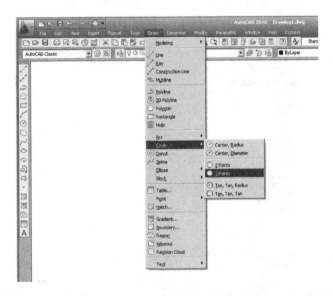

Figure 1.7 – AutoCAD 2010 cascading drop-down menus

- **Method 3:** Use ***toolbar*** icons to activate the commands

This method has been around since AutoCAD switched from DOS to Windows in the mid 1990s and is a favorite of a whole generation of users, and toolbars are a familiar sight with virtually any software these days. Toolbars contain sets of icons and are organized by categories (ex: "Draw" toolbar, "Modify" toolbar, etc.). You press the icon you want and a command is initiated. One disadvantage to toolbars, and the reason the Ribbon was developed, is that they take up a lot of space and, arguably, are not the most effective way of organizing commands on the screen. You can access toolbars by activating the AutoCAD Classic workspace or to have the Ribbon *and* toolbars, do the following while in Initial Setup Workspace:

1) Type in -toolbar and press Enter.
   o AutoCAD will say: Enter toolbar name or [ALL]:
2) Type in draw and press Enter.
   o AutoCAD will say: Enter an option [Show/Hide/Left/Right/Top/Bottom/Float] <Show>:
3) Type in s for "Show" and press Enter. The Draw toolbar will appear as seen in Figure 1.8.

Figure 1.8 – Draw toolbar

To get more toolbars up, you don't need to go through that procedure again. Simply right click on the new toolbar and a menu will appear as seen in Figure 1.9. Then just select the additional toolbars you want to see. Bring up *Modify* and *Standard*; those are the other two we will need in the beginning. Dock them all off to the side or next to the Ribbon on top.

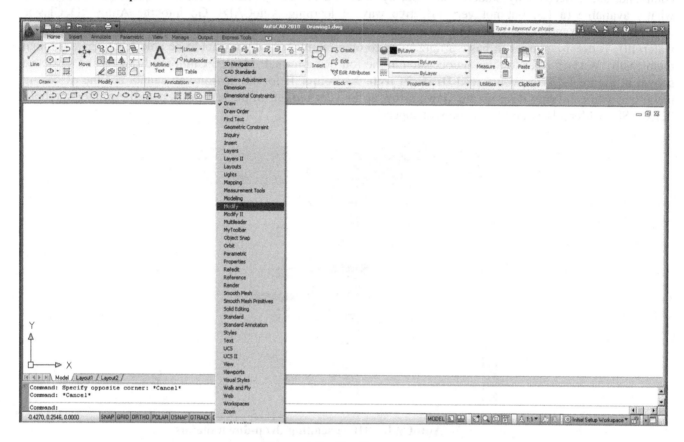

Figure 1.9 – Toolbar menu

- **Method 4:** Use the *Ribbon* tabs, icons and menus

This is the latest and most recently introduced method of interacting with AutoCAD, and follows a new trend in software user interface design started by Microsoft and their new Office 2007 (seen below with Word®).

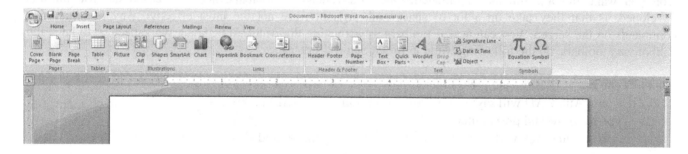

Figure 1.10 – MS Word® Ribbon example

Notice how toolbars have been displaced by "tabbed" categories, where information is grouped together by a common theme.

So it goes with the new AutoCAD. The Ribbon was introduced with the previous release and seems to be here to stay. It is shown again in Figure 1.11.

Figure 1.11 – AutoCAD 2010 Ribbon

Notice what we have here. A collection of tabs, indicating a subject category, is found at the top (Home, Insert, Annotate, etc…) and each tab reveals an extensive set of tools (Draw, Modify, Annotation, etc…). At the bottom of the Ribbon, additional options can be found by using the drop arrows. In this manner the toolbars have been re-arranged in what is, in principle, a more logical and space saving manner. Additionally tooltips appear if you place your mouse over any particular tool for more than a second. Another second will yield an even more detailed tooltip. In Figure 1.12 you will see the Home tab selected, followed by additional options via the drop arrow, and finally the mouse placed over the Polygon command. A few moments of waiting revealed the full tooltip for that command; pressing the icon would have activated the command.

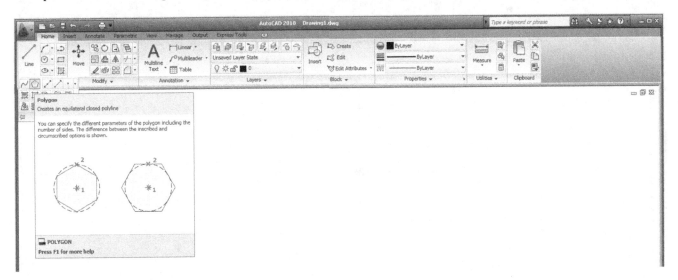

Figure 1.12 – Polygon command and tooltip

Familiarize yourself with the Ribbon by exploring it. It does present some layout advantages and specialized Ribbons can be displayed via other Workspaces. The disadvantage of this new method is that it is a relatively advanced tool that presents many advanced features right away and some confusion is liable to come up for a brand-new user. It is also not ideal for long time users who type, and some veterans in my update classes turn the feature off. The Ribbon is the single biggest change to AutoCAD's user interface and does represent a jump forward in designer/software interaction, but the ultimate decision to use it is up to you.

- Method 5: Use the screen side-menu (*outdated*)
- Method 6: Use a tablet (*outdated*)

These last two methods are rather archaic. Practically no one uses tablets or the screen side-menu, and some younger users wouldn't even know what they are any more, since both have fallen out of favor in the 90s. If you are curious, ask your instructor what a screen side-menu is (accessed through "Options", "Display" tab). A tablet was a piece of digital input hardware with an overlaid command template. It also disappeared a long time ago.

Before we try out our basic commands, let's begin a list of tips. These are a mix of good ideas, essential habits, and time saving tricks that will be passed along to you every once in a while (mostly in the first few chapters). Make a note of them as they are important. Here is the first and most urgently needed one.

> ***TIP #1: The Esc (escape) key in the upper left-hand corner of your keyboard is your new best friend while learning AutoCAD. It will get you out of just about any trouble you get yourself into. If something doesn't look right, just press the Esc key and repeat the command. Mine was worn out learning AutoCAD, so expect to use it often.***

Finally in the interest of trying out the Ribbon, cascading menus, toolbars and typing all at once, go into the AutoCAD Classic workspace and remove all toolbars except the three mentioned (Draw, Modify and Standard). Then using the cascading menus, bring up the Ribbon via **Tools→Palettes→Ribbon**. You should then see the following screen (Figure 1.13), with toolbars docked and even scroll bars removed (ask your instructor how – we will use far more effective *pan* and *zoom* methods soon). Review everything learned thus far and proceed to the first commands!

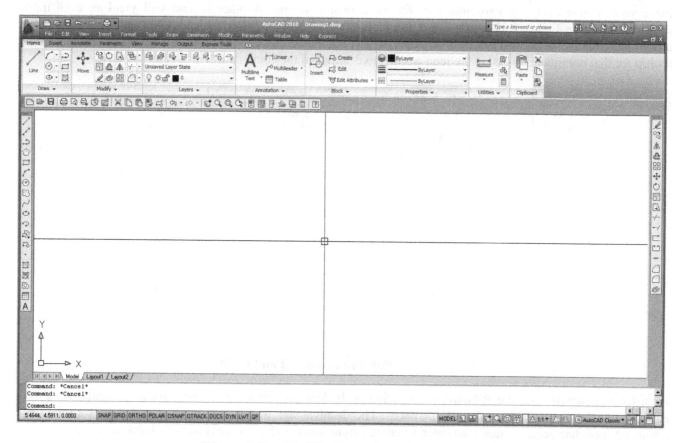

Figure 1.13 – Modified AutoCAD 2010 screen

Sec 1.4 – Practicing the Create Objects Commands

Let's try the commands now one by one. Though it may not be all that exciting, it is extremely important that you memorize the sequence of prompts for each command, as these are really the ABC's of AutoCAD. Remember: if a command isn't working right, or you see little blue squares (they are called GRIPS, and we'll cover them very soon), just press Esc to get back to the Command: status line, at which point you can try it over again. All four methods of command entry will be presented: Typing, Cascading menus, Toolbar icons, and the Ribbon. Alternate each method until you decide which one you prefer. It is perfectly OK to use a hybrid of methods.

➢ **Line**

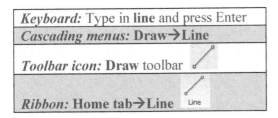

Keyboard: Type in **line** and press Enter
Cascading menus: **Draw→Line**
Toolbar icon: **Draw** toolbar
Ribbon: **Home tab→Line**

Step 1
Begin the line command via any of the above methods
- AutoCAD will say: `Specify first point:`

Step 2
Using the mouse, left-click anywhere on the screen
- AutoCAD will say: `Specify next point or [Undo]:`

Step 3
Move the mouse elsewhere on the screen and left-click again. You can keep repeating Step 2 as many times as you wish. If you're done, click Enter or Esc. You should have a bunch of lines on your screen, either separate or connected together as shown in Figure 1.14.

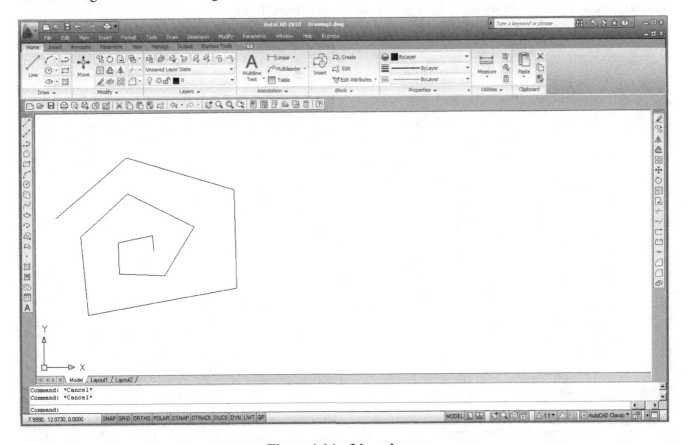

Figure 1.14 – Lines drawn

You don't need to worry about several things at this point. The first is accuracy; we will introduce a tremendous amount of accuracy later in the learning process. The second is the options available in the brackets such as `[Undo]` above; we will cover those as necessary. What is important is that you understand how you got those lines and how to do it again. In a similar manner let's move on to the other commands.

➢ **Circle**

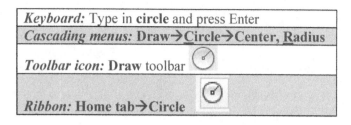

Keyboard: Type in **circle** and press Enter
Cascading menus: **Draw→Circle→Center, Radius**
Toolbar icon: **Draw** toolbar
Ribbon: **Home tab→Circle**

Step 1
Begin the circle command via any of the above methods
- AutoCAD will say: `Specify center point for circle or [3P/2P/Ttr (tan tan radius)]:`

Step 2
Using the mouse, left-click anywhere on the screen and move the mouse out away from that point
- AutoCAD will say: `Specify radius of circle or [Diameter] <1.9801>:`

Step 3
You will notice a circle will form, and will vary in size with the movement of your mouse. The value in brackets in the previous step may also be different; left-click again to finish the circle command. Repeat Steps 1 thru 3 several times, and you should see the following on your screen (Figure 1.15).

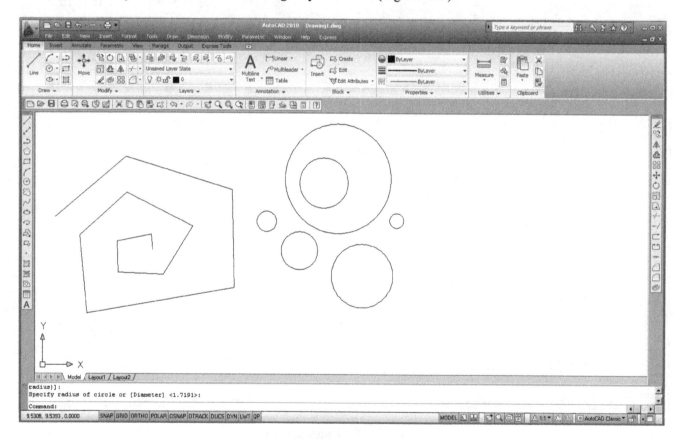

Figure 1.15 – Circles drawn

The method just used to create the circle was called Center, Radius, and you could have specified an exact radius size if you wished, by just typing in a value after the first click (try it). As you may imagine, there are other ways to create circles – six ways to be precise, as seen with the Ribbon and cascading menus (Figure 1.16 and 1.17).

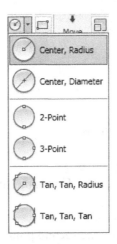

Figure 1.16 – Additional circle options (Ribbon)

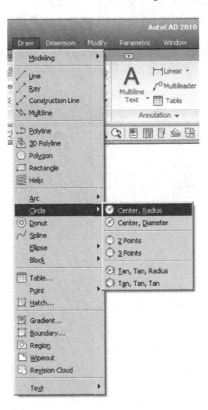

Figure 1.17 – Additional circle options (cascading menus)

Special attention should be paid to the Center, Diameter option! Often students are asked to create a circle of a certain diameter and they inadvertently use radius. This is less of a problem if using the Ribbon or the cascading menus, but you need to watch out if typing or using toolbars, as you will need to press d for [Diameter] before entering a value, otherwise guess what it will be. We will focus much more on these "bracketed" options as the course progresses. *You will get a chance to practice the other circle options as part of the end of chapter exercises.*

> ➤ **Arc**

Keyboard: Type in **arc** and press Enter
Cascading menus: **Draw→Arc→3 Points**
Toolbar icon: **Draw** toolbar 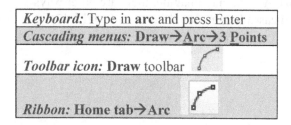
Ribbon: **Home tab→Arc**

Step 1
Begin the arc command via any of the above methods.
Step 2
- AutoCAD will say: `Specify start point of arc or [Center]:`

Take the mouse and left-click anywhere on the screen. This is the first of three points necessary for the arc.
Step 3
- AutoCAD will say: `Specify second point of arc or [Center/End]:`

Click again somewhere else on the screen to place the second point.
Step 4
- Finally AutoCAD will say: `Specify end point of arc:`

Left-click a third and final time somewhere else on the screen to finish the arc.

Practice this sequence several more times to fully understand the way AutoCAD places the arc. Your screen should look something like Figure 1.18.

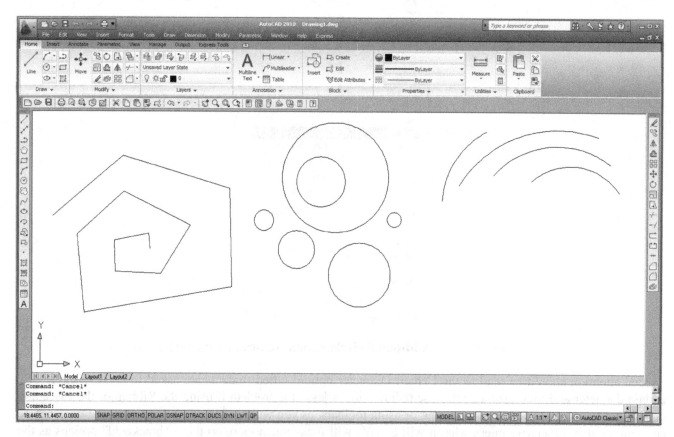

Figure 1.18 – Arcs drawn

The method just used to create the arcs was called 3 Point, and without invoking other options, would be a rather arbitrary (eyeball) method of creating them, which is sometimes just fine for an application. Just as with circles, there are of course other ways to create arcs – 11 ways to be precise, as seen with the Ribbon and cascading menus (Figure 1.19 and 1.20).

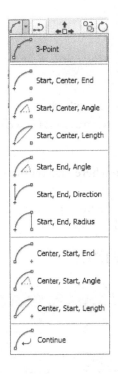

Figure 1.19 – Additional arc options (Ribbon)

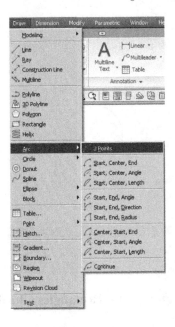

Figure 1.20 – Additional arc options (cascading menus)

Not all of these options are used, and some you will probably never need, but it is worth going over them to know what's available. *You will get a chance to practice the other arc options as part of the end of chapter exercises.*

➢ **Rectangle**

Keyboard: Type in **rectangle** and press Enter
Cascading menus: **Draw→Rectangle**
Toolbar icon: **Draw** toolbar
Ribbon: **Home tab→Rectangle**

Step 1
Begin the rectangle command via any of the above methods
- AutoCAD will say: `Specify first corner point or [Chamfer/Elevation/Fillet /Thickness/Width]:`

Step 2
Left-click, and move the mouse diagonally somewhere else on the screen.

Step 3
- AutoCAD will say: `Specify other corner point or [Area/Dimensions/Rotation]:`

Left-click one more time to finish the command.

You screen should look more or less like Figure 1.21:

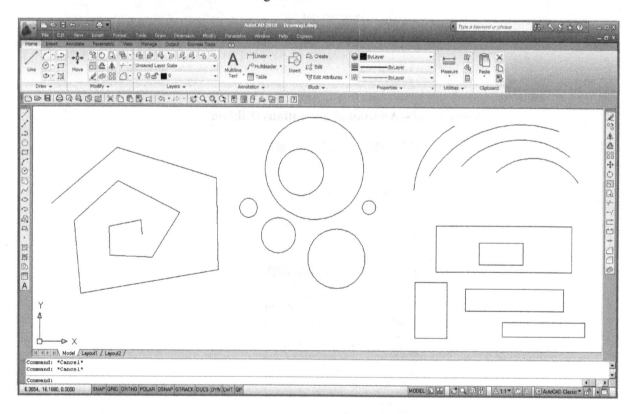

Figure 1.21 – Rectangles added

There are of course more precise ways to draw a rectangle. In Step 3, you can press d for `Dimensions` and follow the prompts to assign length, width and a corner point (where you want it) to your rectangle. Try it out, but as with all basic shapes drawn so far, do not worry too much about sizing and accuracy, just be sure to understand and memorize the sequences for the basic command – that is what's important for now!

So now you've created the four basic shapes used in AutoCAD drawings (and ended up with a lot of junk). Let's erase all of them, and get back to a blank screen. In order to do this we need to introduce Tip #2, which is quite easy to just type. Notice the use of the "e" abbreviation; we will see more of this as we progress.

> *TIP #2: To quickly erase everything on the screen, type in "e" for "erase", press Enter, then type in "all". Then press Enter twice. After the first Enter all the objects will be dashed, indicated they were selected successfully. The second Enter finishes the command by deleting everything. Needless to say, the above command sequence is useful only in the early stages of learning AutoCAD, when you do not intend to save what you are drawing. You may want to conveniently forget this particular tip when you begin to work on a company project! For now it's very useful, but we will learn far more selective erase methods later.*

So let's put all of this to good use. Create the drawing shown below. Use the rectangle, line, circle and arc commands. Don't get caught up trying to make everything line up perfectly, just do a quick sketch. Then you can tell everyone that you learned how to draw a house in AutoCAD within the first half-hour. Well, sort of.

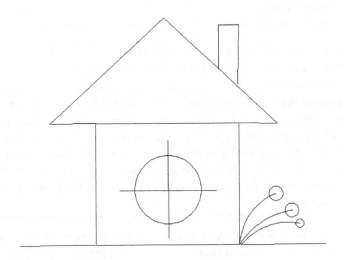

Figure 1.22 – **Simple house sketch**

At this point it's necessary to mention the Zoom and Pan commands. These are very easy if you have a mouse with a center wheel (as most PCs do these days).

Sec 1.5 - View Objects

Zoom – Put your finger on the mouse wheel and turn it back and forth without pressing down. The house you just created will get smaller (roll back) or larger (roll forward) on the screen. Remember though, the object remains the same; you are just zooming in and out.

Pan – Put your finger on the mouse wheel, but this time also press and hold. A hand symbol will appear. Now move the mouse around while keeping the wheel depressed. You will be able to Pan around your drawing. Remember again, the drawing is not moving, you are moving around it.

Regen – This is the easiest command you will learn. Just type in **regen** and press Enter. The screen will refresh. One use for this is if you are panning over and AutoCAD won't let you go further as if you hit a wall. Simply regen and proceed.

While the mouse method is easiest, there are other ways to zoom and pan. One alternative is using toolbars, and part of the Standard toolbar (which you should have on your screen) has some zoom (magnifying glass) and pan (hand) icons for this purpose as seen in Figure 1.23.

Figure 1.23 – Zoom and Pan icons

Explore the icon options on your own, but there is one particular zoom option which is critically important, called Zoom, Extents. If you managed to pan your little house right off the screen and can't find it, this will come in handy, and a full tip on Zoom, Extents is presented next.

> *TIP #3: Type in "z" for "zoom", press Enter, then type in the letter "e", then press Enter again. This is called Zoom to Extents and makes AutoCAD display everything you sketched, filling up the available screen space. If you have a middle-button wheel on your mouse then double click it for the same effect. This technique is very important to fit everything on your screen.*

What if you did something that you didn't want to, and wished you could go back a step or two? Well, like most programs, AutoCAD has an undo command, which brings us to the fourth tip.

> *TIP #4: To perform the Undo command in AutoCAD just type in "u", and press Enter. Don't type the entire command – if you do that, more options pop up which we don't really need at this point, so a simple "u" will suffice. You can do this as many times as you want to (even to the beginning of the drawing session). You can also use the standard Undo and Redo arrows shown in Figure 1.24. Yup, AutoCAD has a "Redo"; it was once called the Oops command. Oops still exists, but its complexity has grown and we won't explore it at the moment.*

Now that you have drawn your house, leave it on your screen and let's move on. We now have to go through all the edit/modify commands one by one. Each of them is critical to creating even the most basic drawings, so go through each carefully, paying close attention to the steps involved. Along the way the occasional tip will be added as the need arises.

Just to remind you again: The Esc key will get you out of any mistakes you make, and return you to the basic command line. Be sure to perform each sequence several times to really memorize them. As you learn the first command **erase**, you will also be introduced to 'picking objects' selection method, which will be similar throughout the rest of the commands. Also, keep a close eye on the command line, as it is where you and AutoCAD communicate, and it *is* a two-way conversation!

Finally before you get started, here is another useful tip that students usually appreciate early on to save some time and effort.

> *TIP #5: Hate to keep having to type, pull a menu or click an icon each time you practice the same command? Well, no problem. Simply press the space bar or Enter, and this will repeat the last command you used. There, saved you a few minutes per day for that coffee break!*

Sec 1.6 - Practicing the Edit/Modify Objects Commands

> **Erase**

Keyboard: Type in **erase** and press Enter
Cascading menus: **Modify→Erase**
Toolbar icon: **Modify** toolbar 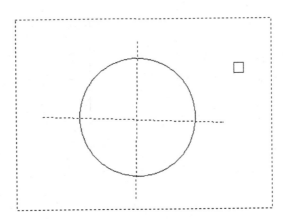
Ribbon: **Home tab→Erase**

Step 1
Begin the erase command via any of the above methods.
- AutoCAD will say: `Select objects:`

Step 2
Select any object from the house in the previous drawing assignment (Figure 1.22) by taking the mouse, positioning it over that object (not in the empty space), and left-clicking once.

Step 3
- AutoCAD will say: `Select objects: 1 found`

The object will become dashed. You've seen this before when we erased everything in Tip #2. It means the objects were selected.

Step 4
- `AutoCAD will ask you again: Select objects:`

Watch out for this step! AutoCAD always asks this in case you want to select more objects, so get used to it.
You're done, however, so press Enter and the object will disappear.

Practice this several times, using the undo "u" command to bring the object back. Make sure it's still there to practice the next few commands on. You can of course select more than one object, as this is the whole point of being asked to select again. As you click each one it will become dashed as seen in Figure 1.24. You can do this until you run out of objects to select; there is no limit. To deselect any objects, hold down the Shift key and click on them.

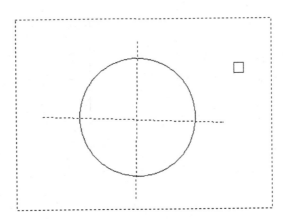

Figure 1.24 – Multiple object selection

> **Move**

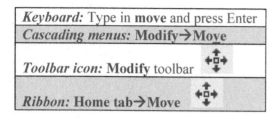

Keyboard: Type in **move** and press Enter	
Cascading menus: **Modify→Move**	
Toolbar icon: **Modify** toolbar	
Ribbon: **Home tab→Move**	

Step 1
Begin the move command via any of the above methods.

- AutoCAD will say: `Select objects:`

Step 2
Select an object by positioning the mouse over that object and left-clicking once.

- AutoCAD will say: `Select objects: 1 found`

The object will become dashed.

- AutoCAD will then ask you again: `Select objects:`

Unless you have more than one object to move, you're done, so press Enter.

Step 3

- AutoCAD will say: `Specify base point or [Displacement] <Displacement>:`

That means left-click anywhere on or near the object to "pick it up"; this is where you will be moving it FROM.

Step 4

- AutoCAD will say: `Specify second point or <use first point as displacement>:`

Move the mouse somewhere else on the screen and left-click to place the object in the new location. This is where you will be moving it TO. Notice that a dashed copy of the object remains in its original location until you complete the command.

Undo what you just did and practice it again. Figure 1.25 shows the move command in progress. The dashed circle will remain there as a "shadow" until you place your circle in its new location.

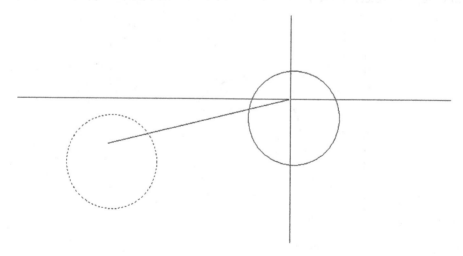

Figure 1.25 – Move command in progress

➢ **Copy**

Keyboard: Type in **copy** and press Enter
Cascading menus: Modify→Copy
Toolbar icon: **Modify** toolbar
Ribbon: **Home tab→Copy**

Step 1
Begin the copy command via any of the above methods.
- AutoCAD will say: Select objects:

Step 2
Select an object by positioning the mouse over that object and left-clicking once.
- AutoCAD will say: Select objects: 1 found

The object will become dashed.
- AutoCAD will then ask you **again**: Select objects:

Unless you have more than one object to copy, you're done, so press Enter.

Step 3
- AutoCAD will say: Specify base point or [Displacement/mOde]
                            <Displacement>:

That means left-click anywhere on or near the object to "pick it up"; this is where you will be copying it FROM.

Step 4
- AutoCAD will say: Specify second point or <use first point as displacement>:

Move the mouse somewhere else on the screen and left-click to copy the object to the new location. This is where you will be copying it TO. Notice that a dashed copy of the object remains in its original location until you complete the command. You can copy as many times as you want.

Undo what you just did and practice it again. Figure 1.26 shows the copy command in progress. The dashed circle will remain there as a "shadow" until you copy your new circles into their new locations.

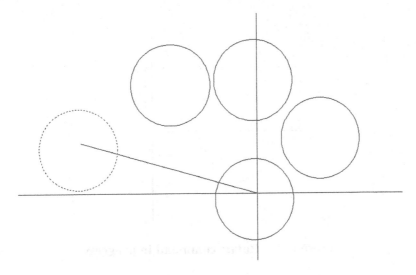

Figure 1.26 – Copy command in progress

➢ **Rotate**

Keyboard: Type in **rotate** and press Enter
Cascading menus: **Modify→Rotate**
Toolbar icon: **Modify** toolbar ↻
Ribbon: **Home tab→Rotate** ↻

Step 1

Begin the rotate command via any of the above methods.

➢ AutoCAD will say: `Current positive angle in UCS: ANGDIR= counterclockwise ANGBASE=0,` and on the next line: `Select objects:`

Step 2

Select any object as before, remembering to press Enter again after selection. So far it is quite similar to erase, move and copy.

• AutoCAD will say: `Specify base point:`

Here this means select the pivot point of the object's rotation (the point *about* which it will rotate). If you selected a circle for your object, try to stay away from the center point, as your rotation efforts may be less than spectacular! Otherwise, click anywhere on or near the object.

Step 4

• AutoCAD will say: `Specify rotation angle or [Copy/Reference] <0>:`

Move the mouse around in a wide circle and the object will also rotate. Notice how the motion gets smoother as you move the mouse farther away. You can click anywhere for a random rotation angle *or* you can type in a specific numerical degree value.

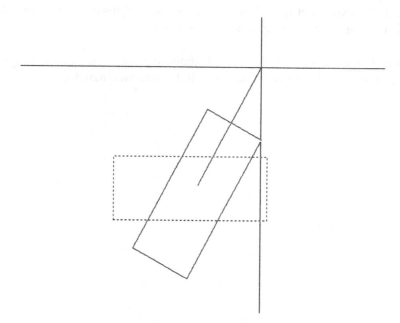

Figure 1.27 – Rotate command in progress

➢ **Scale**

Keyboard: Type in **scale** and press Enter
Cascading menus: **Modify→Scale**
Toolbar icon: **Modify** toolbar ⬜
Ribbon: **Home tab→Scale** ⬜

Step 1
Begin the scale command via any of the above methods.
- AutoCAD will say: `Select objects:`

Step 2
Select any object as before, remembering to press Enter again after the first selection.

Step 3
- AutoCAD will say: `Specify base point:`

This means select the point from which the scaling of the object (up or down) will occur. For now click somewhere on or near the object, or directly in the middle of it.

Step 4
- AutoCAD will say: `Specify scale factor or [Copy/Reference] <1.000>:`

Move the mouse around the screen. The object will get bigger or smaller. You can randomly scale it, or enter a numerical value. For example if you want it twice as big, enter "2"; half size will be ".5".

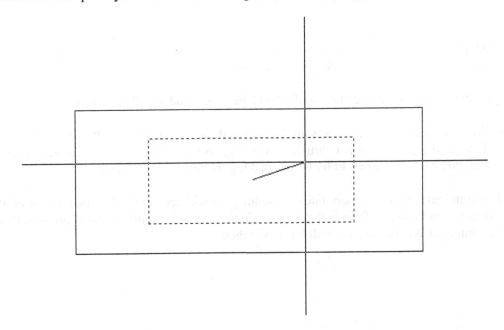

Figure 1.28 – Scale command in progress

> **Trim**

Keyboard: Type in **trim** and press Enter
Cascading menus: **Modify→Trim**
Toolbar icon: **Modify** toolbar -/--
Ribbon: **Home tab→Trim** -/--

In order to practice this command you will first need to draw two intersecting lines, one horizontal and one vertical, similar to a plus sign. Once this is done, go ahead and perform the sequence below.

Step 1
Begin the trim command via any of the above methods.
- AutoCAD will say:
  ```
  Current settings: Projection=UCS, Edge=None
  Select cutting edges...
  Select objects or <select all>:
  ```

Step 2
Using the mouse, left-click one of the lines. You can choose the vertical or horizontal, it doesn't matter. This is your "cutting edge." The other line, the one you did not pick, will be trimmed at the intersection point of the two lines. The cutting edge line will become dashed.
- AutoCAD will say: `Select objects or <select all>: 1 found`
Press Enter.

Step 3
- AutoCAD will say: `Select object to trim or shift-select to extend or[Fence/Crossing/Project/Edge/eRase/Undo]:`

Go ahead and pick anywhere on the line that you did NOT yet select and it will be trimmed.

You can repeat Step 3 as many times as you want, but we're done, so press Enter. Practice this several times by pressing **u** then Enter and repeating Step 1 thru 3. Two things to remember to avoid mistakes: pick the "cutting edge" first, and then don't forget to press Enter before picking the line to be chopped!

You can also do a trim between two or more lines, something we will need in the first major project. Practice both types of trims, as shown in Figure 1.29. In both cases the lines selected as "cutting edges" are now dashed, and the lines about to be trimmed have the cursor (small box) over them.

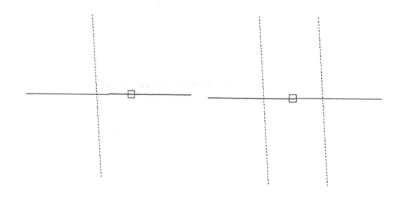

Figure 1.29 – Trim commands in progress

➢ **Extend**

Keyboard: Type in **extend** and press Enter	
Cascading menus: Modify→Extend	
Toolbar icon: **Modify** toolbar	--·/
Ribbon: **Home tab**→**Extend**	--·/

In order to practice this command we will first need to draw two intersecting lines, just as you did for trim, but then use the **move** command to relocate the vertical line directly to the right of the horizontal line, a short distance away, so we can extend the horizontal line into the vertical. Once this is done, go ahead and perform extend.

Step 1
Begin the extend command via any of the above methods.
- AutoCAD will say:
  ```
  Current settings: Projection=UCS, Edge=None
  Select boundary edges...
  Select objects or <select all>:
  ```

Step 2
At this point, left-click the vertical line. This is the TARGET into which you will extend the horizontal line. It will become dashed.
- AutoCAD will say: `Select objects: 1 found`

Press Enter.

Step 3
- AutoCAD will say: `Select object to extend or shift-select to trim or [Fence/Crossing/Project/Edge/Undo]:`

Go ahead and pick the end of the horizontal line that is closest to the vertical line. It is the ARROW that will extend into the vertical line target.

Step 4
You can repeat Step 3 as many times as you want, but we're done, so press Enter. Practice this several times by pressing **u** then Enter and repeating Step 1 thru 3. In Figure 1.30 below, the dashed vertical line is the "target" and the solid horizontal line is the "arrow". The cursor (box) is positioned to extend the arrow into the target by clicking once.

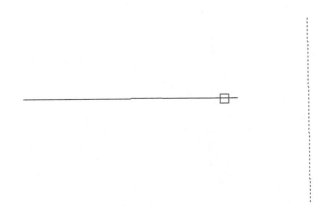

Figure 1.30 – Extend command in progress

> **Offset**

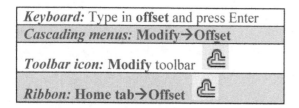

Keyboard: Type in **offset** and press Enter	
Cascading menus: **Modify→Offset**	
Toolbar icon: **Modify** toolbar	
Ribbon: **Home tab→Offset**	

This command is one of the more important ones that you will learn. Its power lies in its simplicity. This command creates new lines by the directional parallel offset concept, where if you have a line on your screen, you can create a new one that's a certain distance away but parallel to the original. To illustrate it, first draw a random vertical or horizontal line anywhere on your screen.

Step 1
Begin the offset command via any of the above methods
*   AutoCAD will say: `Specify offset distance or [Through/Erase/Layer] <Through>:`
Step 2
Enter an offset value (use a small number for now, 2 or 3). Then press Enter.
Step 3
*   AutoCAD will say: `Select object to offset or [Exit/Undo] <Exit>:`
Pick the line by left-clicking it; the line will become dashed.
*   AutoCAD will say: `Specify point on side to offset or [Exit/Multiple/Undo] <Exit>:`
Step 4
Pick a DIRECTION for the line to go, which will be one of the two sides of the original line. You can keep doing this over and over by selecting the new line and then the direction. If you want to change the offset distance, you must repeat Step 1 thru 3.

The significance of offset may not be readily apparent. Many students, especially ones that may have done some hand drafting, come into learning AutoCAD with the preconceived notion that it will just be a computerized version of pencil and paper. While this is true in many cases, in others AutoCAD represents a radical departure. When you begin a new drawing on paper, you literally draw line after line. With AutoCAD, surprisingly little line drawing occurs. A few are drawn, and then the offset command is employed throughout. Learn this command well, you will need it.

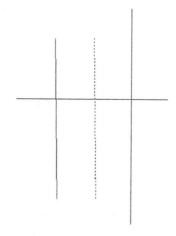

Figure 1.31 – Offset command in progress

➢ **Mirror**

Keyboard: Type in **mirror** and press Enter	
Cascading menus: Modify➔Mirror	
Toolbar icon: **Modify** toolbar 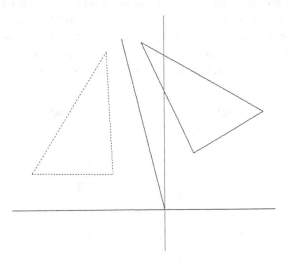	
Ribbon: **Home tab➔Mirror**	

This command, much as the name implies, creates a mirror copy of an object over some plane. It is surprisingly useful because often just copying and rotating is not sufficient or takes too many steps. To learn this command you have to think of a real-life mirror, with you standing in front of it. The original object is you, the plane is the door with the mirror attached (that can swing on a hinge), and the new object is your mirrored reflection. To practice this command, draw a triangle by joining 3 lines together as seen in Figure 1.32.

Step 1
Begin the mirror command via any of the above methods.
- AutoCAD will say: `Select objects:`

Step 2
Select all three lines of the triangle and press Enter.
- AutoCAD will say: `Specify first point of mirror line:`

Step 3
Click anywhere near the triangle and move the mouse around. You will notice the new object appears to be anchored by one point. This is the swinging mirror reflection.
- AutoCAD will say: `Specify second point of mirror line:`

Step 4
Click again near the object (make the second click follow an imaginary straight line). You will notice the new object disappear. Don't panic; rather check what the command line says.
- AutoCAD will say: `Delete source objects? [Yes/No] <N>:`

Step 5
The above just asked you if you want to keep the original object. If you do (the most common response) then just press Enter and you're done. Figure 1.32 illustrates Step 3.

Figure 1.32 – Mirror command in-progress

> **Fillet**

Keyboard: Type in **fillet** and press Enter
Cascading menus: **Modify→Fillet**
Toolbar icon: **Modify** toolbar ⌐
Ribbon: **Home tab→Mirror** ⌐

Fillets are all around us. This is not just an AutoCAD concept, but rather an engineering description for a rounded edge on a corner of an object or at an intersection of two objects. These edges are added to avoid sharp corners for safety or other technical reasons (such as a weld). By definition, a fillet needs to have a radius greater than zero but in AutoCAD a fillet CAN have a radius of zero, allowing for some useful editing as we will see soon. To illustrate this command, first draw two perpendicular lines that intersect.

Step 1
Begin the fillet command via any of the above methods.
- AutoCAD will say:
  ```
  Current settings: Mode = TRIM, Radius = 0.0000
  Select first object or [Undo/Polyline/Radius/Trim/Multiple]:
  ```
Step 2
Let's put a radius on the fillet. Type in **r** for Radius and press Enter.
- AutoCAD will say: `Specify fillet radius <0.0000>:`
Enter a small value, perhaps .5 or 1. Press Enter.
Step 3
- AutoCAD will say: `Select first object or [Undo/Polyline/Radius/Trim/Multiple]:`
Select the first (horizontal) line, somewhere near the intersection. It will become dashed.
Step 4
- AutoCAD will say: `Select second object or shift-select to apply corner:`
Select the second (vertical) line, somewhere near the intersection. The lines will have a fillet added to them of the radius you specified.

These are the steps involved in adding a fillet with a radius. You can also do a fillet with a radius of zero, so repeat steps 1 thru 4, entering **0** for the radius value. This is very useful to speed up trim and extend. Intersecting lines can be filleted to terminate at one point, and if they are some distance apart, they can be filleted to meet together. All this is possible with a radius of zero. Try it. Figure 1.33 summarizes everything described so far. The "1" and "2" indicate a suggested order of clicking the lines for the desired effect.

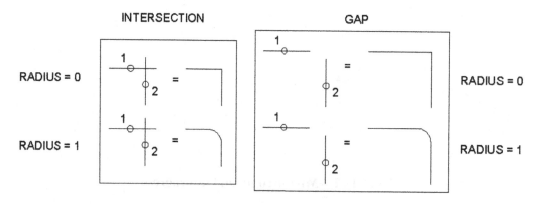

Figure 1.33 – Fillet summary

Well, this is it for now. We have covered the basic commands. These commands form the basis of what we call basic CAD theory. All 2D Computer-Aided Design programs need to be able to perform these functions in order to work, though other software may refer to them differently; i.e.: "Displace" instead of "Move", or "Join" instead of "Extend". Be sure to memorize the above and practice the commands until they become second nature. They are truly the ABC's of basic computer drafting, their importance cannot be overstated, and you need to master them to move on.

Sec 1.7 - Selection Methods

Before we talk about the accuracy tools available to you, we need to briefly cover a more advanced type of selection method called Window/Crossing. You may have already seen this by accident: When you click randomly on the screen without selecting any objects, and drag your mouse to the right or left, a rectangular shape forms. It's either dashed or solid (depending on settings, these shapes may have a green or blue fill) and goes away once you click the mouse button again or press Esc. This happens to be a powerful selection tool.

- **Crossing:** A crossing appears when you click and drag to the *left* of an imaginary vertical line (up or down), and is a *dashed* rectangle. *Any object it touches will be selected.*
- **Window:** A window appears when you click and drag to the **right** of an imaginary vertical line (up or down), and is a *solid* rectangle. *Any object falling completely inside it will be selected.*

Below is drawn a picture of what both the Crossing and the Window look like. Practice selecting objects in conjunction with the erase command. Draw several random objects, start up the **erase** command and instead of picking each one individually (or doing "erase, all"), select them with either the crossing or window and make a note of how each functions. This is a very important and useful selection method.

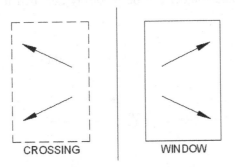

Figure 1.34 – Crossing/Window

Sec 1.8 - Drawing Accuracy – Part 1

Ortho (F8)

So far, when you created lines, they were not perfectly straight. Rather, you may have lined them up by eye. This of course isn't accurate enough for most designs, so we need a better method. Ortho (which stands for "Orthographic") is a new concept that we will now introduce that allows you to draw perfectly straight vertical or horizontal lines. It is not a command, but rather a condition you impose upon your lines prior to starting to draw them. To turn on the Ortho feature just press the **F8** key at the top of your keyboard, or click to depress the **ORTHO** button at the bottom of the screen. Then begin drawing lines. You will see your mouse cursor constrained to only vertical and horizontal motion. Ortho is usually not something you just set and forget about, as your design may call for a varied mix of straight and angled lines, so don't be shy about turning the feature on and off as often as necessary, and sometimes right in the middle of drawing a line.

Sec 1.9 - Drawing Accuracy – Part 2

OSNAPs

While drawing straight lines, we would generally like those lines to connect to each other in a very precise way. So far we have been eyeballing their position relative to each other, but that isn't accurate, and is more reminiscent of sketching, not serious drafting. Lines (or for that matter all fundamental objects in AutoCAD) connect to each other in very distinct and specific locations (two lines are joined end-to-end, for example). Let's develop these points for a collection of objects.

ENDpoint

These are the two locations at the ends of any line. Another object can be precisely attached to an endpoint. The symbol for it is a square as illustrated next.

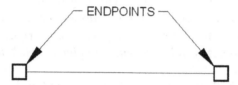

MIDpoint

This is the middle of any line. Another object can be precisely attached to a midpoint. The symbol for it is a triangle as illustrated next.

CENter

This is the center of any circle. Another object can be precisely attached to a center point. The symbol for it is a circle as illustrated next.

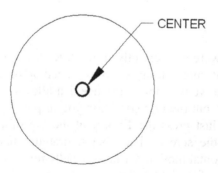

QUADrant

These are the 4 points at the North (90°), West (180°), South (270°), and East (0°, 360°) quadrants of any circle. Another object can be precisely attached to a quadrant point. The symbol for it is a diamond as illustrated next.

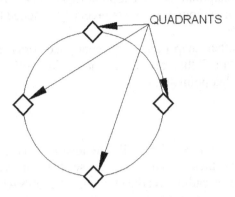

INTersection

This is a point that is located at the intersection of any two objects. You may begin a line or attach any object to that point. The symbol for it is an "X" as illustrated next.

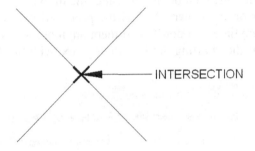

PERPendicular

This is an OSNAP point that allows you to begin or end a line perpendicular to another line. The symbol for it is the standard mathematical symbol for a perpendicular (90°) angle as illustrated next.

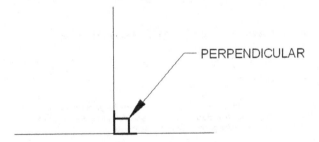

These are the six fundamental OSNAP points we will mention for now. There are others of course such as Node, Extension, Insertion, Tangent, Nearest, Apparent Intersection, and Parallel, but those are used far less often, and with the exception of Near and Node, they will not be discussed in detail.

So how do you use the OSNAPs that we developed so far? There are two methods. The first is to type in the OSNAP point as you need it. For example, draw a line. Then begin drawing another line, but before clicking to place it, type in: "end" and press Enter. AutoCAD will add the word "of", at which point move the cursor to roughly the end of the first line. The snap marker for endpoint (a square) will appear along with a tool tip stating what it is. Click to place the line, and now your new line is precisely attached to the first line.

In the same manner try out all the other snap points we mentioned (draw a few circles to practice Quad and Center). In all cases begin by typing "line" then one of the snap points followed by Enter. Then bring the mouse cursor to the approximate location of that point and click.

Sec 1.10 - OSNAP Drafting Settings

As you may imagine, there is a faster way to use OSNAP. The idea is to have these settings always running in the background so you don't need to type them in anymore. To set them up you need to type in **osnap** or select **Tools→Drafting Settings...** from the cascading overhead menu. The OSNAP dialog box will appear. First, make sure the Object Snap tab is selected from the four available, then select the "Clear All" button, and finally check off the six settings that we mentioned previously, and press OK.

So now you have set the appropriate settings, but you still need to turn them on. For that we need to mention another F-key. If you press **F3** or press the **OSNAP** toggle button at the bottom of the screen, the OSNAP settings will become active. Go ahead and do this, then begin the same line drawing exercise, and see how much easier it is now as the snap settings appear on their own. This is the preferred way to draw, with them running in the background. Feel free to press F3 anytime you don't need them anymore; OSNAP is also not something that you may want all the time. Figure 1.35 is the Drafting Settings dialog box, with the discussed OSNAP points checked.

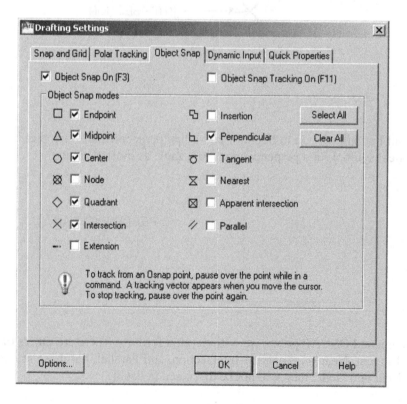

Figure 1.35 – OSNAP dialog box

Chapter 1

Summary

You should understand and know how to use the following concepts and/or commands before moving on to Chapter 2:

- **Create Objects**
  - Line
  - Arc
  - Circle
  - Rectangle

- **Edit/Modify Objects**
  - Erase
  - Move
  - Copy
  - Rotate
  - Scale
  - Trim
  - Extend
  - Offset
  - Mirror
  - Fillet

- **View Objects**
  - Zoom
  - Pan
  - Regen

- **AutoCAD Drawing Environment**
  - Drawing Area, Command Line, Cascading Menus, Toolbars, Ribbon

- **Selection Methods**
  - Direct object selection
  - Select: all
  - Crossing
  - Window

- **Drawing Accuracy**
  - Ortho (F8)
  - Osnap (F3)
    - ENDpoint
    - MIDpoint
    - CENter
    - QUADrant
    - INTersection
    - PERPendicular

Additionally, you should understand and know how to use the following tips introduced in this chapter before moving on to Chapter 2:

TIP #1: The Esc (escape) key in the upper left-hand corner of your keyboard is your new best friend while learning AutoCAD. It will get you out of just about any trouble you get yourself into. Mine was worn out learning AutoCAD, so expect to use it often

TIP #2: To quickly erase everything on the screen, type in "e" for "erase", press Enter, then type in "all". Then press Enter twice. After the first Enter all the objects will be dashed, indicated they were selected successfully. The second Enter finishes the command by deleting everything.

TIP #3: Type in "z" for "zoom", press Enter, then type in the letter "e", then press Enter again. This is called Zoom to Extents and makes AutoCAD display everything you sketched, filling up the available screen space. If you have a middle-button wheel on your mouse then double click it for the same effect. This technique is very important to fit everything on your screen.

TIP #4: To perform the Undo command in AutoCAD just type in "u", and press Enter. Don't type the entire command – if you do that, more options pop up which we don't really need at this point, so a simple "u" will suffice. You can do this as many times as you want to (even to the beginning of the drawing session). You can also use the standard Undo and Redo arrows shown in Figure 1.24.

TIP #5: Hate to keep having to type, pull a menu or click an icon each time you practice the same command? Well, no problem. Simply press the space bar or Enter, and this will repeat the last command you used.

Chapter 1

Review Questions

Answer the following based on what you learned in Chapter 1.

1) List the four **Create Objects** commands covered.

2) List the ten **Edit/Modify Objects** commands covered.

3) List the three **View Objects** commands covered.

4) What are the four main methods to **issue a command** in AutoCAD?

5) What is the difference between a selection **window** and **crossing**?

6) What is the name of the command that forces lines to be drawn perfectly **straight** horizontally and vertically? What is the **F-key** that activates it?

7) What are the six main **osnap** points covered? Draw the snap point's **symbol** next to your answer.

8) What command brings up the **osnap** dialog box?

9) What is the **F-key** that turns the **osnap** feature **On/Off**?

Chapter 1
Exercises

Exercise #1 – Draw the following sets of lines (in bold on left) and practice the 2-Point, 3-Point, and the Tan, Tan, Tan circle options. You may use the Ribbon, cascading menus or typing. In all cases use OSNAPs to connect the circle precisely to the drawn lines. The result you should get is shown in dashes on the right.
(Difficulty level: Easy, Time to completion: 5-10 minutes)

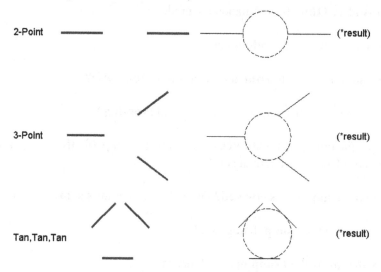

Exercise #2 – Draw an arc using the *Start, Center, End* option (a commonly used one) as seen below. First draw the three lines shown in bold on the left, then the arc. Be sure to use OSNAPs. The result you should get is shown in dashes on the right. *(Difficulty level: Easy, Time to completion: <5 minutes)*

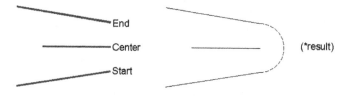

Exercise #3 – Draw a simple door using the *3-Point* arc method. First draw, copy, rotate and position the lines as seen in Steps 1 thru 3, then add the arc as shown in Step 4 (use Ortho and OSNAPs throughout). You may have to try it a few times to get it to look right. Finally, erase the bottom line.
(Difficulty level: Easy, Time to completion: <5 minutes)

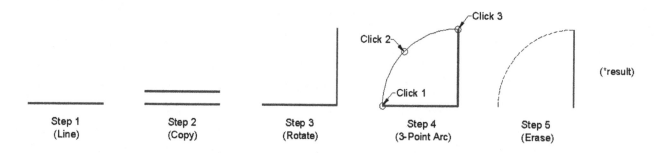

Exercise #4 – Draw the following sets of electrical connection shapes. You will only need the basic create and modify commands, including rectangle, circle, line, offset, fillet and mirror, as well as OSNAPs and Ortho. Specific sizes are not important, but make your drawing look similar, and use accuracy in connecting all the pieces. *(Difficulty level: Easy, Time to completion: 10 minutes)*

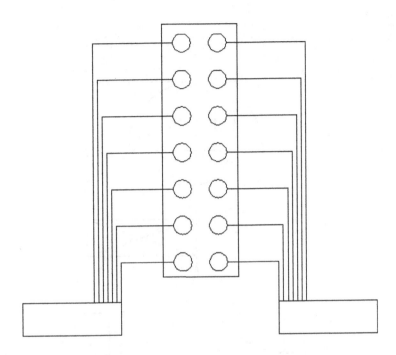

Exercise #5 – Draw the following house elevation using what you learned. You will only need the basic create and modify commands. The overall sizing can be approximated; however, use Ortho (no crooked lines!) and make sure all lines connect using OSNAPs. *(Difficulty level: Intermediate, Time to completion: 15 minutes)*

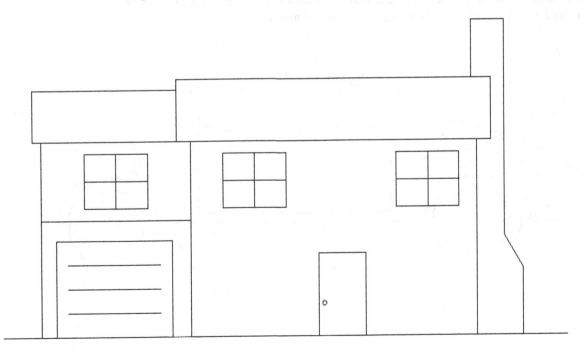

Exercise #6 – Draw the following sets of shapes. You will only need the basic create and modify commands, but you will use almost every one of them, including multiple cutting edges trim. The overall sizing is completely arbitrary, and your design may not look exactly like this one, but be sure to draw all the elements you see here, and use OSNAP precision throughout. *(Difficulty level: Intermediate, Time to completion: 15 minutes)*

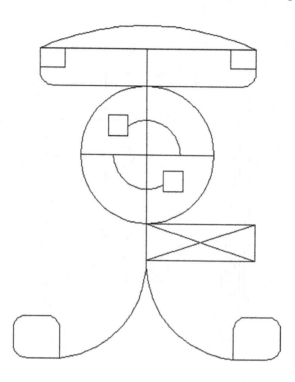

Exercise #7 – Draw the following side view of a car using what you learned. You will only need the basic create and modify commands. The overall sizing (including multiple fillets needed on most corners) can be approximated; however, use accuracy and make sure all lines connect using OSNAPs.
(Difficulty level: Intermediate, Time to completion: 20 minutes)

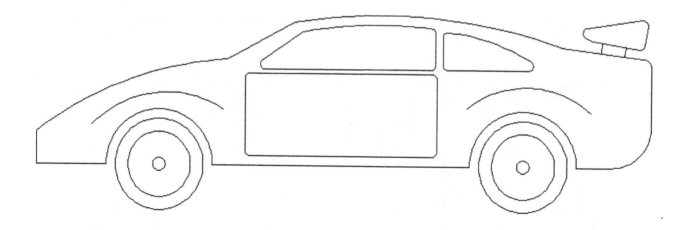

CHAPTER 2

AutoCAD Fundamentals
Part II

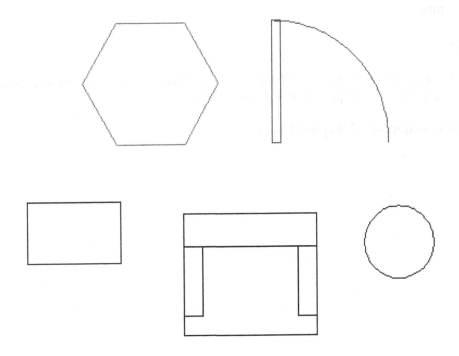

Chapter 2

Learning Objectives

You have learned a great deal in the first chapter, in fact almost everything you need to get started drafting a design. This chapter will serve as a "clean up" and conclude the basic introduction to AutoCAD, as there are just a few more topics to get to. Here we will discuss the following:

- Using Grips
- Setting Units
- Using Snap and Grid
- Understanding the Cartesian coordinate system
- Distance Entry Tools:
  - Direct Distance Entry
  - Relative Distance Entry
  - Dynamic Distance Entry
- Inquiry commands:
  - Area
  - Dist
  - List
  - ID
- Explode command
- Polygon command
- Ellipse Command
- Chamfer command
- Templates
- Setting Limits
- Save
- Help Files

By the end of the chapter you will have learned the necessary basics to begin drafting an architectural floor plan, after an introduction to layers in Chapter 3 you will begin to do so.

Estimated time for completion of chapter: 2 hours

Sec 2.1 - Grips

Let's begin Chapter 2 with Grips, a natural progression from learning the OSNAP points. These are relatively advanced editing tools, but chances are you already made them appear by accident by clicking an object without typing a command first. It's OK, press Esc and they will go away. But what are these mysterious blue squares? Take a close look, after learning OSNAPs, these should look familiar.

Grips are control points, or locations on objects where you can modify that object's size, shape or location by activating these points. (They are quite similar to the "handles" that you'll find in Photoshop® or Illustrator® if you've ever used those applications). Try them out by clicking any line, circle, arc or rectangle. It will become dashed and the grips will pop up. Select a grip and click on it again. It will become red in color. This means it's active (or hot). Now move the mouse around. The object will change in one of several ways. A line or circle can be moved around if the center grip is selected. A circle can be scaled up or down in size if the outer quadrants are selected. Arcs can be made larger, or length increased. Other grips such as the endpoints of a line or corners of a rectangle will change (or stretch) the shape of that object if activated and moved.

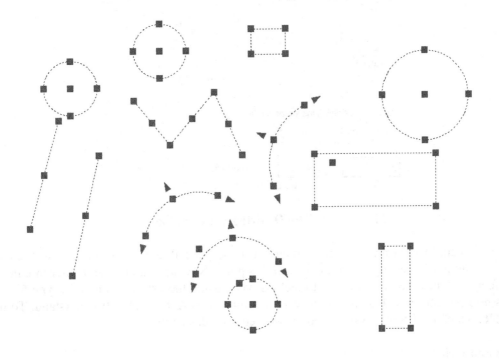

Figure 2.1 – Grips on Circles, Rectangles, Arcs and Lines

Sec 2.2 - Units and Scale

We have not yet talked about any sort of drawing units such as feet, inches or centimeters. When you practiced the offset command you entered a value such as 2 or 3, but what did it mean? Well, as you inadvertently discovered, the default unit system in AutoCAD (until you change it) is a simple decimal system of the type we use in everyday life. However it is unit-less, meaning that the "2" or "3" that you entered could have been literally anything including feet, inches, yards, meters, miles or millimeters. Think about it, it's an easy and powerful idea, but it is up to you to set the frame of reference and stick to it. So if you are designing a city and "1" is equal to a mile, then "2" is two miles and ".5" is half a mile. AutoCAD doesn't care; it gives you a system, and lets you adapt it to any situation or design. Then you just draw in real life units or as we say in CAD, 1-to-1 units. This is the Golden Rule of AutoCAD. **Always draw everything to real life size**. Do not scale objects up or down.

A decimal system of units is OK for many engineering applications, but what about architecture? Architects prefer their distances broken down to feet and inches because it is always easier to understand and visualize a distance of 22'-6" than 270". To switch to architectural units type in **units** and press Enter, or select **Format→Units...** from the cascading menus, and the following dialog box will appear (Figure 2.2).

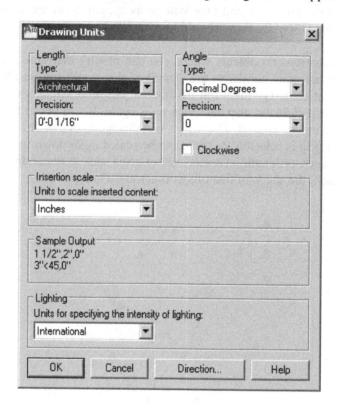

Figure 2.2 – The Drawing Units dialog box

Simply select Architectural from the drop-down menu at the upper left and ignore the rest of the options in the dialog box; there is no need to change anything else at this point. Now as you draw and enter in units they will be recognized as feet or inches. To enter in 5 feet (such as when using the **offset** command), type **5'**. To enter in 5 inches, there is no need for an inch sign, just type in the number **5**; AutoCAD will understand. To mix feet and inches type in **5'6**, which is of course **5'-6"**; there is no need for a dash either.

Sec 2.3 – Snap and Grid

These two concepts go hand in hand. The idea here is to set up a framework pattern on the screen that restricts the movement of the crosshairs to predetermined intervals. That way drawing simple shapes is easy because you know that each jump of the mouse on the screen is equal to some preset unit. Snap is the feature that sets that interval, but you need the grid to make it visible. One without the other does not make sense, and usually they are set equal to each other. Snap and grid are not used that often in the industry, but it's a great learning tool and we will be doing some drawings at the end of this chapter using them. It also helps knowing about this feature as it is often inadvertently activated by pressing the Snap or Grid buttons at the bottom of the screen.

To set SNAP:

Type in **snap**
- AutoCAD will say: `Specify snap spacing or [ON/OFF/Aspect/Style/Type] <0'-0 1/2">:`
Type in **1**, setting the snap to 1 inch.

To set GRID:

Type in **grid**
- AutoCAD will say: `Specify grid spacing(X) or ON/OFF/Snap/Major/aDaptive/Limits/Follow/Aspect]<0'-0 1/2">:>:`
Type in **1**, setting the grid to 1 inch.

The snap and grid are now set, though you may have to zoom to extents (Tip #3) to see everything. Now try drawing a line, and you will notice right away that the cursor "jumps" according to the intervals you set. **F9** will turn snap on/off, while **F7** will take care of grid, and you can always toggle back and forth using the drawing aids buttons at the bottom of the screen.

Figure 2.3 is a screen shot of snap and grid in use while drawing a simple shape. Notice the faint grid dots visible in the background. The cursor will only jump from dot to dot, not in the spaces in between.

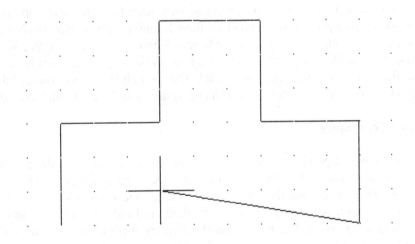

Figure 2.3 – Figure drawn with Snap and Grid

Sec 2.4 - Cartesian coordinate system

AutoCAD, and really all CAD programs, rely on the concepts of axes and planes to orient the drawings in space. In 2D drafting, you are always drawing somewhere on AutoCAD's X-Y plane, which is made up of the intersections of the X and Y axes. When you switch to 3D, the Z axis is made visible to allow drawing into the third dimension. The X, Y and Z axes are part of the Cartesian coordinate system, and we need to briefly review this bit of basic math, and then illustrate exactly why it's helpful to know it. Figure 2.4 is a picture of the 2D coordinate system, made up of the X and Y axis and numbered as shown.

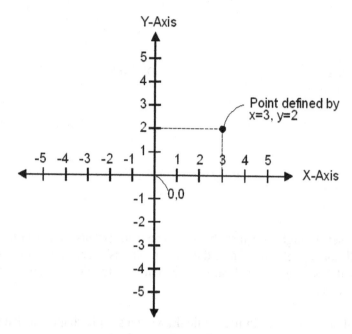

Figure 2.4 – The Cartesian coordinate system

The system shown above is important not only because you may want to know where in space your drawing is, but also you can create objects of a given size by entering their X and Y values. Moreover, we will be referring to this system often for a variety of other construction needs and distance entries, including inserting drawings into the point 0,0 , so review it to familiarize yourself with it again, as likely you may not have seen this in a while. Next we will introduce the various distance entry tools. Whether you choose to use manual distance inputs or the newer "dynamic" methods, knowing the Cartesian coordinate system will help you understand what is going on.

Sec 2.5 - Distance Entry Techniques

In AutoCAD there are a number of precise ways to enter distances and angles, and thereby create initial layouts. These techniques come in two flavors: the older "manual" method, where you enter data via keyboard, and the newer "dynamic" method, where you enter the values on the screen via what is known as a "heads-up" display. Most students will probably gravitate toward the newer method, and indeed it offers many advantages, not the least of which is faster data input. We will consider this the primary method in this text, and present the manual method only as an alternate, and in at least in one case, it is a bit more straightforward.

> *Direct Distance and Angle Entry*

Erase anything that may be on your screen and turn off all drawing aids including snap, grid and anything else you may have on. Now press the DYN drawing aid as seen in Figure 2.5.

Figure 2.5 – Dynamic input

This dynamic input allows you to specify distances and angles directly on the screen. Begin drawing a line and you will see the following display (Figure 2.6).

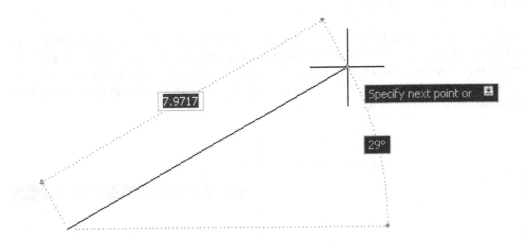

Figure 2.6 – Dynamic input while drawing a line

Notice what AutoCAD is allowing you to do. The "7.9717" value in Figure 2.6 is surrounded by a yellow box, meaning it is editable. All you have to do is enter a desired value, so enter a small value such as 5, but do *not* press Enter yet! What if you wanted to change the angle of the line? Simply press the Tab key and tab over to the "29" degree value, and enter something else, perhaps a 45. Then, finally press Enter, and your line 5 units long at 45 degrees is done. The line command will continue of course and you can repeat this as much as needed.

Manual method: You can do the same exact thing without the DYN input. Simply click the line anywhere, point the mouse in the direction you want the line to go and enter a value on the keyboard. If all lines are at right angles (90°) to each other, then leave Ortho on and "stitch" your way around by just pointing and clicking. This is a very useful method after you do a basic room/building survey and return to the office to enter the data in to AutoCAD. You can also enter data one click at a time by following the coordinate points. Begin the line command, then instead of clicking, type in the following values, pressing Enter after each one:

0,0   0,2   2,2   2,4   4,4   4,2   6,2   6,0   then "c" for "close"

These are the x, y values for the lines to follow, and if you typed correctly, then the following shape should emerge on your screen:

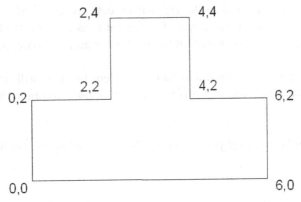

This is a rather slow method of entering data, but it is good for practice and to really understand the Cartesian points. We will not be using this regularly in actual design work.

> *Relative Distance and Angle Entry*

Relative distance frees you from the 0,0 point and allows you to draw any rectangular shape of any size, anywhere on the screen. With dynamic input, this makes little difference as seen next. To draw a 5 × 3 rectangle, make sure DYN is on, begin the command and click anywhere to start the rectangle. You will see two fields to enter the values for the height and width (Figure 2.7). The upper right corner of the rectangle isn't visible because the smaller crosshairs still obscure part of it.

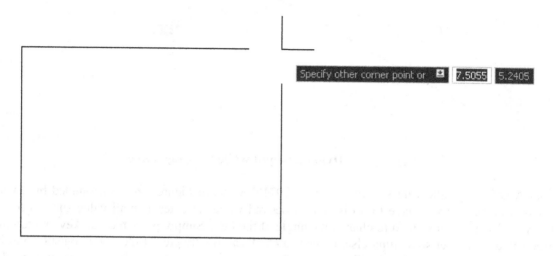

Specify other corner point or   7.5055   5.2405

Figure 2.7 – Dynamic input while drawing a rectangle

As before, enter a new value in the first text field (5) and Tab over to the next field, entering the other value (3). As you press Enter, the 5 × 3 rectangle will be set.

Manual method: You can do the same exact thing without the DYN input, and it may even be a bit faster, but you do have to learn a new trick. The "@" sign is what frees you from 0,0 so you will need to enter it prior to the values. Turn off the DYN, and begin the rectangle command as before by clicking anywhere. Then type in the following: @5,3 and press Enter. The same 5 × 3 rectangle will appear. This manual method is sometimes preferred in this one type of shape creation.

Now that you know about the "@" symbol and what it does, you can also draw lines at an angle using the manual method, something that was briefly ignored in the manual Direct Distance Entry discussion. You will however need to add a new symbol, "< ", to indicate the angle. The format is as follows: "@Distance<Angle". Therefore to create a line that is 10 units long at 45°, you will start the line command and type in: @10<45.

As you can see, the dynamic input is preferred and recommended, but it is still good to know about the manual methods if you just need a quick shape drawn, or if you are stuck on a computer with an older AutoCAD – still very much a possibility!

Dynamic input, meanwhile, can be used for just about anything, including the basic commands you have learned thus far as described next.

> ***Other uses of DYN***

While most users turn on DYN when they need to enter data for basic line work and shapes, its use extends out to just about anything you work on. Shown in Figure 2.8 is DYN with circle and arc, two other major shapes.

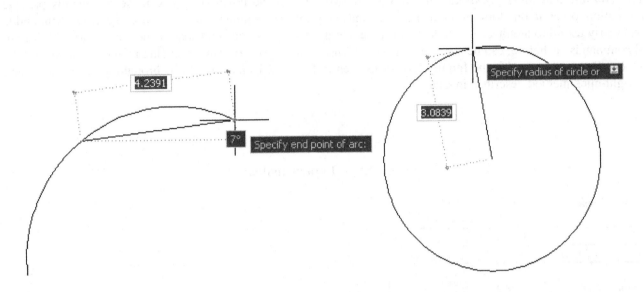

Figure 2.8 – Dynamic input while drawing an arc or circle

You may have noticed also that there is a down arrow in some of the "heads up" menus. You can use the down arrow on your keyboard to drop the menu down and reveal other menu choices, which of course mimic what is available in the parenthesis on the command line. Figure 2.9 is one example with the offset command.

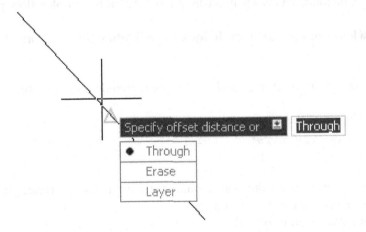

Figure 2.9 – Dynamic input with extra menus on offset

It is up to you if you would like to use the DYN feature throughout your lessons in AutoCAD; in fact it is recommended in some situations! However, it is not good for getting a point across in a textbook, as this information tends to add clutter and lower clarity in screen shots, obscuring the main message, so you won't see it much from here on.

Sec 2.6 - Inquiry Commands

Our next topic in AutoCAD fundamentals is the inquiry commands set. These have been upgraded in AutoCAD 2010, but the fundamental inquiry tools remain *area*, *distance*, *list* and *ID*. You can also now measure the *radius*, *angle*, *volume* and *mass properties*, but those last two will not be covered. All of these commands provide information to you in one form or another and are indispensable in a sophisticated drawing. The *area* command is particularly useful to architects and interior designers to get areas of rooms and spaces and we'll start with it first. All commands, as before, can be accessed through the usual four ways; typing, cascading menus, toolbars and the Ribbon. The toolbar used is the Inquiry toolbar as seen in Figure 2.10 and should be brought up on screen using the right-click method described in Chapter 1.

Figure 2.10 – Inquiry toolbar

> **Area**

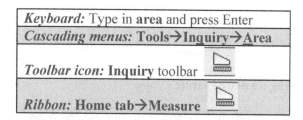

Area calculations are performed in two ways: either on an object or more commonly point-by point (such as the area of a room). Draw a rectangle of any size, and we will go over both methods, although with a rectangle you would only use the `Object` method, saving the point-by-point for a more complex floor plan.

Point-by-Point: This is when you have an area defined by individual lines (not an object) to measure.

Step 1
Make sure your units are set to "Architectural", and you have a rectangle on your screen.
Step 2
Begin the area command via any of the above methods.
- AutoCAD will say: `Specify first corner point or [Object/Add area/Subtract area]`
  `<Object>:`

Step 3
Activate your OSNAPs (F3), and click your way around the corners of the rectangle until the green shading covers the entire area to be measured (four clicks total).
- AutoCAD will give you several lines of:
  ```
  Specify next point or [Arc/Length/Undo]:
  Specify next point or [Arc/Length/Undo]:
  Specify next point or [Arc/Length/Undo/Total] <Total>:
  Specify next point or [Arc/Length/Undo/Total] <Total>:
  ```
Step 4
When done, press Enter, and you will get a total for area and perimeter. Mine said:
```
Area = 67.9680, Perimeter = 34.2044
```
Yours will almost certainly say something different.

Object: This is when you have an object (not an area defined by individual lines) to measure.

Step 1
Begin the area command via any of the above methods.
- AutoCAD will say: `Specify first corner point or [Object/Add area/Subtract area]`
  `<Object>:`

Step 2
Press the letter **o** for `Object`.
- AutoCAD will say: `Select objects:`

Step 3
Select the square and you will get the same area and perimeter measurements.

Practice the radius and angle measurements on your own. They're quite straight forward in process; you will start up the command, select the lines of the rectangle for the angle, getting the info seen in Figure 2.11a, or you will select the circle (need to obviously draw one for that purpose), getting the info seen in Figure 2.11b.

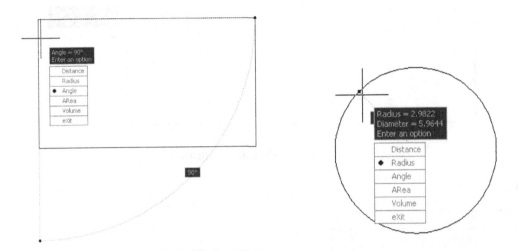

Figure 2.11 a & b – Angle and Radius measurements

➢ **Distance**

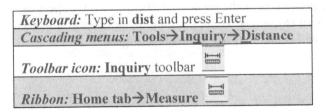

Keyboard: Type in **dist** and press Enter
Cascading menus: **Tools→Inquiry→Distance**
Toolbar icon: **Inquiry** toolbar
Ribbon: **Home tab→Measure**

The distance command will get you a distance between any two points. Although we will practice on the corners of our rectangle, the command will work on any two points, as selected by a mouse click, even empty space.

Step 1
Make sure your units are set to "Architectural", a rectangle is on your screen and your OSNAPs are activated.

Step 2
Begin the dist command via any of the above methods. (*see note below Step 4*)
- AutoCAD will say: `Specify first point:`

Step 3
Select a corner of the rectangle.
- AutoCAD will say: `Specify second point or [Multiple points]:`

Step 4
Select the other corner across the rectangle. AutoCAD will then give you some info. Mine said:
```
Distance = 13.5521,  Angle in XY Plane = 0,  Angle from XY Plane = 0
Delta X = 13.5521,  Delta Y = 0.0000,   Delta Z = 0.0000
```
Yours will say something different.

*NOTE: There is an important difference in procedures if you use the icons or the Ribbon. DYN will activate and you will get a graphical representation of the distance as seen in Figure 2.12.

Figure 2.12 – Distance command

➢ **List**

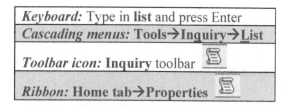

Keyboard: Type in **list** and press Enter
Cascading menus: **Tools→Inquiry→List**
Toolbar icon: **Inquiry** toolbar
Ribbon: **Home tab→Properties**

The list command gives you info about not just the geometric properties of the object but also its CAD properties such as what layer it is on; the subject of the next chapter. It is an extremely useful command that clues you in on what you are looking at (whether a beginner or an expert, you will sometimes ask yourself this question).

Step 1
Make sure you have something on your screen to investigate.
Step 2
Begin the list command via any of the above methods.
- AutoCAD will say: `Select objects:`

Select your object (it will become dashed) and press Enter. The command line will pop up in mid-screen. Look it over and close it (X – upper right) when done. Here's what mine said about my rectangle in Figure 2.13.

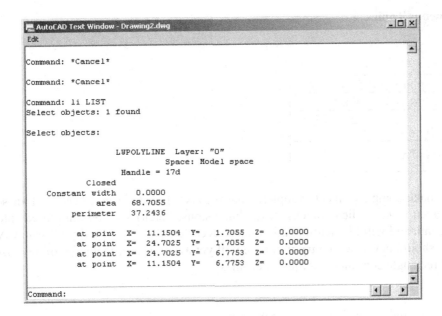

Figure 2.13 – List command

> **ID**

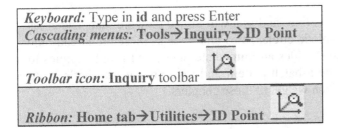

This command is the simplest of all. It tells you the coordinates of a point in space, such as the intersection of two lines. This is useful for inserting blocks (to be covered later), and just plain old figuring out the general location of a design, something that will be needed for Xrefs way ahead in advanced chapters.

Step 1
Make sure you have something on your screen to ID.
Step 2
Begin the ID command via any of the above methods.
  • AutoCAD will say: `Specify point:`
Step 3
Using OSNAPs select a point, such as the corner of a rectangle. AutoCAD will give you the X, Y, Z coordinates. Here is what I got; you will get different values.

`X = 11.1504      Y = 6.7753      Z = 0.0000`

This concludes the section on inquiry commands. We're now on the home stretch, and will explore some commands that didn't quite fit in anywhere else.

Sec 2.7 – Miscellaneous Topics

> **Explode**

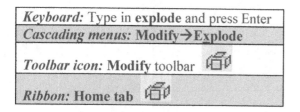

Keyboard: Type in **explode** and press Enter
Cascading menus: **Modify→Explode**
Toolbar icon: **Modify** toolbar
Ribbon: **Home tab**

Explode is used to make complex objects simpler. That means a rectangle will become four separate lines, and other objects we haven't yet studied (blocks, text, dimensions, hatch) will be simplified into their respective components. This command will be mentioned in almost every chapter, as virtually all AutoCAD entities can be exploded and many should never be! To try it out we will yet again need a rectangle of any size. Make sure you are using the actual rectangle command, not just four lines.

Step 1
Begin the explode command via any of the above methods.
- AutoCAD will say: `Select objects:`

Step 2
Pick the rectangle and it will become dashed.
- AutoCAD will say: `1 found`

Press Enter and the rectangle will be exploded. Click any of its lines and you will notice they are independent as seen in Figure 2.14 a & b. Keep in mind that you cannot explode circles, arcs and lines as those are already the simplest objects possible. There is a way to reverse explode (besides an immediate undo), but it only applies to simple compound objects such as a rectangle. If you explode a hatch pattern and save your work, that's it, no going back! This is why you will see appropriate warnings when we get to those chapters.

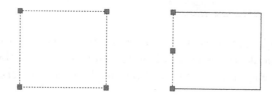

Figure 2.14 a & b – Unexploded and exploded rectangle side by side

> **Polygon**

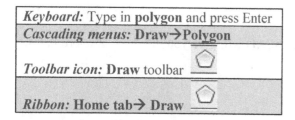

Keyboard: Type in **polygon** and press Enter
Cascading menus: **Draw→Polygon**
Toolbar icon: **Draw** toolbar
Ribbon: **Home tab→ Draw**

The polygon command is useful and we will need it in later chapters for a project. It is also easy to do and can be introduced early on, though it is not nearly as important as the primary four objects (line, arc, circle and rectangle), and due to that it wasn't introduced in the first chapter.

A polygon, strictly speaking, is any shape with three or more sides to it. Of course in those cases when it has three sides (triangle), you can draw it with existing techniques. Same thing with four sides (square or rectangle). However, what about five sides or more, such as pentagon or hexagon? Here a polygon command is needed, as it's a real chore to draw these shapes by just using the line command (just try figuring out all the correct angles).

There is, however, a complication in drawing polygons that you may not have with simpler shapes. With a rectangle you specify the height and width and that's it, but how do you do that with, let's say, a pentagon? Here's a picture of one (Figure 2.15). What is the height? What is the width? It's not an easily measured shape.

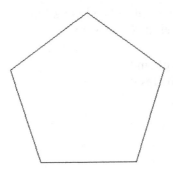

Figure 2.15 – Basic pentagon

The solution that drafters came up with a long time ago (many of AutoCAD's techniques are just adaptations of pencil and paper board-drafting methods used for decades), is to enclose the shape in a circle of a known radius or diameter.

You can also enclose the circle inside the polygon. With this technique it's easier to specify a size – you just say "I need a pentagon that fits inside a circle of 10in. diameter". This is where the terms Inscribed and Circumscribed come from. Figure 2.16 is a diagram of two pentagons both inscribed *in* and circumscribed *about* a circle – note the tangencies. In one case the points of the polygon touch the circle; in the other case its sides touch it.

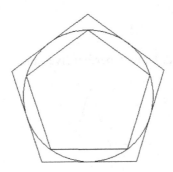

Figure 2.16 – Inscribed and Circumscribed pentagon

Let's go through the steps in making this polygon.

Step 1
On a blank screen draw a circle with a radius of 5. Note that we don't need this step to create a polygon, but for your first try, it's a great help in visualizing what you're doing.
Step 2
Begin the polygon command via any of the above methods.
- AutoCAD will say: `Enter number of sides <4>:`

Step 3
Enter the number 5 for five sides (making it a pentagon), and press Enter.
Step 4
- AutoCAD will say: `Specify center of polygon or [Edge]:`

With your OSNAPs on, click to select the center of the circle.
Step 5
- AutoCAD will say: `Enter an option [Inscribed in circle/Circumscribed about circle] <I>:`

Step 6
Now that you know what these terms mean, go ahead and press Enter to select "I" for "`Inscribed in circle`" (the default value). You can just as easily select "C" as well.
Step 7
- AutoCAD will say: Specify radius of circle:

Enter the value 5. Here's a screen shot of the result:

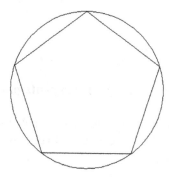

You can now erase the circle, since it was just a guide. In the future you won't need it, just click to place the polygon anywhere on the screen you need to. Go ahead and practice this command again, but this time selecting a different number of sides (six or more) and also using the circumscribed option. This new command will lead us into the first challenge exercise.

 ### *Drawing Challenge*

Try this exercise based on what you learned. Draw a perfect five-sided star such as the one shown below:

Think about everything we have talked about up to this point. Only four commands are needed to create the star. Take a few minutes to try to draw it. If you can't get it, don't despair, about a third of my students through the years typically didn't get it right away either. The answer is at the end of Chapter 8.

➢ **Ellipse**

The ellipse command is easy to do and will be needed when we draw furniture in Chapter 4, so we'll include it here. The command can be executed in one of two ways: *Center* and *Axis, End*. You can also create a partial ellipse by defining just a section of it.

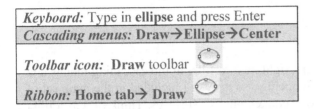

Keyboard: Type in **ellipse** and press Enter
Cascading menus: **Draw→Ellipse→Center**
Toolbar icon: **Draw** toolbar
Ribbon: **Home tab→ Draw**

Step 1
Begin the ellipse command via any of the above methods.
- AutoCAD will say: `Specify axis endpoint of ellipse or [Arc/Center]:]:`
Click anywhere on the screen.

Step 2
- AutoCAD will say: `Specify other endpoint of axis:`
With Ortho on (not required but good to have), click again somewhere off to the left or right of the first point. Then move the mouse back in toward the center as the ellipse is formed as seen here:

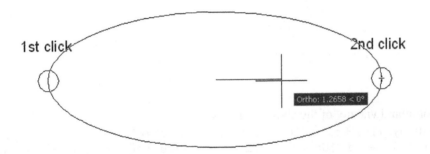

Step 3
- AutoCAD will say: `Specify distance to other axis or [Rotation]:`
Click anywhere again to complete the ellipse. You can also specify a rotation angle along the major axis (the line between click 1 and 2). This will be as if the ellipse is rotating away (or toward) you.

For a variation on this, try the center version of the ellipse, which is essentially similar in execution. The final version of the command (Elliptical Arc) involves creating the ellipse, followed by tracing out how much of it you want to show as seen here:

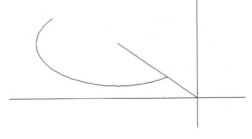

> ➢ **Chamfer**

Keyboard: Type in **chamfer** and press Enter	
Cascading menus: **Modify→Chamfer**	
Toolbar icon: **Modify** toolbar	
Ribbon: **Home tab→ Draw**	

You may have already seen or tried the chamfer command as it sits right next to fillet in the cascading menus and the Modify toolbar. It is not as critical as fillet, and not used as frequently, so it wasn't included in the main set of commands, but we'll introduce it now. A chamfer is a beveled edge connecting two surfaces. Think of it as a fillet that is angled instead of round. Chamfers may have a 45° corner, but this is not required; AutoCAD allows for asymmetrical chamfers.

To try out the command, draw two perpendicular lines similar to the fillet exercise in Chapter 1, as shown below.

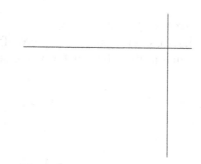

Step 1
Begin the chamfer command via any of the above methods.
- AutoCAD will say: (TRIM mode) Current chamfer Dist1 = 0.0000, Dist2 = 0.0000
  Select first line or [Undo/Polyline/Distance/Angle/Trim/mEthod/Multiple]:

No distance has been entered yet, and any attempt at chamfer will result in a chamfer radius zero. Press **d** for Distance.

Step 2
- AutoCAD will say: Specify first chamfer distance <0.0000>:

Enter a small value such as 1.

Step 3
- AutoCAD will say: Specify second chamfer distance <1.0000>:

Here the 1 is not needed again so just press Enter.

Step 4
- AutoCAD will say: Select first line or [Undo/Polyline/Distance/Angle/Trim/mEthod/ Multiple]:

Select the first line.

Step 5
- AutoCAD will say: Select second line or shift-select to apply corner:

Select the second line

The chamfer will be created as seen in Figure 2.17. Zoom in if needed to see it. If the chamfer fails it's possible that you selected values too large for the available lines. Another way to create a chamfer is using length of one line segment together with an angle, using the Angle option. Experiment with both.

Figure 2.17 – Chamfer

> **Templates**

Templates have a bit of a mystery for a beginner AutoCAD student. What exactly are they? While templates can get complicated in what they contain, all they really are is a file with some information embedded in it, and all the proper settings set, but no actual design. So it's a blank page with some intelligence built in and everything set up and ready for you to draw. These templates can be saved under a *.dwt extension, but you really don't even need to concern yourself with this. What is important is being able to open a new file and knowing what you are looking at. Using the cascading menu **File→New…** you will default to the following (Figure 2.18). Open up the highlighted *acad.dwt* and you have a blank file, end of story!

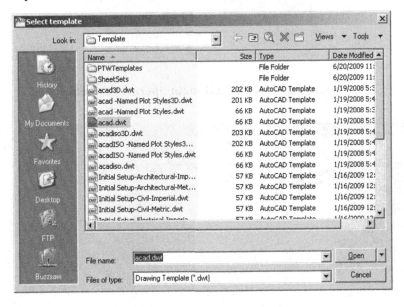

Figure 2.18 – acad.dwt template

> **Limits**

This is yet another topic that causes confusion. Limits are not technically necessary, but many textbooks persist in having you set them up. All they are is an indication of the size your drawing area is preset to. Remember, however, that it is limitless, so all limits refer to is the area you can work on without a regen. If you draw a shape that is way beyond your limits, you simply regenerate the screen (Zoom→Extents) and the limits move to just beyond this new shape, so setting them is rather redundant. If you would like to set them anyway, just type in **limits**, press Enter and set the lower left corner to 0,0 and then the upper right corner to perhaps 100', 50' or anything else you wish.

➢ **Save**

Needless to say, you need to save your work; AutoCAD will eventually lock-up, crash and die. If not, Windows will, or there will be a power failure, and there is never a guarantee of a save, so it's just a good habit to get into. Starting in the next chapter you will need to start saving your work. To do so, you can type in **save** and press Enter or press the graphic of a disk at the top of the screen. Another way is to drop down the arrow located next to the big red "A" at the upper left of the screen (Figure 2.19) and choose Save or Save As. You will then get the usual save dialog box (Figure 2.20) from which you should know how to proceed.

Figure 2.19 – AutoCAD main operations menu

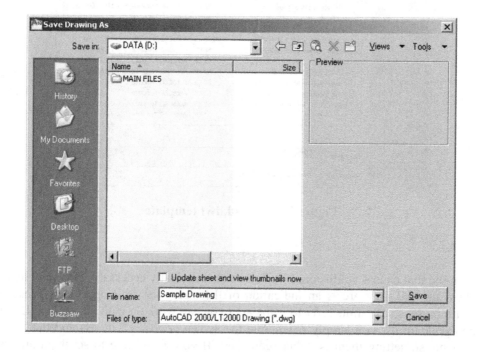

Figure 2.20 – Save Drawing

> ➢ **Help Files**

Finally we get to our last topic, the Help Files. AutoCAD, like any software, has extensive help files built in and indeed that is one way (though not recommended) to attempt to learn the software. A better application of Help is to explore it once you know how to use AutoCAD and need assistance with a particular aspect of a command. Tool tips that appear with every command (unless you turn them off) in various levels of detail also can be considered a part of the help files.

To access Help, simply press F1. It is context sensitive and will go to whatever command was active at the time of pressing F1. Figure 2.21 is a screen shot of the main page (with no commands active). The best way to use it is via the Index, where you search the command on the left and tab through Concept, Procedure and Quick Reference on the right. You can also just look up the commands alphabetically in Contents. To close help you will need to "X it out"; it won't go away on its own.

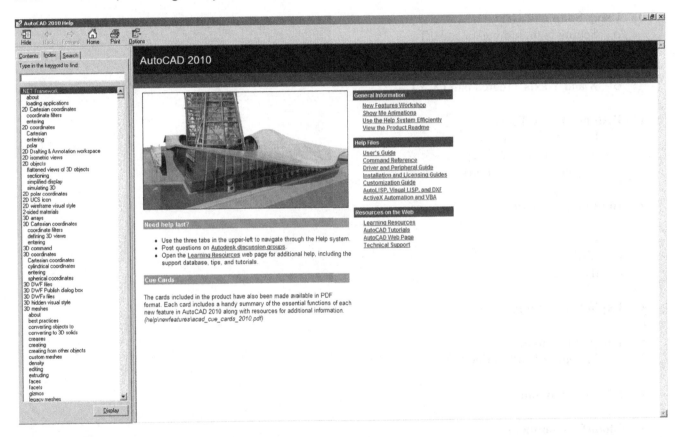

Figure 2.21 – AutoCAD 2010 Help Files

Review everything presented so far; in Chapter 3 we will introduce the layers concept, and put all of the preceding together to begin drawing a floor plan. Remember that learning AutoCAD is cumulative, and you will need everything we have covered so far, no more and no less, to successfully be able to deal with the upcoming material, so make sure you understand it all before moving on.

Chapter 2

Summary

You should understand and know how to use the following concepts and/or commands before moving on to Chapter 3:

- **Using Grips**
  - Activating them, Esc to get out

- **Setting Units**
  - Architectural and Engineering (Decimal) units

- **Using Snap and Grid**
  - Setting both, turning on and off

- **Understanding the Cartesian coordinate system**
  - X and Y axis, ordered pair (X,Y)

- **Distance Entry Tools**
  - Direct Distance Entry
  - Relative Distance Entry
  - Dynamic Distance Entry

- **Inquiry commands**
  - Area
  - Dist
  - List
  - ID

- **Explode command**

- **Polygon command**
  - Circumscribed/Inscribed

- **Ellipse command**

- **Chamfer command**

- **Templates**
  - acad.dwt

- **Setting Limits**

- **Save**

- **Help Files**
  - F1

Chapter 2
Review Questions

Answer the following based on what you learned in Chapter 2.

1) How do you activate *Grips* on AutoCAD's geometry?

2) How do you make *Grips* disappear?

3) What are *Grips* used for?

4) What is the command to bring up the **units** dialog box?

5) The **snap** command is always used with what other command?

6) Correctly draw a **Cartesian coordinate system** grid. Label and number each axis appropriately, with values ranging from -5 to +5.

7) On your **Cartesian Coordinate System** grid above, locate and label the following points: (3,2) (4,4) (-2,5) (-5,2) (-3,-1) (-2,-4) (1,-5) (2,-2)

8) What are the two main types of **distance entry**?

9) How do you use **Dynamic Input**?

10) List the four primary **Inquiry** commands we covered

11) What command will simplify a rectangle into four separate lines?

12) What is the difference between an **Inscribed** and a **Circumscribed** circle in the polygon command?

13) What is a **chamfer** as opposed to a **fillet**?

14) What template should you pick if you want a new file to open?

15) What is the idea behind limits?

16) Why should you save often?

17) What is the graphic symbol for save?

18) How do you get to the Help Files in AutoCAD?

Chapter 2

Exercises

Exercise #1 – Open a new file, set units to architectural and set your Snap and Grid both to 1". Then draw the following shape. Keep in mind the spacing between each dot is now 1".
(Difficulty level: Easy, Time to completion: 5 minutes)

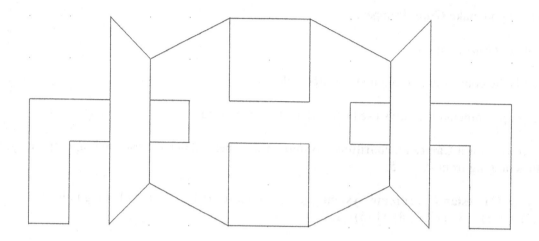

Exercise #2 – Open a new file and set units to Architectural. Then draw the following figure using the Dynamic Input (DYN) drawing aid. When done, use the area command to find its area.
(Difficulty level: Easy, Time to completion: 5 minutes)

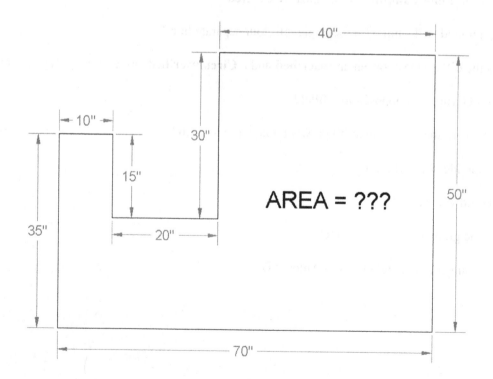

Exercise #3 – Open a new file and set units to Architectural. Draw the following shapes using the manual Relative Distance Entry (the @ sign) method. The first number is always the X, the second is the Y. Repeat the exercise using the DYN method instead.
(Difficulty level: Easy, Time to completion: 5 minutes)

24" × 36", 17.5" × 3", 12" × 12", 3" × 40"

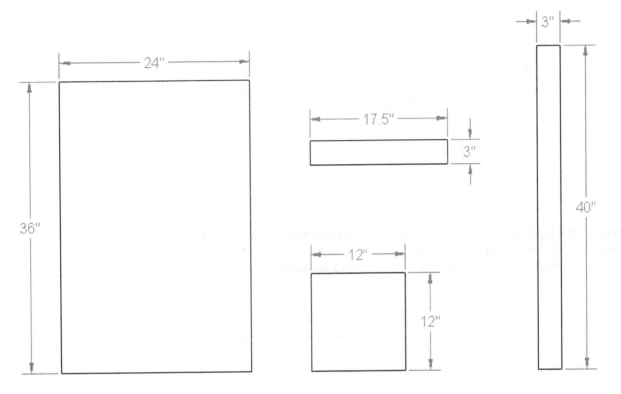

Exercise #4 – Open a new file and set units to Architectural. Draw the following figure using basic circle, line, offset and fillet commands.
(Difficulty level: Easy, Time to completion: 5 minutes)

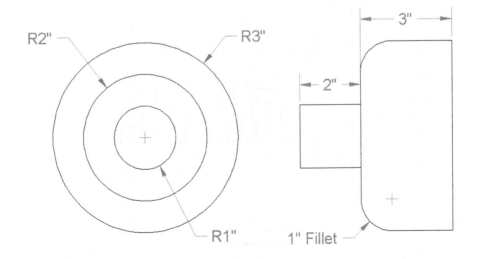

Exercise #5 – Open a new file and leave the units as Decimal. Draw the following simplified nut and bolt diagram. You have all the needed information, but may have to carefully project some parts of the side view. You will need the polygon, circle, line, trim, chamfer and other commands.
(Difficulty level: Intermediate, Time to completion: 10-15 minutes)

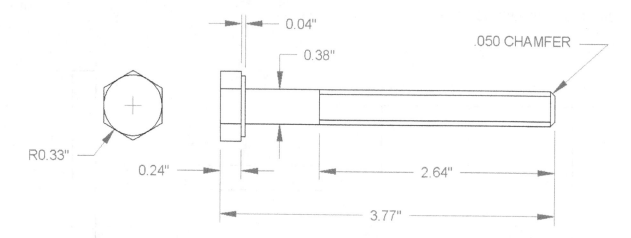

Exercise #6 – Open a new file and set units to Architectural. Draw the following bookshelf. The spacing of the shelves and the size/rotation of the books can be anything you want. All dimensions are for reference only.
(Difficulty level: Intermediate, Time to completion: 15 minutes)

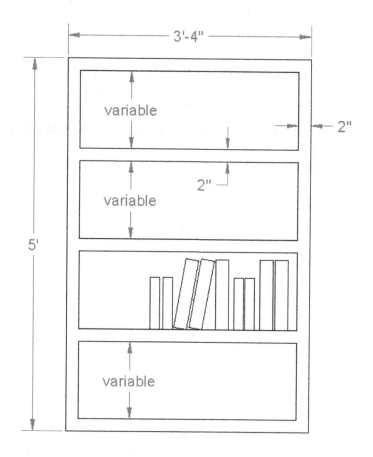

LEVEL 1

CHAPTER 3

Layers, Colors, Linetypes and Properties

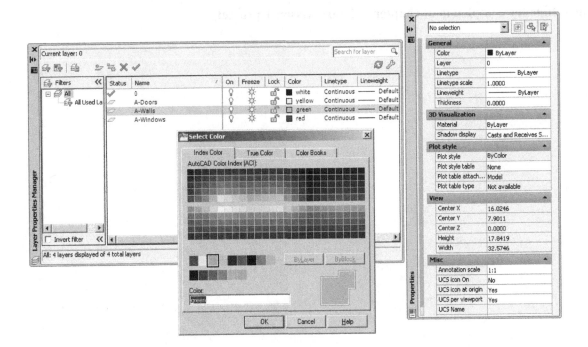

Chapter 3

Learning Objectives

Layers are an essential concept and need to be introduced before a serious drawing is attempted. In this chapter we will introduce layers and discuss the following:

- What are layers?
- Why use them?
- Creating and deleting layers
- Making a layer current
- Assigning colors:
  - Index color
  - True color
  - Color books
- Layer Freeze/Thaw
- Layer On/Off
- Layer Lock/Unlock
- Loading and setting Linetypes
- Properties Palette
- Match Properties
- Layers Toolbar

Here you will have learned all the remaining tools to begin creating a floor plan which we will start at the end of the chapter.

Estimated time for completion of chapter: 3 hours (lesson + project)

Sec 3.1 – Introduction to Layers

Layers will be our first major topic after the previous chapters of basics. It is a very important concept in AutoCAD, and no designer or draftsman would ever consider drawing anything more complex than a few lines without making use of them.

> ➢ **What are layers?**

"Layers" is a concept that allows you to group drawn geometry in distinct and separate categories according to similar features or a common theme. This allows you to exercise control over your drawing by (among other things) applying properties to those groups such as assigning colors, or making them visible or invisible for clarity.

So, for example, all the *walls* of a floor plan will exist on the "Walls" layer, all the *doors* on the "Doors" layer, and *windows* on the "Windows" layer and so on. You can think of layers as transparency sheets in an Anatomy textbook, with each sheet showing a unique and distinct category of data, (skeleton, muscle, circulatory system, etc.) but when viewed all at once show the complete picture of a person. Similarly in AutoCAD there can be any number of layers (often the problem is using too many) with all of them holding their respective data and combining into the complete design.

> ➢ **Why use them?**

As stated above, the key word here is "control". Each layer can be assigned certain properties such as color or linetype, or it can be manipulated to make it visible or invisible, as well as locked or unlocked to prevent editing. If all your geometry was lumped together on one layer, as some beginners do, then whatever you do to one set of objects, you will do to all; certainly not what you want.

Let's take a look at the Layers Properties Manager. Open a new file and type in **layer** or just **la**, and here's what you'll see (Figure 3.1).

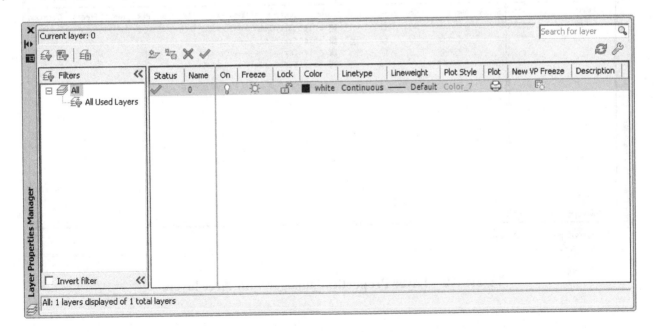

Figure 3.1 - AutoCAD 2010 Layer Properties Manager

Let's take a look at all the layer features one by one.

> **Creating and deleting layers**

Each new layer needs to be initially typed in according to what you need or the company standard. To do that you need to highlight (click on) the layer 0 grey band. When it turns blue, simply press Enter once and the blue ribbon bar will jump to another line and create Layer 1. Another way to get a new layer is to press a button at the top of the layer box (two icons to the left of the red "X") that looks like a sheet of paper with a sparkle next to it. Either way, once you do that, just type in the layer name followed by Enter. Some helpful tips:

- If you noticed that you made a spelling mistake after you already pressed Enter, you can just click back into the layer name once or twice until the cursor enters edit mode and retype.

- Do not attempt to change the "0" layer's name. It is the default layer and cannot be deleted or renamed.

- If you want to delete a layer (perhaps an accidentally made one), simply highlight the layer, click the red "X" and then Apply at the bottom right of the layer box. If the layer contains nothing it will be deleted.

Following the above procedure create three layers:

- A-Doors
- A-Walls
- A-Windows

Here's what you should see:

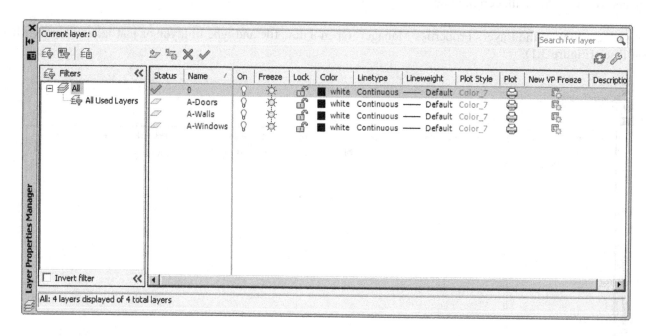

Figure 3.2 – Layer Properties Manager with three layers

➢ **Making a layer current**

This is simply the active layer that you are able to draw on at any given moment. You are always on *some* layer (even if you're not immediately sure which one exactly). So logically if you are about to draw a wall, you should be in the A-Walls layer and so on. To specify what layer to have as current either click the green check mark button (to the right of the red "X") or **double-click** the layer name itself. Go ahead and make A-Walls current.

➢ **Assigning layer colors**

Assigning colors to layers is very important, and should be done as you create them or soon after. The reasons to have color are two-fold. On the most basic level, a colorful drawing is more pleasant to look at on the screen, but more importantly it is easier to interpret and work with. If, for example, your walls are green, and you keep this in mind, then your eye will be able to quickly see, and assess the design. Likewise, yellow doors will be easy to spot. This "color separation" greatly aids in drafting. Imagine a scenario where your building floor plan is made up of all white lines; you wouldn't be able to tell one line from another. As that drawing grew more complex, they would all melt into one monochrome mess! Finally, colors will be assigned pen thicknesses (an advanced topic), and will contribute to more professional looking output. We will address this in Level 2.

To assign colors, let's turn our attention back to layer dialog box. Highlight, by clicking once, layer A-Walls. Now follow the blue ribbon across until you see a box with the word "white" next to it. It will be under the "Color" column. Click it, and the Select Color dialog box pops up as seen in Figure 3.3.

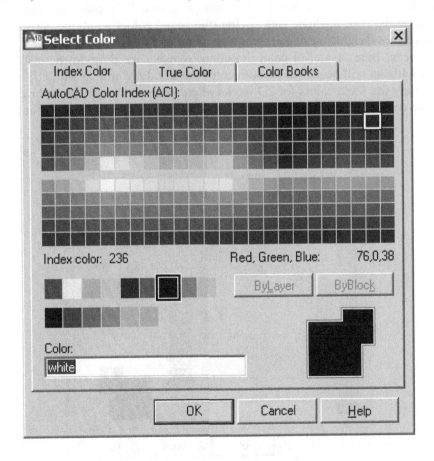

Figure 3.3 –Select Color, Index Color

> **Index color**

Let's take a tour of this Index Color tab. The two large 24 × 5 stripes of colors you immediately see at the top is the **AutoCAD Color Index (ACI)** and it contains 240 colors. Don't be too quick in selecting one of those right away. Take a look below; there are two smaller color stripes. The rainbow-like stripes #1 through #9 are the ones we're most interested in. If you take a close look, they are the "primary" colors and the 240 colors above them are just different shades of the primaries. You should always use up those nine colors first; they are all very distinct from each other, and only then move on to the others. Also this makes setting pen weights easier later down the road, as you won't have to scroll through dozens of colors to find the ones you used.

The colors from left to right are Red→ Yellow→ Green→ Cyan (not Aqua)→ Blue→ Magenta (not Maroon), then white and two shades of grey, #8 and #9. The color stripe below that are just additional shades of grey and we won't need them for now. So finally pick a color for the A-Walls layer (green is what many architects use for walls), click it, then press Enter, and you are taken back to the Layer dialog box to assign more colors to more layers. Yellow is recommended for A-Doors and red for A-Windows.

> **True color**

There's more to colors of course. Go back to the Select Color dialog box. Now click the second tab called True Color. Here's where some experience in graphic design may come in handy. The main color window you are looking at now contains over 16 million colors, and should be familiar to you if you used Photoshop®, Illustrator®, PaintShop® or any of those types of design programs. You can now move your mouse around selecting colors based on the HSL method. That is Hue (the actual colors – moving across), Saturation (moving up or down), and Luminance (brighter or darker – moving slider up or down next to main window). You may also use the RGB (Red, Green, Blue) model of color selection by using the drop-down selector at the upper right. Whichever color you settle on will be shown by both color models. Play around with all the buttons to get a feel for them. Designers rarely use these colors for 2D work (where you want objects to be bright and visible), but do pick these more subtle colors for realistic-looking 3D models. Figure 3.4 is a screen shot of what this tab looks like.

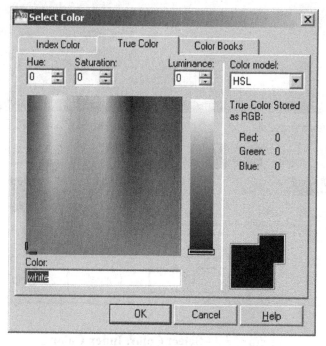

Figure 3.4 – Select Color, True Color

➤ **Color books**

Here's where things get even fancier. Color books are Pantone colors. If you are an architect or an interior designer, you already know what these are. If you're shaking your head and have never heard of Pantone, then don't worry about it, chances are you won't need to use them. Simply scroll through the palettes and select a color. Figure 3.5 is a screen shot of the Color Books tab.

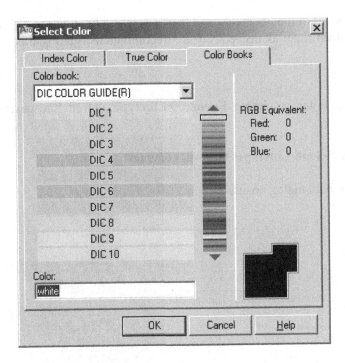

Figure 3.5 – Select Color, Color Books

➤ **Layer Freeze/Thaw and On/Off**

Here we come to a very important part of the layers dialog box. Freeze and Thaw are opposites of the same command, one that will make a layer disappear from view. The command will freeze or thaw the layer for the entire drawing. Pick a layer, highlight it, follow the band across to the "Freeze" column and underneath you will find a yellow "sun" icon. If you click it, that layer will "freeze" and the icon will become a blue "snowflake". Clicking again will undo the action. So what does freeze mean? It will make the layer (or rather all the objects on that layer) disappear from view, but remember, they are NOT deleted, only invisible. You will use this feature routinely in drawings to control what is displayed. The On/Off feature represented by the light-bulb (yellow to blue and back) is essentially the same as Freeze/Thaw, with some subtle (and minor) details so we won't go into them. You can use either method; though stay with Freeze/Thaw just to keep things consistent.

➤ **Layer Lock/Unlock**

This layer command, represented by the lock, does exactly that; it locks the layer so you can't edit or erase it. The layer remains visible, but untouchable. The unlocked layer's icon is blue and looks "unlocked," while a locked layer's lock will turn yellow and change to a "locked" position. Play around with all the buttons we mentioned thus far to get a feel for them. When done, make sure everything is reset (not locked or frozen) and click the "X" in the upper corner of the layer box to close it.

Sec 3.2 – Introduction to Linetypes

Linetypes are the different lines that come with AutoCAD. As a designer, architect or engineer you may need a variety of lines to convey different ideas in your design. Just like in hand drafting, where you created Dashed, Hidden, Phantom and other line types to show cabinets, demolition work or hidden geometry in part design, in AutoCAD you load the lines and use them as necessary.

The idea here is to load them all in the beginning and then assign the linetypes to either a layer or sometimes just a few items. You can also load them as needed, but it's recommended getting this out of the way so you have them at your fingertips.

Step 1
Type in **linetype** and press Enter or choose **Format→Linetype...** from the cascading menu. The Linetype Manager will pop up.
Step 2
Press **Load...**. The **Load or Reload Linetypes** box will come up.

Figure 3.6 is a screen shot of both dialog boxes.

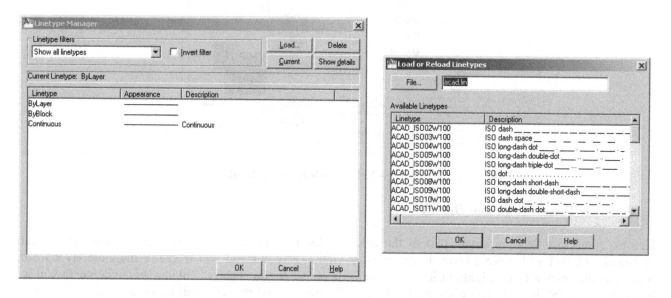

Figure 3.6 – Linetype Manager and Load/Reload Linetypes

Step 3
Position your mouse in the white empty space between the two columns (Linetype and Description) of the Load/Reload Linetypes box and right-click. The **Select All / Clear All** menu will pop up.
Step 4
Press **Select All** and every linetype in the left column will turn blue (selected).
Step 5
Press OK (and possibly a **Yes to All** as well) and the dialog box will disappear. All the linetypes are loaded now so press OK again in the Linetype Manager and that's it.

We will now use the loaded linetypes and assign a linetype to a layer.

Step 1
Go back to layers dialog box.
Step 2
Create a new layer called *Hidden* and let's assign a linetype to it (the "hidden" linetype naturally).
Step 3
Make sure the Hidden layer is highlighted with the blue bar and move your mouse to the right until it matches up with the header category Linetype. It should say Continuous right now.
Step 4
Click the word "Continuous" and the "Select Linetype" (Figure 3.7) dialog box will come up with all your linetypes pre-loaded. If we didn't load them earlier, the box would have been practically empty and we would have had to do the loading at that point.
Step 5
So now all we have to do is scroll up and down the list to select the appropriate linetype (Hidden in this case), select it, and then press OK.

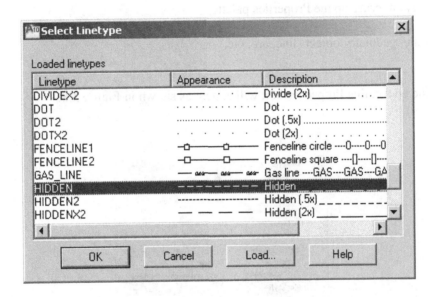

Figure 3.7 – Select Linetype

Make a note of the important linetypes for future reference: Center, Dashed, Divide, Hidden and Phantom are some of the key ones. Ignore the ones named ACAD_ISO… for the simple reason that their names are not descriptive enough to recognize at a glance. Also make a note of why there are three types of the same linetype (such as DASHED, DASHED2 and DASHEDX2). By checking the appearance carefully, you'll see that the scale is different. Finally let's get back to the layers dialog box. Now if you set your HIDDEN layer to Current, then everything you draw will now be a dashed line.

We'll conclude with one final important note on linetypes. If you cannot see the linetype, and it appears as a solid line when it should be something else, chances are the scale is too small. Type in **ltscale** (linetype scale) and enter a larger number. This is mentioned again as a reminder in a future tip.

Sec 3.3 – Introduction to Properties

Sometimes there is a need to change objects from one layer to another, (perhaps because of an error in originally placing it on the correct layer, or a later decision to change to another layer). There may also be a need to change the color, linetype or other properties of objects. AutoCAD has a collection of tools to change properties of objects, with the three most common methods being the Properties palette, the Match Properties tool and the Layer toolbar. We will explore each in order.

> **Properties Palette**

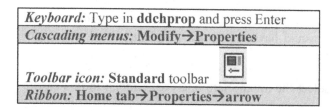

Keyboard: Type in **ddchprop** and press Enter
Cascading menus: **Modify→Properties**
Toolbar icon: **Standard** toolbar
Ribbon: **Home tab→Properties→arrow**

There are two other ways to bring up the Properties palette:

- Double-click any geometry object (line, arc, etc.) or
- Press Ctrl+1

Once you do any of the above, this Properties box will appear as shown in Figure 3.8.

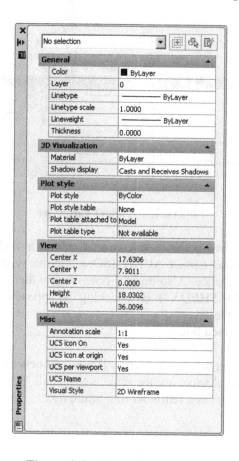

Figure 3.8 – Properties palette

Look at the palette in detail from top to bottom. There are a number of categories (General, View, etc.), and a number of features that can be modified in those categories (Color, Layer, Linetype, etc.). Also there are two columns, a grey one on the left and a mix of grey (inactive) and white (active) fields on the right. To use this palette first draw several lines on your screen that are each on a different layer. Then follow the steps below.

Step 1
First make sure you have selected one or more items to change. They will become dashed. If you selected two different items, only those properties that can be changed commonly to both of them will be available.
Step 2
In the Properties box, find the category you want changed; color for example.
Step 3
Click on the right-hand column directly across from that property.
Step 4
A drop-down arrow will show. Click it and view the choices within that category.
Step 5
Select the different property that you want the new objects to have, such as a new color.
Step 6
Close the Properties box (the "X" in the upper corner) and press Esc once or twice and the objects will have that property. The Properties toolbar is a quick and easy way to change a number of properties but there is another way shown next.

> **Match Properties**

This is a really great way to very quickly match up the properties of one object with the properties of another. The procedure is simple. Start the command (methods in table below) then select the first object whose properties you like, and then click on another object that you want to have those properties.

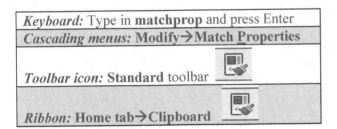

Keyboard: Type in **matchprop** and press Enter	
Cascading menus: **Modify→Match Properties**	
Toolbar icon: **Standard** toolbar	
Ribbon: **Home tab→Clipboard**	

Match Properties will match almost anything: layers, colors, linetypes, linetype scales, text sizes, hatch patterns and many more. Finally we have the layers toolbar, which has its own method of layer changes, and deserves its own separate discussion.

> **Layers toolbar**

The layers toolbar is shown in Figure 3.9. Add it to your visible toolbars on the screen, and use it throughout the remainder of the chapters.

Figure 3.9 – Layers toolbar

To change layers using this toolbar simply select whatever you would like to change to another layer (what you selected will become dashed), then drop down the layer menu in the toolbar and pick the new layer you want the highlighted geometry to go on, and it will change. You may have to press Esc to deselect the layer(s) afterwards.

This toolbar is also useful for freezing, turning off and locking layers as seen in Figure 3.10. Also it is a great visual reference to determine what layer is current, which will of course be the layer that is being shown. The toolbar also has some useful icons, though the only one we covered so far was the one all the way to the left that brings up the Layer Properties Manager.

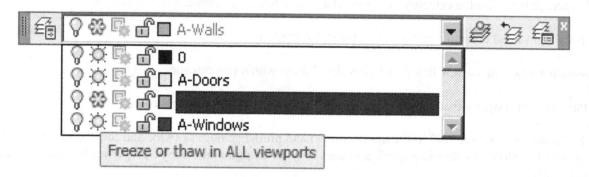

Figure 3.10 – **Layers toolbar in use**

This is it for basic layers and associated topics. You should know the following:

- What layers are and why they are needed
- How to open the layer dialog box
- How to enter a new layer, delete it and change its name
- How to make a layer current
- How to assign a color to a layer, and all the color options
- How to Freeze/Thaw, Off/On and Lock/Unlock the layers
- How to select and set Linetypes
- How to use the Properties dialog box
- How to use Match Properties
- How to use the layers toolbar

What we haven't covered yet:

- The left side of the Layer Properties Manager (Filters and Layer States) – to be covered in Level 2
- Several other Layer features – Lineweights, Plot Style, Plot, etc...

For now it is time to put everything to good use and draw a full apartment floor plan. Review everything thus far and begin the assignment on the following page.

In Class Drawing Project - Floor Plan Layout

This will be your first major drawing assignment. You will now get a chance to put together everything you learned so far and apply it to a realistic drawing situation. Although this floor plan is very simple compared to major architectural or engineering projects created in the industry, it still features components of a professional drawing, just fewer of them, so you don't get overwhelmed right away. Follow all instructions carefully and proceed slowly. Speed will come later when all the commands become second nature. Our goal is to draw the floor plan shown in Figure 3.11.

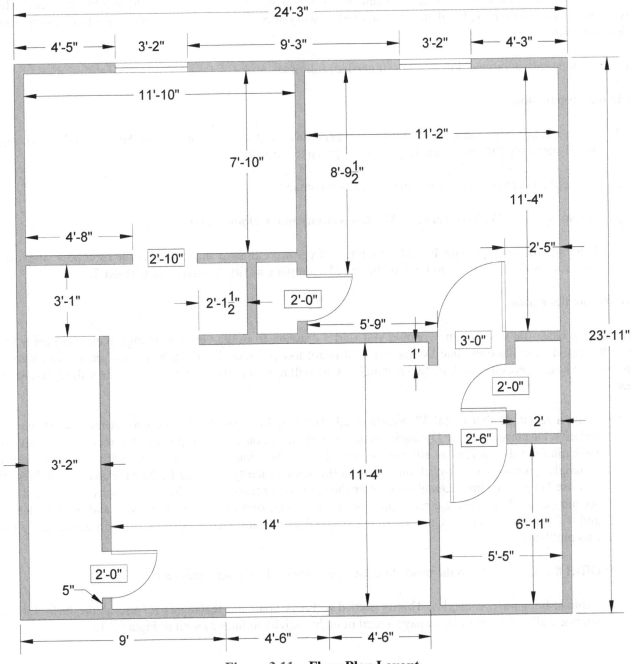

Figure 3.11 – Floor Plan Layout

It is a simple one-bedroom apartment on the second floor (hence no apparent door to get in; you walk up the stairs from the first floor on the left). The kitchen is on the upper left, bedroom on the upper right, bathroom on the lower right and the living room right in the middle. There are three closets (one over the stairs), three windows and five doors (kitchen has no door). The dimensions are only for you to look at and use, not draw in yet as we haven't learned them yet. The numbers in the boxes are the door opening sizes. All walls are 5in thick.

Outlined below is a complete discussion on how to start drawing the floor plan, intended for those students that are studying AutoCAD by themselves without an instructor. If you are taking a class your instructor will assist you but should essentially follow the same format. Once you learn all the steps involved and build confidence, aim to draw it again in less than 60, then 30 and finally less than 10 minutes, starting from a blank screen. This may sound unreasonable now, but that is the required industry speed and you will be able to get to that quite soon with some practice.

Outline of steps with additional information (if needed) follows.

Basic file preparation:

1) Open a brand new AutoCAD file using the *acad.dwt* template. Save it as *YourName_AptProject1.dwg*. We'll start from the very beginning again for practice purposes.

2) Set your units to Architectural using the **units** command.

3) Set up layers: **A-Walls (Green)**, **A-Windows (Red)** and **A-Doors (Yellow)**.

4) Load all your linetypes (we'll need them later). If you were to save and close this file right now, and use it later for another project, then it would be called a template as first explained in Chapter 2.

Starting the floor plan:

Now that the basic file is set up, how would you start a drawing from the very beginning? Well there are a few ways. We could certainly draw line by line, but that is not too efficient. Think back to your set of tools learned thus far. The best approach is to draw two rectangles. One will represent the outer walls and one will represent the inner.

5) Draw a rectangle that is **24'-3"** wide and **23'-11"** tall. Be careful! Use the correct method of manual Relative Distance Entry or Dynamic Input; your choice. You can even just use the rectangle command, indicating the dimensions as allowed by one of the sub-options. If you do it manually, you will start a rectangle, press Enter, left-click anywhere in the screen then type in @24'3,23'11 exactly as written. If you use DYN, start the rectangle and enter the values on screen, using Tab to hop between the first and second values. If you use the rectangle dimension option, start the rectangle, press **d** and enter its length and width when prompted. Once again, make sure the units are Architectural or you will immediately run into problems.

6) Offset the rectangle **5"** to the inside to create the inner walls (Offset command).

7) Explode both wall rectangles. This is needed to break apart the rectangles for use as the basis for the interior walls. You should now have a total of eight individual lines as seen in Figure 3.12.

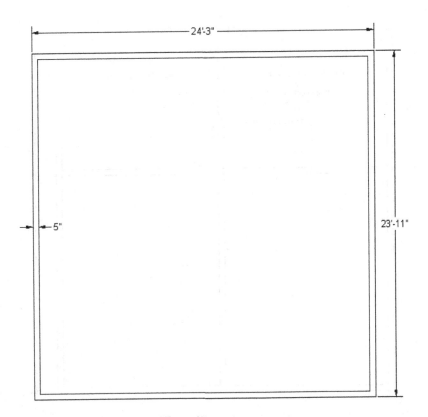

Figure 3.12 – Floor Plan, Step 7

Drawing the inner wall geometry:

As mentioned in Chapter 1, one of the biggest misconceptions beginners have about drafting in AutoCAD is that they will spend lots of time drawing line after line. That expectation is natural, after all in hand drafting that is exactly what you would do - pick up the pencil and T-square and draw lines connecting them in intricate ways to create a design. Not so in AutoCAD. You would still draw lines occasionally, but most of the time you would use the **offset** command to create new geometry such as in this case here. For example, we will use the inner main walls as basis for the room walls and offset as needed.

8) Offset inner left vertical wall 11'-10" to the right to create part of the kitchen wall. Then right away offset 5" for thickness of wall. Repeat with inner top horizontal wall (offset 7'-10") to get the lower part of kitchen. Offset 5" right away for thickness. Figure 3.13 illustrates this graphically.

9) Fillet the two sets of walls to create a sharp corner. This is where the fillet command's advantage really shows. Follow it carefully and select the two inner wall lines first, then the outer.

10) Create a 2'-10" opening for the kitchen door. Once again use the offset command. Offset the inner wall 4'-8" to the right as shown on the plan. Then offset again 2'-10" for the door itself. Then trim out an opening by selecting *all* the lines using a crossing, pressing Enter, and clicking the lines you don't need. You will later use this technique to create the windows the same way.

11) Trim out the extra line where each inner wall meets the main wall. Figure 3.14 shows you what you should see so far (be sure to ask your instructor if something isn't working out).

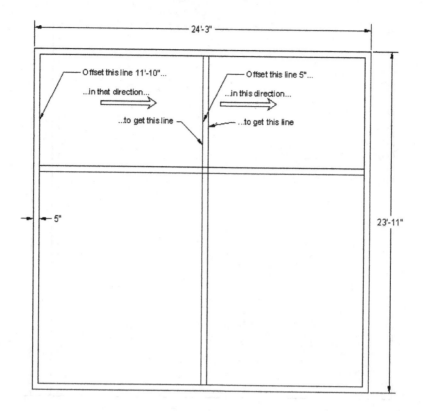

Figure 3.13 – Floor Plan, Step 8

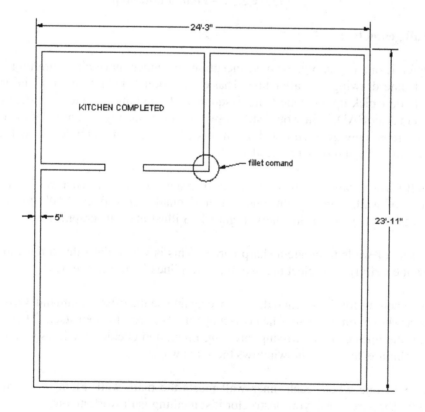

Figure 3.14 – Floor Plan, Step 11

Drawing the rest of the inner walls:

If you have understood the previous steps, then you should have a basic idea of what is needed to draw the rest of the floor plan. It really is mostly similar. You will of course have to get creative on occasion, but just stop and think of the basic tools that you have available to you. If it seems overly complicated, you are doing it wrong. After the kitchen go on to do the bathroom in the lower right corner, then the bedroom and stairs. Here are some additional tips to help you along:

- Create the openings for the doors and windows, but do the actual doors and windows last. Be sure to use the correct layers!
- To create a door, draw a rectangle of the appropriate size (all doors will be 2 inches thick), put it in its proper place and add an arc for the door swing.
- To create windows draw a rectangle with a line through it. Position into the window opening.

Some additional discussion on how to create the doors and windows is presented next.

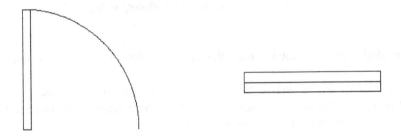

Figure 3.15 – Door and Window Symbols

The easiest way to create a door is to draw a rectangle of a needed size. The door sizes will vary somewhat throughout the floor plan but the thickness will always be 2". Start the rectangle command and click once where the door will go (use OSNAP accuracy). Then relative to that point determine what to type in using Relative Distance Entry or Dynamic Input. Be careful as negative numbers may be involved.

So in this example here, a 36" door is created by placing a rectangle at the upper right corner of the wall and typing in **@-2,36**. Make a note of why it is a negative 2. If you are using Dynamic Input, you still have to use negative numbers; you just type on screen and get out of needing the @ sign, but the rest applies.

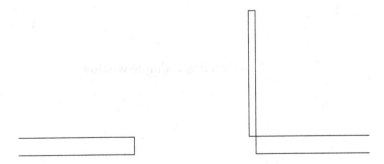

To create the door swing, you can carefully position an arc or as an alternative create a circle based at the hinge point and a radius of 36" as seen in Figure 3.16.

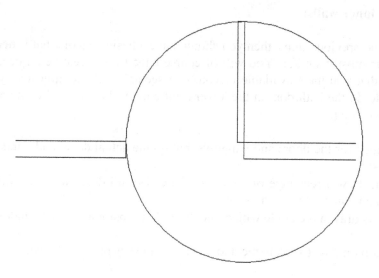

Figure 3.16 – Creating a door swing

Then just trim the unneeded ¾ of the circle by using the top of the door and the left wall as cutting edges.

The window is much easier to do. As stated before, just draw a rectangle in the window opening and add a straight line across, then insert into the opening (Figure 3.17). Once again it bears repeating, for both the doors and the windows, be sure to be on the correct layer for each.

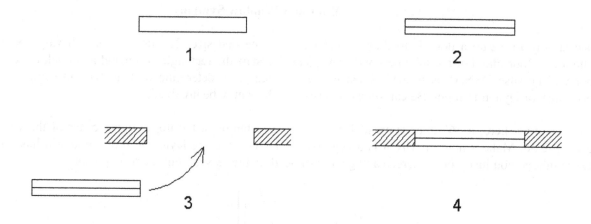

Figure 3.17 – Creating a simple window

Chapter 3

Summary

You should understand and know how to use the following concepts and/or commands before moving on to Chapter 4:

- **Layers**
  - Basic concept and purpose
  - Creating and deleting
  - Current layer

- **Layer Colors**
  - Setting and changing colors
  - Index Color
  - True Color
  - Color Books

- **Layer Freeze/Thaw**

- **Layer On/Off**

- **Layer Lock/Unlock**

- **Linetypes**
  - Loading and selecting

- **Properties Palette**

- **Match Properties**

- **Layers Toolbar**

Chapter 3
Review Questions

Answer the following based on what you learned in Chapter 3.

1) What is a **layer**?

2) What are some of the main reasons for **using layers** in AutoCAD?
(list at least two)

3) What typed command brings up the **layers dialog box**?

4) List two ways to **create** a new layer

5) How do you **delete** a layer?

6) How do you make a layer **current**?

7) What are the three available **color tabs**?

8) What does **freezing** a layer do?

9) What does **locking** a layer do?

10) What are some of the **linetypes** mentioned in the text?

11) What are three ways to bring up the **Properties toolbar**?

12) What are the other two ways to **change properties or layers** that were discussed?

Chapter 3

Exercises

Exercise #1 – In a new file, load all linetypes, then open the layer dialog box and set up the following layers, colors and linetypes. Then lock all E-layers, make A-Walls-New current, and freeze A-Text.
(Difficulty level: Easy. Time to completion: 7-10 minutes)

A-Walls-Demo, Grey, Hidden
A-Walls-New, Green, Continuous
A-Windows-New, Red, Continuous
A-Doors-Demo, Grey, Hidden
A-Doors-New, Yellow, Continuous
A-Shelves, Blue, Hidden
A-Furniture, Magenta, Continuous
A-Text, Cyan, Continuous
A-Dims, Cyan, Continuous
A-Carpeting, Grey, Continuous
E-Outlets-Duplex, Red, Continuous
E-Switches, Red, Continuous
E-Switches-3way, Red, Continuous

Exercise #2 – In a new file, leave units as Decimal, load all linetypes, set up layers M-Bracket, Hidden, and Center and assign them proper linetypes and colors of your choosing. Then draw the following mechanical object.
(Difficulty level: Easy. Time to completion: 12-15 minutes)

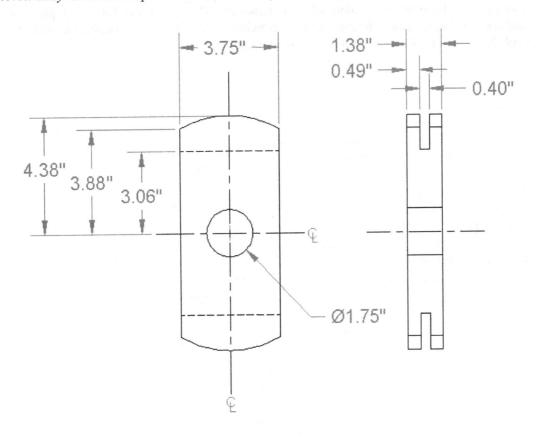

Exercise #3 – In a new file, leave units as Decimal, load all linetypes, set up layers M-Part, Hidden, and Center and assign them proper linetypes and colors of your choosing. Then draw the following mechanical object. All dimensions and text are just for your reference, and the drawing is slightly over-constrained to assist you.
(Difficulty level: Intermediate. Time to completion: 20-25 minutes)

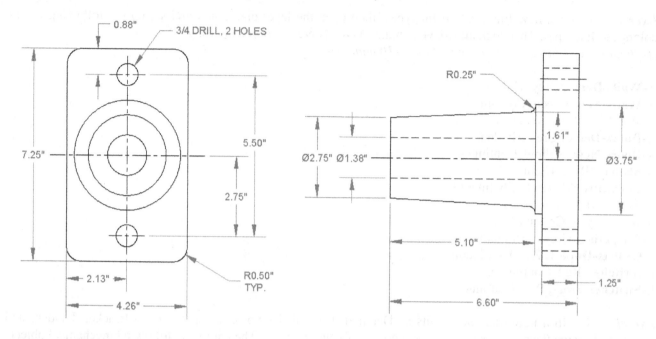

Exercise #4 – In a new file, leave units as Decimal, load all linetypes, set up layers M-Part, Hidden, and Center and assign them proper linetypes and colors of your choosing. Then draw the following pipe assembly. All dimensions and text are just for your reference, and the drawing is slightly over-constrained to assist you.
(Difficulty level: Intermediate. Time to completion: 20 minutes)

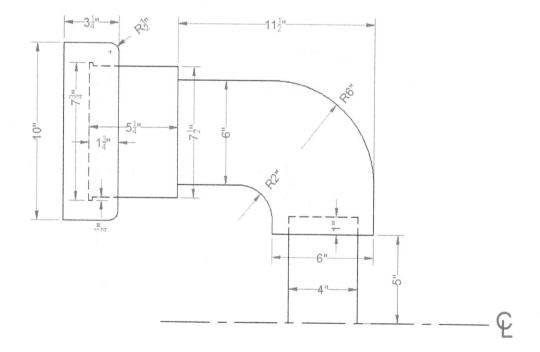

Exercise #5 – In a new file, set up Architectural units, the correct layers (only A-Walls and A-Doors are needed), and draw the following floor plan. Dimensions are only for your reference.
(Difficulty level: Intermediate/Advanced, Time to completion: 60 minutes)

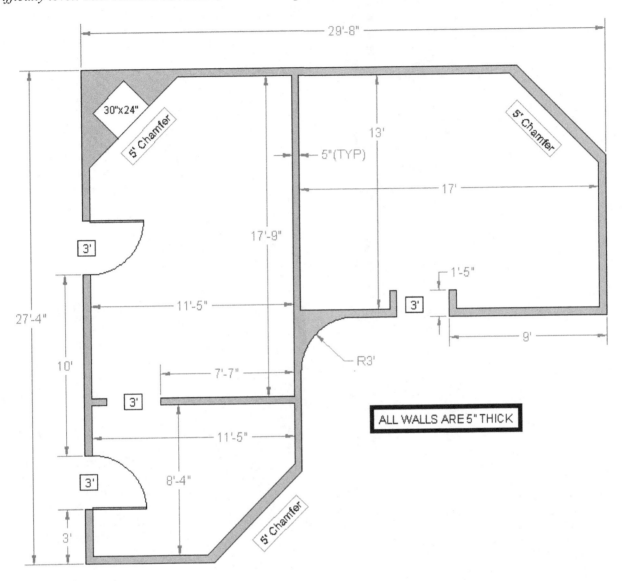

LEVEL 1

CHAPTER

Text, Mtext, Editing and Style

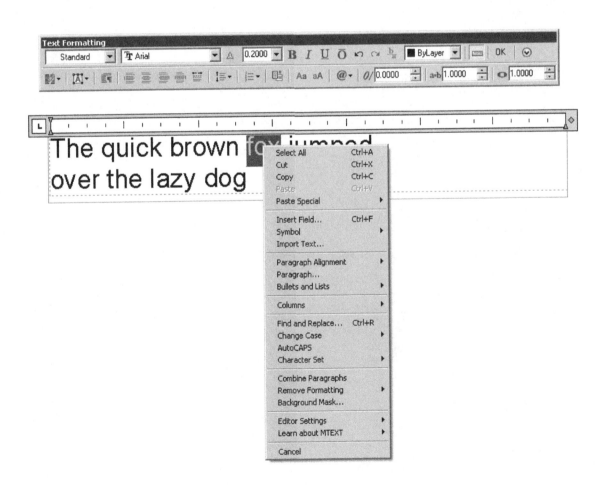

Chapter 4

Learning Objectives

Text allows your design to communicate beyond what just a line drawing can. In this chapter we will introduce Dtext, Mtext, Style and Editing and discuss the following topics:

- Text
- Properties and applications of Text
- Editing all types of text
- Mtext
- Properties and applications of Mtext
- Mtext formatting
- Mtext symbols
- Style
- Spell check

At the end of this chapter you will be able to annotate your floor plan as well as add additional features such as furniture and stairs.

Estimated time for completion of chapter: 2 hours (lesson + project)

Sec 4.1 - Introduction to Text and Mtext

Adding text to your AutoCAD drawing is the next logical step once you learned how to create and edit basic designs. After all, the purpose of most drawings is to describe how to build something or show what a design looks like. Text goes a long way in assisting in this, and along with dimensions is usually the next item to be added to a new design.

Text in AutoCAD comes in two versions: regular **text** (single line text) and **mtext** (multi-line text). We shall look at how to create and edit both types. We'll then take a look at how to choose and set fonts, and conclude the chapter by adding text to the previously designed floor plan.

Sec 4.2 - Text

The first type of text command we'll cover is regular text. This is your basic text creation command, and it creates a text field anywhere you click on the screen, into which you can type whatever you need to. It does have a carriage return that goes to another line upon pressing Enter, so you need to press Enter twice to get out of the field. While an unlimited number of lines of text can be typed, typically regular text is used for only a single line as multiple lines of text are not joined together in paragraph form and cannot be formatted to any extent. Let's summarize:

- Text is used mostly when only one or several lines are needed.
- The text cannot be formatted or any effects added, beyond underlining and few other minor effects.
- Multiple lines of text are not in paragraph form, rather they remain individual lines.

Open a new file, bring up the Text toolbar (Figure 4.1) and let's create a few sample lines.

Figure 4.1 – Text toolbar

Keyboard: Type in **text** and press Enter
Cascading menus: Draw→Text→Single Line Text
Toolbar icon: **Text** toolbar A_I
Ribbon: **Home tab→Multiline Text→** A_I

Step 1
Begin the text command via any of the above methods
- AutoCAD will say:
  ```
  Current text style: "Standard" Text height: 0.2000 Annotative: No
  Specify start point of text or [Justify/Style]:
  ```

Step 2
Left-click anywhere on the screen.
- AutoCAD will say: `Specify height <0.2000>:`

Step 3
Enter a new height if desired, say **1.0**, and press Enter.
- AutoCAD will say: `Specify rotation angle of text <0>:`
You can rotate the text, but there's really no need to, so just type in **0** for now, and press Enter.

Step 4

A text field will open up with a blinking cursor. Go ahead and type something, pressing Enter to go to the next line or twice to finish. Here is what the basic text will look like (Figure 4.2). If yours is a different size, you may have to zoom in or out to fit it nicely on the screen.

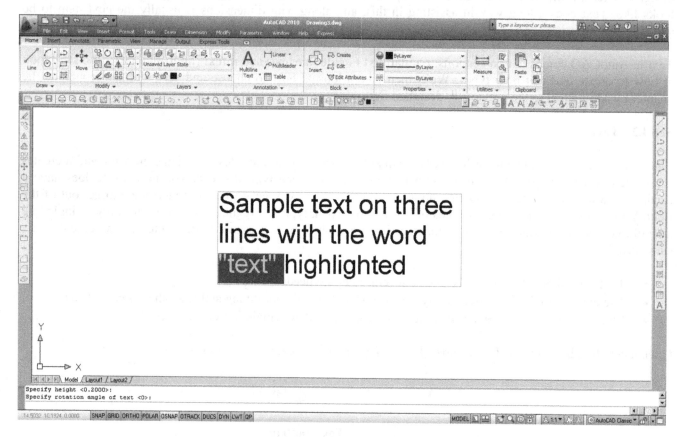

Figure 4.2 – Basic Text

The text is still rather simple; we haven't yet addressed font, sizing, or any effects. Notice that the text is three separate lines. You can move, copy or erase any of the two lines without affecting the other. So how would you edit this text?

> ➤ **Editing Text**

Editing, or changing, what the text says is very simple, and is the same procedure for both text and mtext, so we will go over it right away and you can apply it from now on. The easiest way to edit the text is to *double click* on it, which will open up the text field. This is such an overriding and best method that we won't present a command matrix, as the only other way is to type in **ddedit** (or use the appropriate "Edit…" icon – see Figure 4.3), but this method is rarely used. However, remember ddedit for Chapter 6! It is the only way to change dimension text values, so you *will* need it eventually, just not yet. When you're done editing, just press Enter again.

Figure 4.3 – Edit… icon

Here's a trick with regular text that you won't hear about much anymore. If you want to underline it, type in **%%u** just before the text string you want underlined. This old code dates back to DOS days, and it will automatically and immediately add an underline to your basic text. You can also experiment with **%%c**, **%%p** and **%%d**. See what those three do. None if this is needed with mtext, as discussed next.

Sec 4.3 - Mtext

Mtext is multiline text. In other words, this is the command you use when you anticipate typing in a paragraph as opposed to one or two lines, and more importantly you require some advanced formatting and effects. As of AutoCAD 2010 it even has a spell-checker built in. Many designers use mtext for all their text needs, just in case formatting needs to be applied to even just one word. The only downside is that it takes a few seconds longer to set up and the features make it a more complex command. The idea here is to define an area where your text will go. Once you do that the new paragraph will fit into that area.

Let's summarize the mtext features and try the command.

- Used primarily for writing paragraphs
- Has extensive formatting and editing features
- Has an extensive symbol library and spell-check
- Can accept extensive text importing

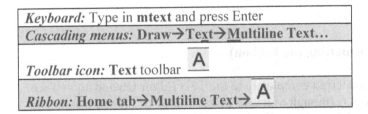

Keyboard: Type in **mtext** and press Enter
Cascading menus: **Draw→Text→Multiline Text...**
Toolbar icon: **Text** toolbar
Ribbon: **Home tab→Multiline Text→**

Step 1
Begin the mtext command via any of the above methods
- AutoCAD will say: `Current text style:   "Standard"  Text height:   0.2000`
                    `Annotative:  No`
                    `Specify first corner`

Step 2
Left-click anywhere on the screen, noticing the "abc" next to the crosshairs. A rectangle with an arrow will appear as shown in Figure 4.4.

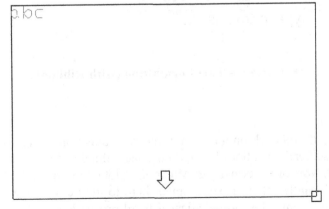

Figure 4.4 – Mtext field

Continue to move your mouse down and across to the right, making the rectangle bigger. This is your text field; make it as large or small as you need it to be to hold all the text. As you're doing this,

- AutoCAD will say:
    ```
    Specify opposite corner or [Height/Justify/Line spacing/Rotation/
    Style/Width/Columns]:
    ```
Click again when you have defined the field.

Here AutoCAD 2010 throws you a bit of a twist. If you are ***not*** using the Ribbon, you will see the text field with a toolset just above (but not attached to it) as seen in Figure 4.5, where you can type in your text, pressing Enter to go on to the next line. To finish up, press OK in the upper right of the toolset or just click anywhere outside the field. We will go over some of the tools in a moment, but what if you have the Ribbon up?

Figure 4.5 – Text Formatting (no Ribbon)

If you are using the Ribbon, the formatting duties are transferred completely to the Text Editor tab and all you see below it is the text field, no toolset, as seen in Figure 4.6 (though you *can* bring it back as shown later). The functionality is essentially the same, just the presentation differs. We will cover mtext's extensive formatting tools in the next section.

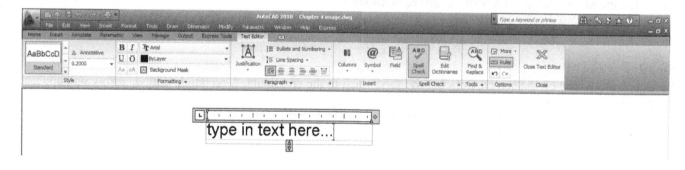

Figure 4.6 – Text Formatting (with Ribbon)

> **Formatting Mtext**

First of all, notice something: if you click on mtext, it remains a paragraph, a big difference from a few lines of regular text. To edit mtext, as already mentioned, you do the same thing as for regular text; that is double click it, use the icon or type in ddedit. Now on to formatting! Mtext has a lot of additional features, and once you input the text, you can modify it significantly. There is some intent here to mimic MS Word's text editing abilities, and although AutoCAD can't really come close to a dedicated word processing program, the available tools are still quite extensive. We will take a look at the non-Ribbon toolset first and cover them again (in less detail since you already will know them) using the Ribbon's Text Editor.

Looking at the Text Formatting tool bar from left to right as shown below, we have:

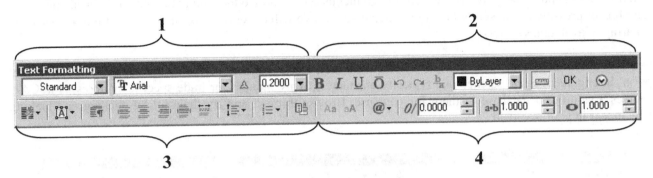

Group 1

This is where you would set *name*, *font* and *size*, although this is generally done using *style*, to be discussed later. You can also make the text annotative, to be covered in Level 2.

Group 2

These are the standard *bold*, *italic*, *underline*, *overline* and *undo/redo* buttons. This is also where you can set *color*, though that is usually done by layer. The ruler icon will turn off the ruler grid, the OK button closes the field, and the down arrow brings up additional *options*, to be covered soon.

Group 3

Here you will find a tool for making *columns* and some of the associated settings, *paragraph justification* tools as well as tools to change *line spacing*, *numbering* and *fields*.

Group 4

Here you will find tools to change from *upper to lower case (and vice versa)*, *symbols* (to be discussed soon), *oblique*, *width factor* and *tracking tools*.

You may be familiar with most of the above if you ever typed a document in MS Word®. Just like Word, you can highlight the text that you want the changes to apply to. Experiment with the buttons to see what they all do. Notice also how Arial is the default font in AutoCAD 2010.

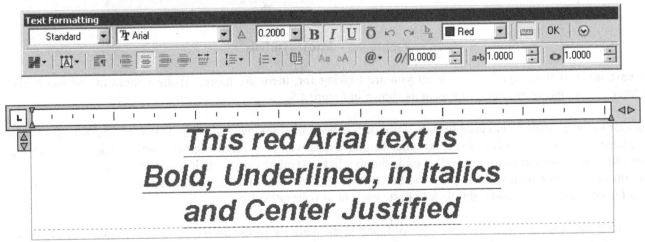

Figure 4.7 – Mtext sample

There is more to the mtext command than just the main toolbar seen above, although it does contain most of the often used commands. There are also two additional menus accessed by either right clicking while inside the mtext box or pressing the @ symbol icon in Group 4 above. We will cover this "@" menu first; it introduces a vast array of available symbols.

Here's what the menu looks like in Figure 4.8. Take a look at the menu closely; this is an interesting and useful option. In engineering and science one typically encounters many industry or trade specific symbols. AutoCAD provides a significant database of them for your use. Browse through and try them all by clicking and making them appear in your text field.

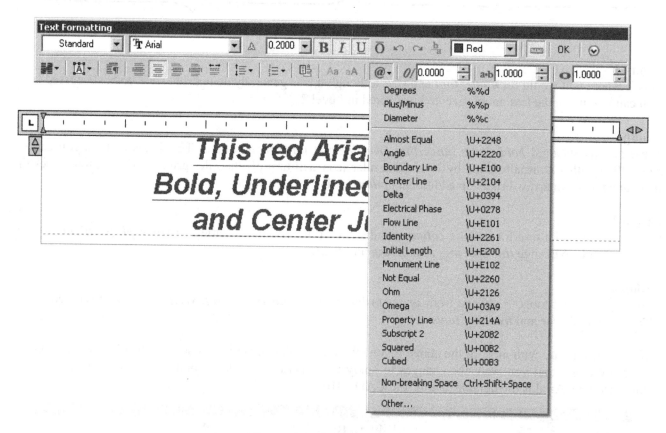

Figure 4.8 – Mtext symbols

In case none of those symbols are what you are looking for, there are more. At the bottom of the menu click "Other…". A *character map* will appear as shown in Figure 4.9.

Similar to the character map used by MS Word, it functions the same way as well. This is basically a database of available letters (in English and several other languages) as well as symbols and characters. The choices change somewhat from font to font. To use this tool, simply click a symbol you want (it will temporarily increase in size so you can examine it closer) and press the Select button. Then minimize the character map box (or close it if you don't need it any more), and right-click to paste back in to your mtext field.

Character maps are generally for symbols not found in the main list of the drop down table (Figure 4.8). If you see the symbol you want there first, use it.

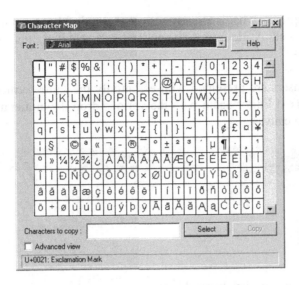

Figure 4.9 – Character Map

To introduce the final menu, right click in the mtext field and you will see the following menu (Figure 4.10).

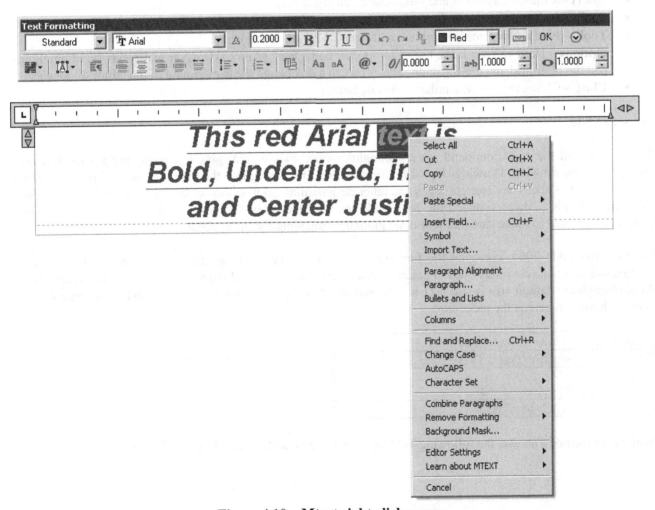

Figure 4.10 – Mtext right-click menu

Some of these menu choices duplicate what is already seen in the main Mtext menu bar, and you can also access all the symbols through here as well. Aside from some cut and paste choices, this menu just presents an alternate way of making option selections.

As mentioned before, Ribbon users are presented with essentially the same formatting tools but in a rearranged manner. Here is the Ribbon's Text Editor again with the various drop-down menus "thumb tacked" (by clicking the tack symbol) on the drawing canvas.

- **Style** – Annotative and size can be changed here.
- **Formatting** – Bold, italics, underline, overline, font, color, background mask (color), oblique, spacing and width factor can be changed here.
- **Paragraph** – Justification, line spacing, bullets and numbering can be changed here.
- **Insert** – Columns, symbols and fields can be changed here.
- **Spell Check** – You can run the spell checker, set dictionaries and settings here.
- **Tools** – Find and replace, import text and AutoCAPS are found here.
- **Options** – Here you can make the editor toolbar come back (AutoCAD never really gets rid of anything), add the ruler grid and use the undo/redo. You can also change the background of the editor to opaque.
- **Close** – Closes the editor, similar to the OK button.

Sec 4.4 - Style

The idea behind the style command is very straightforward. Pick a font, give it a name and a size, and use it throughout your drawing. Drawings typically use only one font throughout the main design, and perhaps another fancier one in the title block area for logos and other designations, so it makes sense to set one style and stick to it. You have already perhaps changed the font while learning the mtext command, but that was only for THAT instance. You need to make sure that font is set globally – meaning for the entire drawing.

Prior to AutoCAD 2009, changing the font right away was a necessity as the default font (Simplex) was an unattractive one that was rarely used in practice. With AutoCAD 2010 and the previous release, the default font is Arial (though the default size is still 0.2), so you may want to stay with this popular font and not change it. If you need to, however, here is the procedure.

Keyboard: Type in **style** and press Enter	
Cascading menus: **Format→Text Style…**	
Toolbar icon: **Text** toolbar 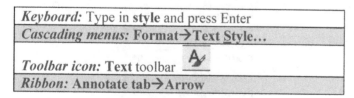	
Ribbon: **Annotate tab→Arrow**	

Whichever method you use, the following Text Style dialog box will appear (Figure 4.11).

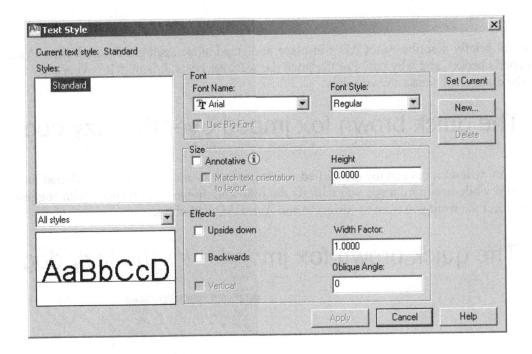

Figure 4.11 – **Text Style**

Taking a look at this dialog box, let's say Arial isn't what we want here, and we would like to set a RomanS font that is 6" high instead. To do this, we need to go through the following three steps.

Step 1
Press the **New...** button and type in the name and size of the font: **RomanS_6**
Step 2
Pick **romans.shx** from the **Font Name:** drop down menu.
Step 3
Highlight the 0.0000 in the **Height** field and type in **6** (no need for the inch symbol). Press **Apply** and **Close**.

All text that you type from now on will be RomanS 6" font. To create another font just repeats these steps. To size your font up and down, you can also create new font styles, but in this case it may be more practical to just use the scale command.

The additional options in the style dialog box are not used that often but review them just in case. They are listed below: All effects can be previewed in real time in the Preview box in the lower right.

- **Upside down** – Flips the text upside down
- **Backwards** – Flips the text backwards
- **Vertical** – Stacks the text vertically
- **Width Factor:** - Widens the text if (>1) narrows text if (<1)
- **Oblique Angle:** - Leans the text to the right if positive (+) and to the left if negative (-)
- **Annotative** – An advanced topic for Level 2

Sec 4.5 – Spell Check

This section will briefly describe AutoCAD's in-place and stand-alone spell-checking abilities. Yes, AutoCAD does have a spell-checker, and why not? For the price the software costs, it better! It is very easy to use, so let's try it out on the misspelled phrase shown below:

The quick brown fox jmped over the lazy dog

The first thing you will notice is that the misspelled word (jumped) is underlined by a red dotted line. If you right click on the word while in editing mode you will get the following in-place spell check with suggestions (Figure 4.12). Simply select the word you want to correct to and AutoCAD will oblige.

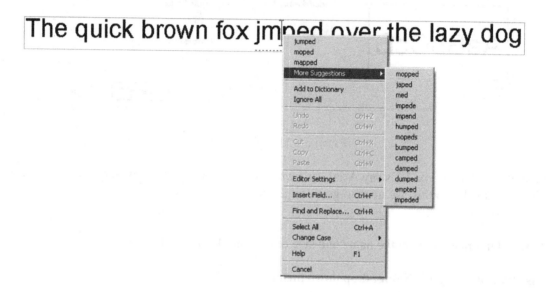

Figure 4.12 – In-place spell checker

Now what if you already have a significant amount of text (perhaps from an import) and need to spell-check it? Then you can use the stand-alone utility as described next.

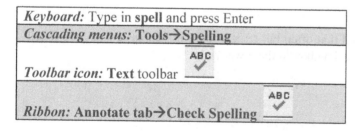

Keyboard: Type in **spell** and press Enter	
Cascading menus: **Tools→Spelling**	
Toolbar icon: **Text** toolbar	
Ribbon: **Annotate tab→Check Spelling**	

Using any of the above methods, the Check Spelling dialog box will appear as seen in Figure 4.13.

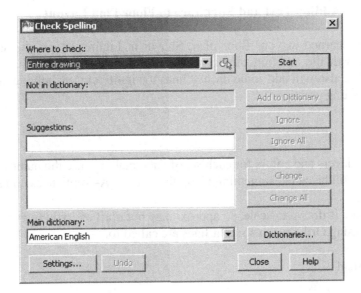

Figure 4.13 – Check Spelling

Using the drop down menu of the spell dialog box, you can select specific paragraphs or lines of text, or just press Start to initiate spell checker for all the text in the entire drawing.

In either case, the following will appear (Figure 4.14).

Figure 4.14 – Check Spelling initiated

Much like the MS Word spell checker, you can run through some options such as ignoring the word, changing it or adding it to the dictionary so it is not flagged again (good for names and acronyms). There is even a variety of languages available. When done, spell checker will tell you and you can close it.

In Class Drawing Project – Adding Text and Furniture to Floor Plan Layout

Let's now apply what you learned to your floor plan. Shown in Figure 4.15 is the same floor plan you have worked on in Chapter 3, with the addition of text in the rooms and closets as well as new features such as furniture and appliances. We also will add some dashed lines to signify closet shelving and of course the furniture and stairs for additional drafting practice.

Some general tips:

- Make up your own layers and colors for each set of data, making sure the names are logical! For example, kitchen appliances will be called something along the lines of A-Appliances and so on.

- None of the furniture is drawn to scale, so approximate but draft everything correctly; connecting all lines with OSNAP and using Ortho when straight lines are called for.

- Save your work often!

Also some specific drawing tips:

- All kitchen furniture on the left is formed using basic rectangles, drawn to no particular size. To create the range top burners with the four even circles, you need to draw one as shown in the following sequence using two circles and two lines, then mirror it using midpoints of the range itself as seen next.

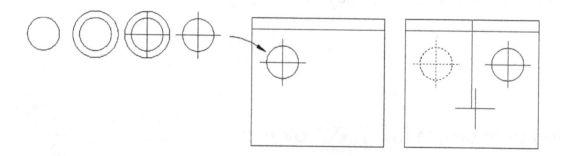

- To create the table you could use a rectangle with a small guideline sticking out of the left vertical side. That could be the anchor for the second point of the arc. Then a mirror and an explode command will allow you to erase the unneeded parts as seen next.

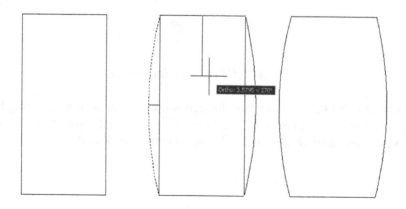

- To create the stairs you will need to draw one line at the top of the stairs and then offset it some value (perhaps 10"). To create the original line you will need to use the NEAr OSNAP point. This new point (not introduced with the original six) simply shadows a line or any object and allows you to begin a new object anywhere along that perimeter. We do not generally leave NEAr running along with the other ones as it makes using them more difficult; rather, just type it in or use the toolbar as needed. An hourglass figure will appear, just click wherever you want to start the stair line. The other side will be connected via the more familiar PERPendicular OSNAP. Below is a screen shot of NEAr in action.

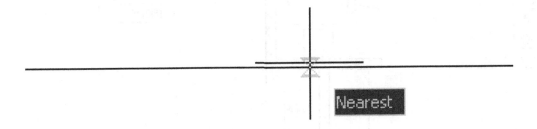

- Draw rectangles around each of the room text; we will need this later for hatch.

- The bathtub is done using the basic ellipse command, introduced in Chapter 2.

Finally we have two more tips. Tip #6 has to do with ltscale, something briefly mentioned in Chapter 3. Tip #7 shows you how to get rid of the UCS icon, something many students ask about.

TIP #6: If you draw your Hidden lines for the closet shelving, and they look solid not dashed, then the problem may be incorrect linetype scaling, a common error to watch out for. Type in LTSCALE, press Enter and input a new value (a larger one since your floor plan is bigger than the default original space of 8.5" x 11"). Do this a few times until you get it to look just right.

TIP #7: Want to get rid of that pesky Universal Coordinate System icon that is hanging out in the drawing area? It's somewhere near the lower left-hand corner of your screen, looks like a big "L" and has a "Y" and an "X" on its tips. To make it go away, type UCSICON, press Enter and then type in "OFF". Of course if it doesn't bother you and you actually like it there, then ignore this whole tip.

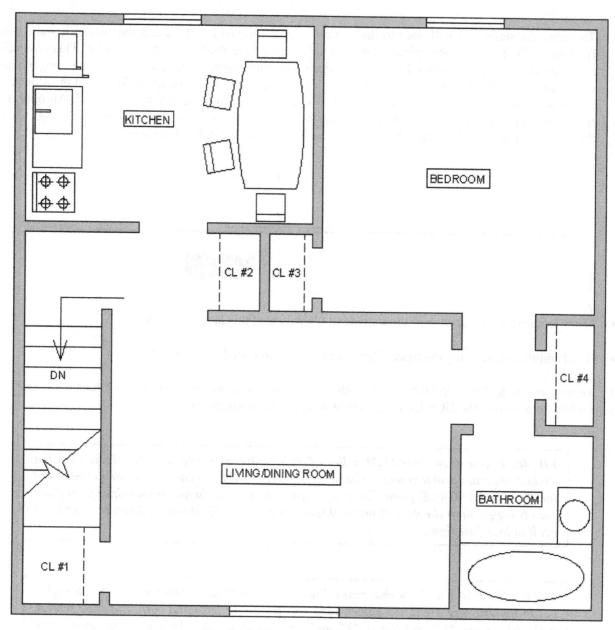

Figure 4.15 – Floor Plan Layout #2

Chapter 4

Summary

You should understand and know how to use the following concepts and/or commands before moving on to Chapter 5:

- **Text**
  - Creating basic Text

- **Mtext**
  - Creating basic Mtext

- **Mtext formatting options**
  - Bold
  - Underline
  - Italics
  - Justification
  - Symbols
  - Other

- **Editing text**
  - double click
  - ddedit

- **Style**
  - Font Name
  - Font Type
  - Font Size
  - Width Factor
  - Oblique Angle

- **Spell checker**

Chapter 4

Review Questions

Answer the following based on what you learned in Chapter 4.

1. What are the two types of **text** in AutoCAD? State some **properties** of each, and describe when you would likely use each type.

2. What are the two methods to **edit text** in AutoCAD?

3. What feature allows you to add extensive symbols to your mtext?

4. What are the three essential features to input when using **Style**?

5. What is **width factor**?

6. What is **oblique angle**?

7. How do you initiate **spell check**?

8. What is the new OSNAP point introduced in this chapter?

9. How do you get rid of the UCS icon?

10. What can you do if you cannot see the dashes in a Hidden linetype?

Chapter 4

Exercises

Exercise #1 – Open a blank file and create a new style called RomanS_2, choose font: RomanS, height: 2" and width factor: .88. Then create the following table using basic linework (distances between horizontal lines are 4"). Finally, fill it in with the data shown using your choice of Text or Mtext commands.
(Difficulty level: Easy. Time to completion: 20 minutes)

		LIST OF MAIN PARTS		
ITEM	QTY	DESCRIPTION	SIZE DWG	DWG. NUMBER
1	1	SKID ASSEMBLY	A1	83006
2	1	STRAINER ASSEMBLY	A1	83008
3	1	GENERATOR ASSEMBLY	A1	59470–001
4	1	GENERATOR ASSEMBLY	D1	59470–002
5	2	POWER SUPPLY (OIL IMMERSED)	B1	58717–001
6	1	POWER SUPPLY (OIL IMMERSED)	B1	58717–002
7	1	LOCAL CONTROL PANEL ASSEMBLY	D1	83017–001
8	2	LOCAL CONTROL PANEL ASSEMBLY	A1	88017–002
9	1	INSTRUMENT AIR ASSEMBLY	A1	83022
10	1	AIR PUMP	A1	86584

Exercise #2 – Using the same font as in Exercise #1, but with a width factor of 1.0, draw the following piping symbol legend. All symbols are drawn approximately and not to scale. Add text as shown.
(Difficulty level: Intermediate. Time to completion: 40-60 minutes)

SYMBOL LEGEND

VENT

BALL VALVE

CHECK VALVE

3–WAY SOLENOID VALVE

SOLENOID ACTUATOR

DIAPHRAGM ACTUATOR

PNEUMATIC ACTUATOR.

NEEDLE VALVE NORMALLY OPEN

REGULATOR VALVE,
SELF CONTAINED, INTERNAL

DIAPHRAGM SEAL

BLOWER/MOTOR

DRAIN, FLOOR

INTERLOCK LOGIC

TURBINE ELEMENT

TANK CLOSED TOP FLAT BOTTOM

REDUCER

INSTRUMENTS SHARING COMMON HOUSING

SCREEN

INSTRUMENT AIR 20 PSI

FILLED CAPILLARY

ELECTRICAL LINE, POWER/CONTROL

INSTRUMENT ELECTRICAL LINES (SIGNAL DIGITAL)

INSTRUMENT ELECTRICAL LINES (SIGNAL ANALOG)

GROUND EARTH

Exercise #3 – Open a new file, set up architectural units, create the proper layers/colors and draw the following floor plan. Dimensions are only for your reference.
(Difficulty level: Intermediate. Time to completion: 60-90 minutes)

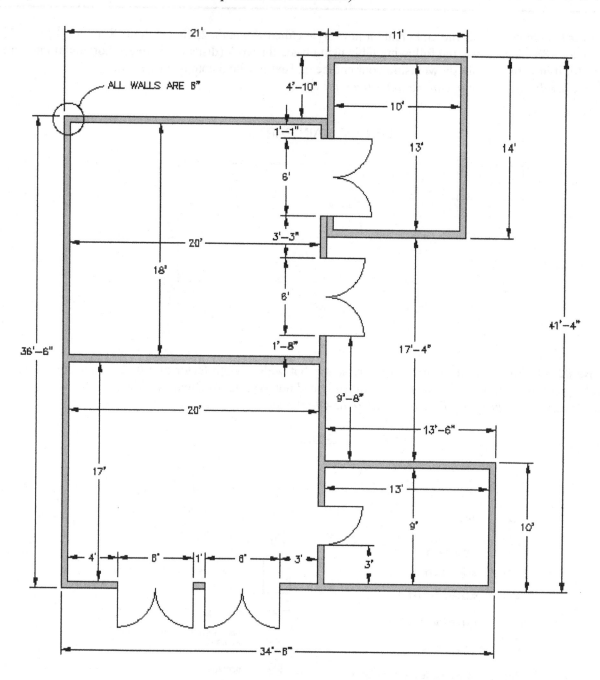

LEVEL 1

CHAPTER  5

Hatch Patterns

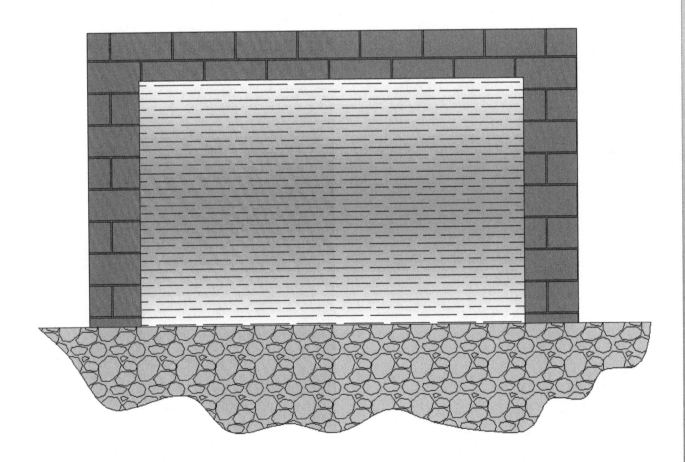

Chapter 5

Learning Objectives

Hatch patterns allow you to indicate a variety of surfaces, from wood to concrete, on your design. In engineering they are also used to indicate cross section surfaces. In this chapter we will introduce the Boundary Hatch and discuss the following:

- Hatch fundamentals
- Picking patterns
- Picking area or objects to hatch
- Adjusting scale and orientation
- Working with hatch patterns
- Gap tolerance
- Gradients and solid fills

At the end of this chapter you will be able to add hatch patterns to your floor plan.

Estimated time for completion of chapter: 2 hours (lesson + project)

Sec 5.1 - Introduction to Hatch

This is a chapter on the concept of Boundary Hatch or just Hatch for short. This is simply a tool for creating patterns and adding fills to your design. These are not merely decorative; rather they serve an important function of indicating types of material, ground and floor covering, and cross sections. As such, this topic is quite important, and is the next logical concept to master in AutoCAD.

Hatch patterns are not an AutoCAD invention. In the days of hand drafting there was also a need to visually indicate what sort of material one was looking at. Architects and engineers created easy to understand repeating patterns that somewhat, if not closely, resembled the actual material or at least gave you a very good idea of what it was. Just a few of the uses of hatch patterns are listed below:

Architecture:
- *Brick* – An important and popular pattern to render exterior and (sometimes) interior of buildings. AutoCAD has numerous brick patterns available.
- *Herringbone, Parque* – Important for flooring designations.
- *Honeycomb* – Insulation designation.

Civil Engineering:
- *Concrete, Sand, Clay, Earth, Gravel* – All used in designating surfaces in civil and site plan design.

Mechanical Engineering:
- *Various ANSI diagonal patterns* – Used to designate the visible inside of an object cut in cross section.

With AutoCAD, creating these patterns is easy, as they are all pre-drawn and saved in a library that is called up anytime you use the command. Often a designer can purchase or download additional patterns if the basic ones are inadequate. You can even make your own and save them for future use, although this is a tedious process and is only briefly covered in the appendix.

Sec 5.2 - Hatch Procedures

At the most basic level, AutoCAD's hatch command requires only four steps:

- *Step 1 - Pick the hatch pattern you want to use.*
- *Step 2 - Indicate where you want the pattern to go.*
- *Step 3 - "Fine tune" the pattern by adjusting scale and angle (if necessary).*
- *Step 4 - Preview the pattern, and accept it if OK.*

Memorize the four steps and their sequence; this will help you when looking at the hatch command dialog box.

Before we get into the details, it goes without saying that you first need something to hatch. The command will not work on a blank screen or a bunch of unconnected random lines. You need a closed area (although later on we will violate that rule), and the easiest to do is a circle or rectangle. Open a new file and draw either of the two shapes. Follow the steps carefully as described next. All advanced hatch functions flow from these basics.

- ***Step 1 - Pick the hatch pattern you want to use***

Keyboard: Type in **hatch** and press Enter
Cascading menus: Draw→Hatch...
Toolbar icon: **Draw** toolbar 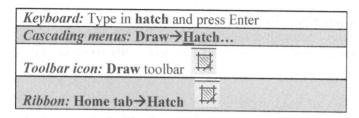
Ribbon: **Home tab**→**Hatch**

Begin the hatch command using any of the above methods. The following Hatch and Gradient dialog box will appear (Figure 5.1). At the bottom right of the dialog box is an arrow that expands or collapses additional options. Make sure it is expanded as seen in the figure.

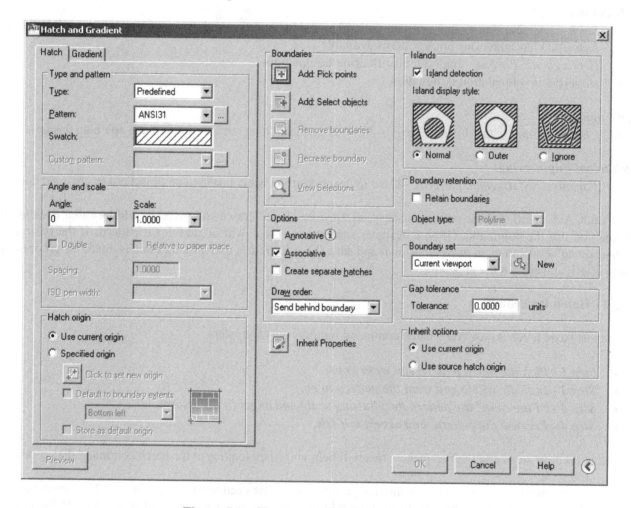

Figure 5.1 – Hatch and Gradient dialog box

Now select the pattern. The choices are listed in the upper left of the dialog box under the heading **Type and pattern**. Below that are three fields called **Type**, **Pattern:** and **Swatch:**. Leave **Type** as predefined and move down to the next one. If you use **Pattern:** then you will select the patterns *by name*, if you pick **Swatch:** patterns are selected *visually*, a much better option. Click on the diagonal lines to the right of **Swatch:** (or the little box with the three dots to the right of **Pattern:**) and a new box called the Hatch Pattern Palette appears (Figure 5.2).

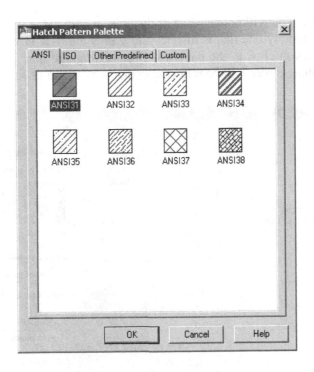

Figure 5.2 – Hatch Pattern Palette

Take a look through the tabs – ANSI, ISO, Other Predefined, and Custom. We will have a use for some of the ANSI patterns, not so much for the ISO ones, and Custom is where the custom-defined patterns would go if you were inclined to create them. Click on the "Other Predefined" tab and this is what you will see (Figure 5.3).

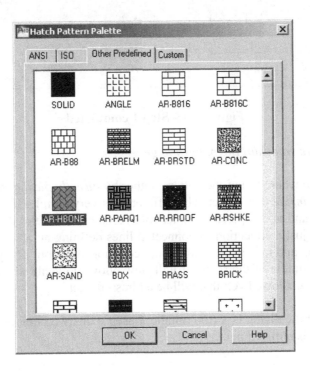

Figure 5.3 – Other Predefined tab

Scroll up and down the patterns. You will notice some of the ones mentioned at the start of this chapter. Go ahead and pick one that you like, preferably one that is distinct and stands out; AR-HBONE is a good choice, and will be used in this example. When you click it, the pattern will be highlighted blue. Go ahead and click OK. We're done with Step 1. The new pattern choice will be reflected in the swatch area as seen in Figure 5.4.

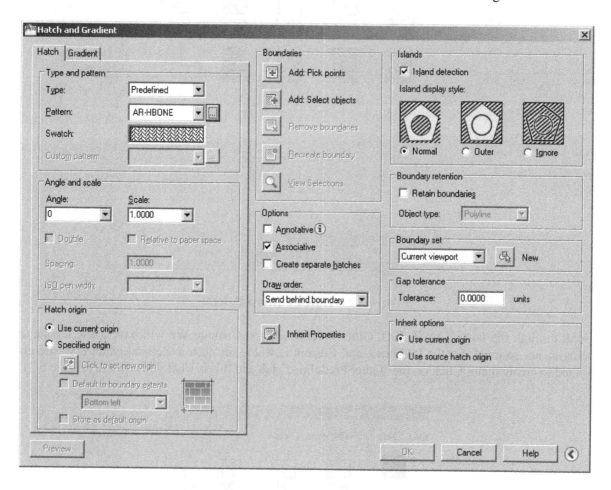

Figure 5.4 – Step 1 completed

- *Step 2 - Indicate where you want the pattern to go*

There are two ways to indicate where to put the pattern, either by *directly picking* the object that will contain the pattern, or by *picking a point inside* that object. The difference is very simple. If the object actually is an object made of joined-together lines and is one piece (such as your rectangle or circle) then it can be picked directly. If not, and the "object" is really just a collection of connected lines defining an area, then the best way is to pick a point in the middle of that area; and indeed this is the more common situation. The pattern will then behave like a bucket of paint spilled in the middle of the room. It will flow outward evenly and stop only when it hits a wall. Once again, no holes or gaps are allowed yet; that will be addressed later.

In the "Boundaries" section of the dialog box (top middle) observe the two choices just discussed and pick either one. The AutoCAD procedure is outlined next.

Add: Pick points
If you pick this option AutoCAD will say:
```
Pick internal point or [Select objects/remove Boundaries]:
```
Click somewhere *inside* the object and AutoCAD will say:
```
Selecting everything...
Selecting everything visible...
Analyzing the selected data...
Analyzing internal islands...
```
Press Enter and you will return back to the hatch dialog box with Step 2 completed.

OR

Add: Select objects
If you pick this option AutoCAD will say:
```
Select objects or [picK internal point/remove Boundaries]:
```
Click on the *object* itself (not the empty space) and press Enter. You will return back to the hatch dialog box with Step 2 completed.

Successfully picking a boundary can sometimes be tricky business. The existence of gaps, no matter how small, can be a source of occasional frustration to designers, and until just recently it used to be difficult to tell where those gaps were hiding in order to fix them. As of AutoCAD 2010, red circles now appear where the gaps are, aiding you in locating them, but this all can be avoided in the first place by not doing sloppy drafting (i.e.: not connecting lines together properly). It is a sign of advanced AutoCAD skills when a complex area hatches right the first time!

In cases where boundary selection isn't successful, complex areas can often be broken down to smaller ones by using lines to divide the area into smaller pieces and hatch the non problematic portions first, though this is a last resort. A few releases ago AutoCAD added the Gap Tolerance command that ignores gaps up to a preset limit. This is a great tool that we will cover shortly, along with the entire gap issue in general, but it is somewhat of a "band-aid" and still doesn't address the underlying sloppy drafting, but merely allows you to get away with it.

Remember "Garbage in = Garbage out", and it is essential to master the fundamentals of AutoCAD drafting early on. The hatch command is an "early warning" to students. If they are having problems using it smoothly, they need to go back and refine their basic drafting (linework and accuracy) skills.

- *Step 3 - "Fine tune" the pattern by adjusting scale and angle (if necessary)*

You're almost done and at this point could probably just press OK and finish the hatch pattern; however, press Preview instead. It is found at the bottom left of the dialog box, and is generally a good habit to get into. What you will see is your hatch pattern (if it's the right size) with the object's border in dashed lines (Figure 5.5).

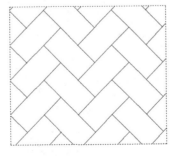

Figure 5.5 – Bhatch preview

If everything looks right (as it does in this example) AutoCAD will say what to do next:

```
Pick or press Esc to return to dialog or <Right-click to accept hatch>
```

Go ahead and right-click or press Enter.

Sometimes, however, the pattern is too big or too small. You can usually tell it is too big by either not being able to see it, or seeing a small part of it. If it is too small the problem is worse because you may not be able to finish the pattern at all. Instead, AutoCAD will tell you the following:

```
Hatch spacing too dense, or dash size too small. Pick or press Esc to return
to dialog or <Right-click to accept hatch>:
```

Nothing will appear. In either case press Esc. once and you will be returned to the dialog box. Under the "Angle and scale" change the scale to a larger or smaller number (by typing it in) and Preview again. You may need to do this several times until the hatch pattern is just right. In the same manner, you may adjust the angle of the entire pattern if desired. Angle and scale is shown in Figure 5.6.

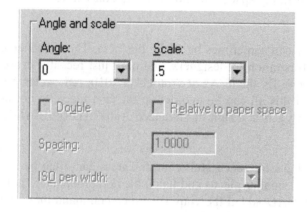

Figure 5.6 – Angle and scale

- ***Step 4 - Preview the pattern, and accept it if OK***

Finally you're done! Do one more Preview and press OK. Here is what the sample hatch pattern looked like. Yours may or may not be similar depending of course on what pattern and scale you selected, as well as what shape you used.

Figure 5.7 – Final hatch pattern

Sec 5.3 – Working with Hatch Patterns

In this short section we will mention some of the additional tools and concepts that pertain to hatch patterns. First of all, the most basic question at this point is how do you edit a hatch pattern after you created it? It's very easy; to edit a hatch pattern simply double click it. Make sure you get one of the lines of the pattern, not just empty space. You will be taken back to the main dialog box of Figure 5.1 to change whatever you need to.

➤ Exploding Hatch Patterns

This may be one of the worst things you can do. Never explode hatch patterns; depending on size they will turn into hundreds of pieces that cannot be put back together again in the same editable form. Reasons designers give for exploding hatch patterns (making them fit, trimming them, moving the containment border, etc.) can all be addressed with additional hatch tools.

➤ Hatch Pattern Layers and Colors

Hatch patterns should be on a generic A-Hatch layer or on whatever layer name best describes what the pattern represents (A-Carpet, M-Cross-Sec, C-Gravel, etc.). Choose any color you want for the hatch patterns, though you shouldn't go with anything too bright as it may overwhelm the design. Shades of grey are common for cross sections in mechanical design, as are various shades of rusty orange for brick. As a side note, hatch patterns can be easily erased by the usual erasing methods. The patterns can also be moved, copied, rotated and mirrored, though there is rarely a reason to do so.

➤ Advanced Hatch Topics

Except for Gap Tolerance, we won't cover the rest of the options in too much detail, but will summarize some of them in an overview below. Feel free to explore on your own; this is an important part of learning AutoCAD and will greatly enhance your confidence with the software.

Hatch Origin
This allows you to begin hatch patterns from a precise location (origin) as opposed to a random fill-in. AutoCAD will choose where the origin on the pattern itself is and you get to choose where on the object that origin will be aligned to. Try it out.

Options
- *Annotative* – Will not be covered until Level 2.
- *Associative* – Keep this checked. This simply means that if you move the border, the hatch will move with it; very useful in case of future updates.
- *Create separate hatches* – If there are multiple separate areas to hatch, this option will keep them separate; sometime useful, sometimes not.
- *Draw order* – This is exactly what it sounds like; it gives you options in how you want to stack the hatch and its boundary. Check out the options in the drop down menu.
- *Inherit Properties* – This option lets you borrow the features of another hatch pattern and use them in a new pattern you are creating, saving you some effort and time of picking out a pattern from scratch.

Islands
This gives AutoCAD guidance in how to interpret boundaries inside boundaries (i.e., a chair in the middle of your floor). It simply tells AutoCAD to ignore them, or if not, how to deal with them. Try drawing another rectangle inside your first one and running through the options represented by the pentagon shapes.

Gap Tolerance

This is a simple to use and popular new tool that was mentioned earlier. Simply set the tolerance to a value that will be big enough to span anticipated holes or gaps in your design and hatch as usual. AutoCAD will remind you of what you're doing via the following alert seen in Figure 5.8 (notice the red circles mentioned earlier to indicate a gap).

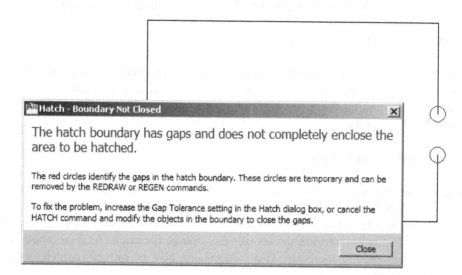

Figure 5.8 – **Boundary Not Closed alert**

Close the alert and press Esc. to continue. Then, returning to the main dialog box, set a value for the gap tolerance and re-hatch the area. You will get another warning (which can be eliminated from future occurrence by checking off "Always perform my current choice") as seen in Figure 5.9.

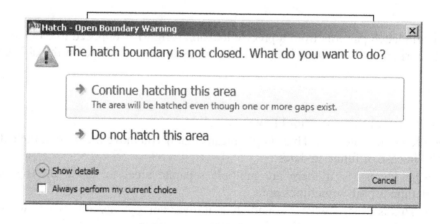

Figure 5.9 – **Open Boundary Warning**

Select *Continue hatching the area* choice and AutoCAD will temporarily close the gap to continue the hatch command. After successful completion, the gap will show again, but the hatch will remain as seen in Figure 5.10. Some designers have used the command to ignore door openings when hatching carpeting in floor plans. You can try this as well on your floor plan later.

Figure 5.10 – Bhatch with Gap Tolerance

Sec 5.4 – Gradient and Solid Fill

At the top of the Hatch and Gradient dialog box click the Gradient tab. The left side of the box will change while the right-hand side will stay essentially the same as shown in Figure 5.11. The gradient option is just a solid fill with some flair. You can choose the desired pattern, in one or two colors, and adjust the shade, tint, orientation and angle of the gradient. The rest of the steps are similar; pick point or select object, preview and done!

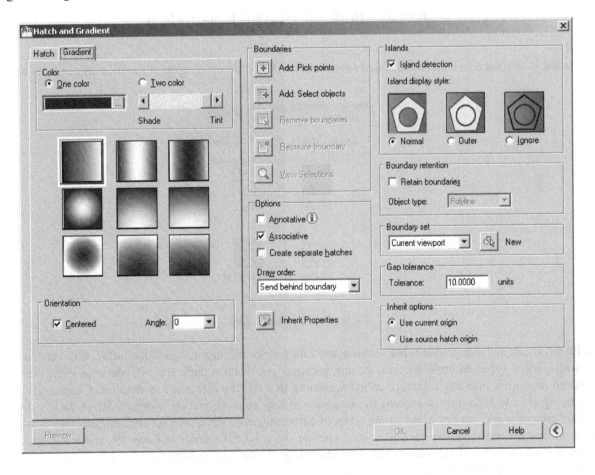

Figure 5.11 – Gradient tab

Shown in Figure 5.12 is a possible use for gradient in rendering a brick wall. A brick pattern was created and a brick-colored fill on top of each other. Note that you may need to use **Tools→Draw Order→Bring to Front** from the cascading menu to position the brick above the gradient fill (or the gradient fill behind the brick, depending on what you pick first). Try it!

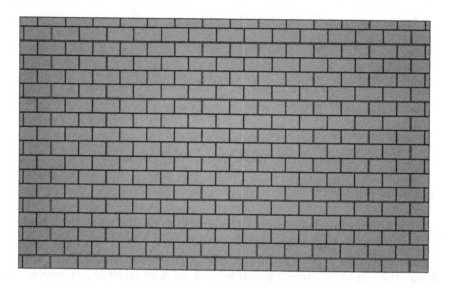

Figure 5.12 – Brick wall using hatch and gradient

Solid Fill

This fill is the first choice in Other Predefined (shown below), and deserves special mention as our final topic.

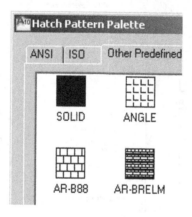

Figure 5.13 – Solid Fill

This solid fill is similar to a gradient, but without any fancy color mixing; it's just one color. It is very handy as a fill for walls, some types of cross-sections or any location you want a dark area. It becomes even more useful when paired up with a lineweight feature called screening that will be discussed in detail in Chapter 20. It has to do with being able to fade out (or screen) the intensity of any color (typically down to 30%), and is perfect for indicating "area of work" on key plans, some types of carpeting, even pavement on civil engineering drawings.
A way to mimic this now without screening is to change the solid fill's color to Grey #9, and a soft grey haze is then seen that is distinct and visible, yet not overwhelming. If you're wondering how the walls were filled in on the apartment drawing, this is how.

In Class Drawing Project – Adding Hatch to Floor Plan Layout

Go ahead and add hatch to the rooms of your floor plan, including the walls. Some tips are listed below.

To add hatch to rooms, freeze the A-Doors layer, and instead draw *temporary* lines closing off the door entrances. After hatching, remove the lines. You may also be able to use gap tolerance set to at least 3 ft, but that is not recommended.

Make sure the hatch patterns are on the proper layers, whatever they may be (carpeting or just hatch_1, hatch_2, etc.; you decide). You don't have to pick the same hatch patterns shown, but make sure they are reasonable and at the proper scale.

To fill in the apartment walls, freeze everything except the A-Walls and the hatch layer. Then use the solid fill, color #9 as described earlier. If you created everything properly, there should not be any gaps or breaks in the wall islands. Check to make sure the windows are drawn correctly, meaning the wall has to have an actual closed gap, into which the window is inserted. Here's what everything should look like *prior to wall hatching (1) and right after (2)*. The apartment is shown in Figure 5.14 on the next page.

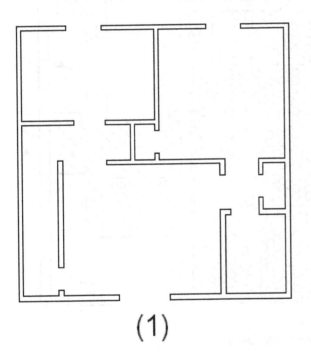

(1)

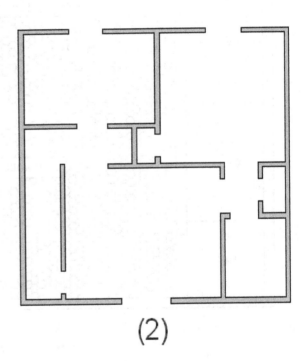

(2)

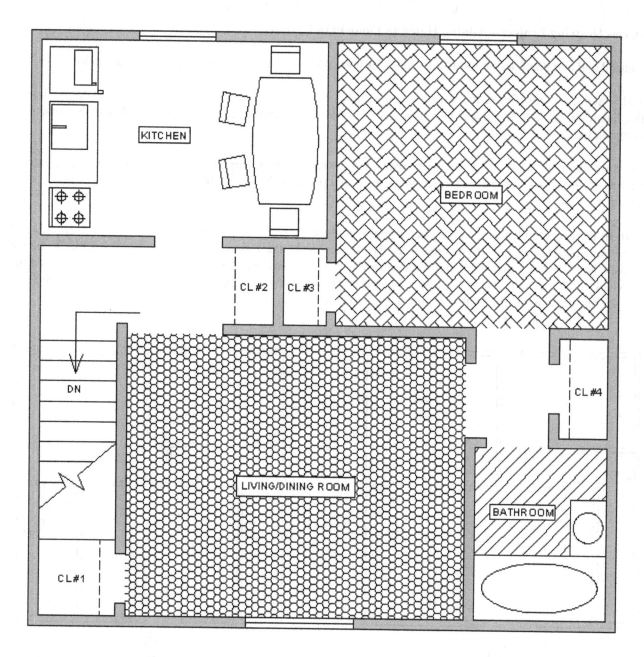

Figure 5.14 – Floor Plan with hatch

Chapter 5
Summary

You should understand and know how to use the following concepts and/or commands before moving on to Chapter 6:

- **Hatch**
  - Picking pattern
  - Selecting area or object
  - Scale and orientation adjustments
  - Preview

- **Editing a Hatch**
  - double-click

- **Additional Options**
  - Islands
  - Inherit properties
  - Associative
  - Draw order

- **Gap Tolerance**

- **Gradients and Fill**

Chapter 5

Review Questions

Answer the following based on what you learned in Chapter 5.

1) List the four steps in creating a basic **hatch pattern**.

2) What are the two methods to select **where to put** the hatch pattern?

3) What commonly occurring problem can prevent a hatch being created?

4) What is the name of a tool recently added to AutoCAD to address this problem?

5) How do you edit an existing hatch pattern?

6) What is the effect of **exploding** hatch patterns? Is it recommended?

7) What is an especially useful pattern mentioned in the chapter to fill in walls?

Chapter 5

Exercises

Exercise #1 – In a new file, create the following shapes and hatch patterns. Sizing and patterns are arbitrary, but try to duplicate what is shown.
(Difficulty level: Easy. Time to completion: 3-5 minutes)

Exercise #2 – In a new file, create the following shapes and hatch patterns. Sizing and patterns are arbitrary, but try to duplicate what is shown.
(Difficulty level: Easy. Time to completion: 3-5 minutes)

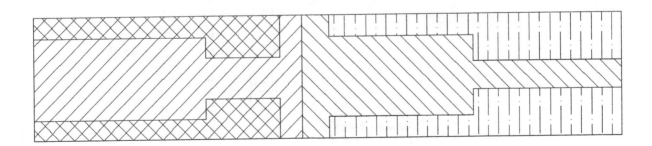

Exercise #3 – Open a new file and set units to Architectural. Create the following stair layout and hatch patterns. Any sizes and dimensions that are not given can be estimated, as can be the scale of the patterns.
(Difficulty level: Intermediate. Time to completion: 20-30 minutes)

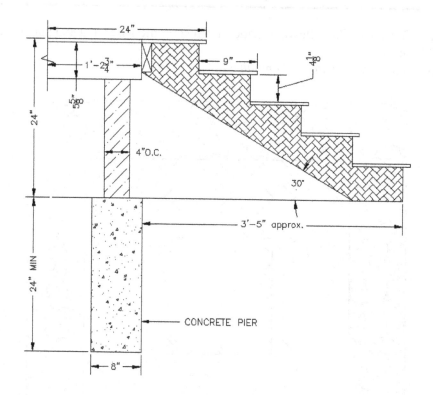

Exercise #4 – Open a new file and set units to Architectural. Create the following layout of a set of cylinders. Dimensions are provided, but sizing is of secondary importance to creating the hatch patterns. Notice how gradients are used to depict curvature inside the pipe cross sections; a good trick to add realism to 2D drawings.
(Difficulty level: Intermediate. Time to completion: 20 minutes)

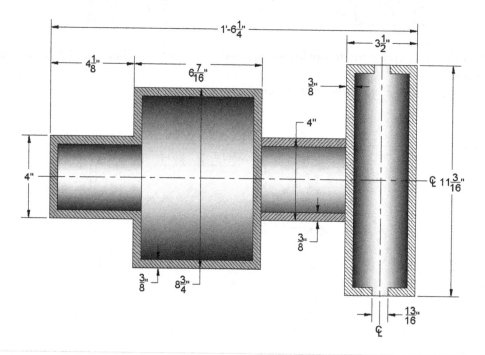

CHAPTER  6

Dimensions

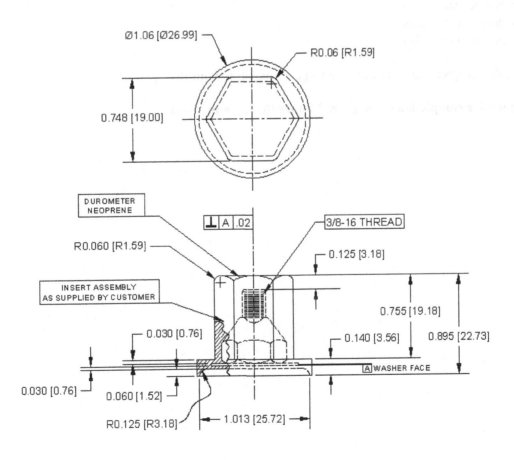

Chapter 6

Learning Objectives

In this chapter we will introduce the first part of AutoCAD's extensive dimensioning capabilities and discuss the following:

- Linear (horizontal and vertical) dimensions
- Aligned dimensions
- Diameter dimensions
- Radius dimensions
- Angular dimensions
- Continuous dimensions
- Baseline dimensions
- Leader and Multileader
- Arc Length dimensions
- Ordinate dimensions
- Jogged dimensions
- ddedit
- ddim
- Dimension units
- Dimension font
- Dimension arrowheads
- Dimension overall size

At the end of this chapter you will be able to add dimensions to your floor plan.

Estimated time for completion of chapter: 3 hours (lesson + project)

Sec 6.1 - Introduction to Dimensions

Dimensioning in AutoCAD is a major topic, one that is so extensive that it is split into two separate discussions. Here we will address the basics of what dimensions are, which ones are available and how to use and edit them. In Chapter 13 we will look at extensive customization and additional options. AutoCAD permits significant sophistication with dimensioning; here we will only cover what is used most often to get you going quickly. In Chapter 13 we will also introduce a brand new addition to the dimensioning family: Parametric dimensioning and the concept of constraints and dimension-driven design.

So what are dimensions? They are simply visible measurements of something for the purposes of conveying information to the audience that will be looking at your design. It is how you describe the size of the design and where it is in relation to everything else. Dimensions can be "natural," meaning they display the actual value, or "forced" when they display an altered value, such as when you dimension an object with a break line. The break line indicates that a chunk was taken out to conserve space, yet the dimension still needs to state the true value – which is entered manually.

Sec 6.2 - Types of Dimensions

The first step in learning dimensions is to know what is available, so you can use the appropriate type in any situation. The primary and secondary dimensions that are available to you are shown next. Commit them to memory, as you may need many, if not most of them, on a complex project.

Primary Dimensions – These are dimensions used most often by the typical architectural or engineering designer:

> ➢ Linear (horizontal and vertical)
> ➢ Aligned
> ➢ Diameter
> ➢ Radius
> ➢ Angular
> ➢ Continuous
> ➢ Baseline
> ➢ Leader.

Secondary Dimensions – These are used less often, or are quite specialized, but still need to be mentioned:

> ➢ Arc Length
> ➢ Ordinate
> ➢ Jogged

Let's take the primary list one by one. Look them over and understand what each is for. Then draw the appropriate shape to practice with, and add the new dimension by following the step by step instructions. We will, as always, use typing, toolbars, cascading menus and the Ribbon. Open a new file and bring up the dimension toolbar (Figure 6.1) adding it to the toolbars already on your screen.

Figure 6.1 – Dimension toolbar

Note an important point if typing dimensions! You will need to type in **dim** and press Enter. This will get you into the dimension subset. The command line will then say `Dim:` and you can type in whatever dimension you want to create (usually abbreviating by only typing the first few letters as shown) and press Enter again. The command will then execute normally. So there is an advantage to using other input methods here rather than typing, though the differences in effort and speed is rather small. Remember also that you will need to press Esc. to get out of this subset back to the regular command line.

> **Linear Dimensions** – These are any dimensions that are strictly horizontal or vertical (Figure 6.2).

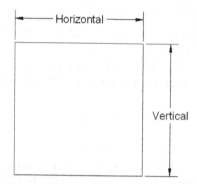

Figure 6.2 – Linear Dimensions

Keyboard: Type in **dim,** press Enter, type in **hor**, press Enter
Cascading menus: **Dimension→Linear**
Toolbar icon: **Dimension** toolbar
Ribbon: **Annotate tab→Dimension→Linear**

Step 1
Create a small to medium size rectangle or square as seen in Figure 6.2 and begin the linear dimension command via any of the above methods.
Step 2
- AutoCAD will say: `Specify first extension line origin or <select object>:`
Turn on your OSNAPs and using ENDpoint, select the left corner of the rectangle/square.
Step 3
- AutoCAD will say: `Specify second extension line origin:`
Again, using ENDpoint, select the right corner of the rectangle/square (we will be doing the *horizontal* one first).
Step 4
The dimension will appear
- AutoCAD will say: `Specify dimension line location or [Mtext/Text/Angle/Horizontal/Vertical/Rotated]:`
Move the mouse up and down and click where you want the dimension to go, usually a short distance from the object. It will set and the value will be shown by AutoCAD saying `Dimension text = 4.4681` (your value will likely be different of course).

Note an important distinction with typing! Instead of setting the value in Step 4, AutoCAD throws in one more step of allowing you to alter the dimension on the spot (forcing it). You can enter a new value, or press Enter to accept the natural one.

The previous linear dimensions are by far the most common in most drawings, and are also the basis for the continuous and baseline dimensions to be covered later on. Be sure to also run through the vertical version of the linear dimension, selecting the upper and lower right (or left) corners of the rectangle/square. Next is the aligned dimension, which gives a true distance of a slanted surface.

> **Aligned Dimension** – This is a dimension that measures a slanted line or object (Figure 6.3).

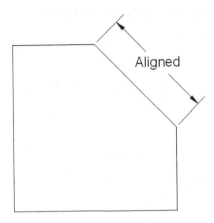

Figure 6.3 – Aligned dimension

Keyboard: Type in **dim,** press Enter, type in **align,** press Enter
Cascading menus: Dimension→Aligned
Toolbar icon: **Dimension** toolbar
Ribbon: **Annotate tab→Dimension→Aligned**

Step 1
Create a small to medium size rectangle or square and chop off a corner as seen in Figure 6.3. Then begin the aligned dimension command via any of the above methods.
Step 2
- AutoCAD will say: `Specify first extension line origin or <select object>:`
Turn on your OSNAPs and using ENDpoint, select the upper left start of the slanted line.
Step 3
- AutoCAD will say: `Specify second extension line origin:`
Again, using ENDpoint, select the lower right end of the slanted line.
Step 4
The dimension will appear.
- AutoCAD will say: `Specify dimension line location or [Mtext/Text/Angle]:`
Move the mouse up and down and click where you want the dimension to go, usually a short distance from the object. It will set and the value will be shown by AutoCAD saying `Dimension text = 2.9255` (your value will likely be different of course).

Again, note an important distinction with typing! Instead of setting the value in Step 4, AutoCAD throws in one more step of allowing you to alter the dimension on the spot (forcing it). You can enter a new value, or press Enter to accept the natural one.

The next two dimensions, diameter and radius, are essentially similar, though we will cover both separately for clarity. Note the primary difference: the symbol indicating diameter (Ø) and the symbol for radius (R) prior to the value. Although you can set this up, by default there is no line crossing the circle half-way (for radius) or all the way (for diameter) so you will have to read the values carefully, and look for the appropriate symbol to know what you are looking at. Also note the location of the values; they are at roughly the 10, 2, 4 and 8 o'clock positions relative to the circle. Don't just stick them anywhere; hand drafting rules still apply in CAD.

➢ **Diameter Dimension** – This is a dimension that measures the diameter of a circle or an arc (Figure 6.4).

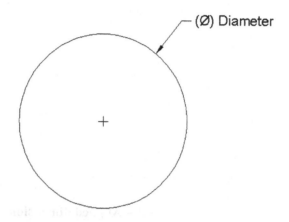

Figure 6.4 – **Diameter dimension**

Keyboard: Type in **dim,** press Enter, type in **dia**, press Enter
Cascading menus: **Dimension→Diameter**
Toolbar icon: **Dimension** toolbar
Ribbon: **Annotate tab→Dimension→Diameter**

Step 1
Create a small to medium size circle as seen in Figure 6.4. Then begin the diameter dimension command via any of the above methods.
Step 2
- AutoCAD will say: `Select arc or circle:`
Pick the circle and the dimension will appear attached to your mouse
Step 3
- AutoCAD will say: `Dimension text = 3.1357` (your value will likely be different of course)
                    `Specify dimension line location or [Mtext/Text/Angle]:`
Position your mouse somewhere at the upper right of the circle (rarely inside, usually outside) at 2 o'clock, or at any of the other accepted positions and click. The value and dimension will set.

Again, note an important distinction with typing! Instead of setting the value in Step 3, AutoCAD throws in one more step of allowing you to alter the dimension on the spot (forcing it). You can enter a new value, or press Enter to accept the natural one.

> **Radius Dimension** – This is a dimension that measures the radius of a circle or an arc (Figure 6.5).

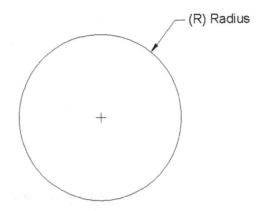

Figure 6.5 – Radius dimension

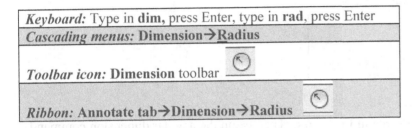

Keyboard: Type in **dim,** press Enter, type in **rad,** press Enter
Cascading menus: Dimension→Radius
Toolbar icon: **Dimension** toolbar
Ribbon: **Annotate tab→Dimension→Radius**

Step 1
Create a small to medium size circle as seen in Figure 6.5. Then begin the radius dimension command via any of the above methods.

Step 2
- AutoCAD will say: `Select arc or circle:`

Pick the circle and the dimension will appear attached to your mouse

Step 3
- AutoCAD will say: `Dimension text = 1.5679` (your value will likely be different of course)
            `Specify dimension line location or [Mtext/Text/Angle]:`

Position your mouse somewhere at the upper right of the circle (rarely inside, usually outside) at 2 o'clock, or at any of the other accepted positions and click. The value and dimension will set.

Again, note an important distinction with typing! Instead of setting the value in Step 3, AutoCAD throws in one more step of allowing you to alter the dimension on the spot (forcing it). You can enter a new value, or press Enter to accept the natural one.

> ➤ **Angular Dimension** – This is a dimension for angles between two lines or objects (Figure 6.6).

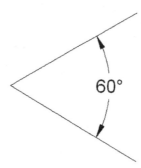

Figure 6.6 – Angular dimension

Keyboard: Type in **dim,** press Enter, type in **ang**, press Enter
Cascading menus: Dimension→Angular
Toolbar icon: **Dimension** toolbar
Ribbon: **Annotate tab→Dimension→Angular**

Step 1
Create two lines spread out at a random angle as seen in Figure 6.6. Then begin the angular dimension command via any of the above methods.
Step 2
• AutoCAD will say: `Select arc, circle, line, or <specify vertex>:`
Pick both lines one after the other.
Step 3
• AutoCAD will say: `Specify dimension arc line location`
`or[Mtext/Text/Angle/Quadrant]:`
Drag the mouse out and move it around, selecting the best position for the new dimension and click when you find a spot you like. Be careful as the supplement of the degree value will show if you move behind the lines.
• AutoCAD will say: `Dimension text = 60` (your value may of course be different).

Again, note an important distinction with typing! Instead of setting the value in Step 3, AutoCAD throws in one more step of allowing you to alter the dimension on the spot (forcing it). You can enter a new value, or press Enter to accept the natural one and position the text.

Angular completes the basic set of new fundamental dimensions. Next we have *continuous* and *baseline*, which are not really new as we will see, but rather an extension (or automation) of the linear dimension learned earlier. We will then conclude with a leader – not so much a dimension but a useful member of that family.

➤ **Continuous Dimensions** – These are a continuous *string* of dimensions (Figure 6.7).

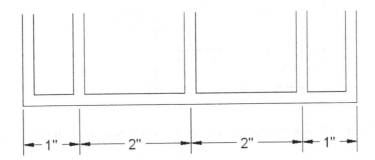

Figure 6.7 – **Continuous dimensions**

A short explanation is in order, as continuous and baseline give new students some initial problems. Continuous dimensions are nothing more than a string of familiar horizontal or vertical ones. The idea here is to create one of those two types of linear dimensions, then start up continuous where you left off and let AutoCAD quickly fill them in as you pick contact points. It's really the same as using linear over and over again, but faster and more accurate since it's automated.

The easiest way to practice the continuous dimensions is to draw a set of squares attached to each other. These will represent a simplified view of building and room walls. Basically draw one random-sized square and copy it, endpoint (lower left corner) to endpoint, until you see the following:

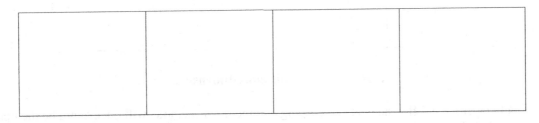

The next step is to draw one horizontal (it can also be vertical) linear dimension as you already have done previously. Locate it about where you would like the entire string to go as seen here:

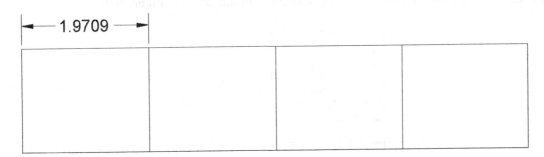

This is the essential first step in creating both continuous and baseline dimensions. Now you can practice continuous as outlined next.

Keyboard: Type in **dim,** press Enter, type in **cont**, press Enter
Cascading menus: Dimension→Continue
Toolbar icon: **Dimension** toolbar
Ribbon: **Annotate tab→Continue**

Step 1

Once you have the squares and one linear dimension, start up the continuous command via any of the above methods.

- AutoCAD will say: `Specify a second extension line origin or [Undo/Select]` `<Select>:`

Step 2

A new dimension will appear. Pick the *next* point (corner) along the string of rectangles and click on it (always using OSNAP points, no eyeballing!)

- AutoCAD will say: `Dimension text = 1.9709` (your value may of course be different).

You can continue this until you run out of objects to dimension; at that point just press Esc. Your result is shown in Figure 6.8.

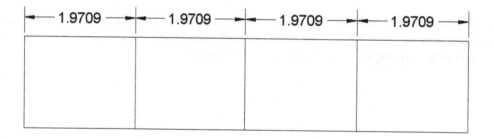

Figure 6.8 – Continuous dimensions

As you are already aware, the difference between typing and the other methods will once again be the opportunity to enter a value other than what is given (force the dimension) when typing. You will then have to repeat the continuous command (in other words re-type **cont**) to carry on.

> **Baseline Dimensions** – These are a continuous *stack* of dimensions (Figure 6.9).

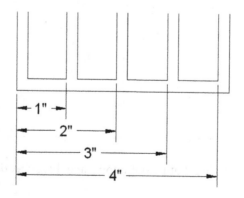

Figure 6.9 – Baseline dimensions

The baseline dimension, as mentioned before, is very similar. The goal is to make a neat stack of evenly spaced dimensions that all start at one point (the *Base* in *Base*line). To begin, erase the previous continuous dimensions (leaving the squares), and once again draw one linear dimension as seen below:

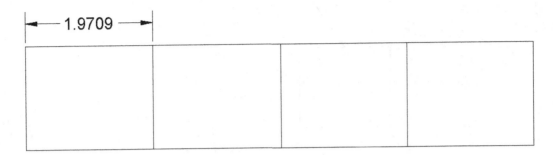

Keyboard: Type in **dim,** press Enter, type in **base**, press Enter	
Cascading menus: **Dimension→Baseline**	
Toolbar icon: **Dimension** toolbar	
Ribbon: **Annotate tab→Baseline**	

Step 1
Once you have the squares and one linear dimension, start up the baseline command via any of the above methods.

- AutoCAD will say: `Specify a second extension line origin or [Undo/Select]` `<Select>:`

Step 2
A new dimension will appear. Pick the *next* point (corner) along the string of rectangles and click on it (always using OSNAP points, no eyeballing!)

- AutoCAD will say: `Dimension text = 3.9417` (your value may of course be different).

You can continue this until you run out of objects to dimension; at that point just press Esc. Your result is shown in Figure 6.10.

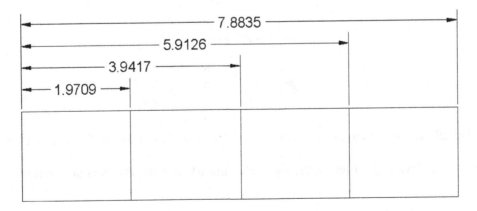

Figure 6.10 – Baseline dimensions

> ➤ **Leader and Multileader** – This is an *arrow and label* combination pointing at something (Figure 6.11).

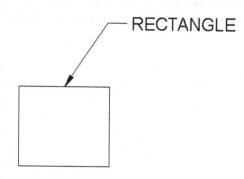

Figure 6.11 – Leader

While a leader isn't a true dimension by definition, it's still a very common and necessary member of the dimension family. The values shown by the leader can be not only numerical, but also text comments from the designer, concerning the part or object it is pointing to.

Leaders are so important that AutoCAD has given them a major rework in recent releases. We will go over the two main options in increasing order of complexity, starting with the basic leader command, followed by the more feature-rich multileader with its add, remove, align and collect options.

Leader command:
Step 1
Draw a small box on your screen, making sure Ortho is off. Type in the command **leader** and press Enter.
- AutoCAD will say: `Specify leader start point:`

Step 2
Click with OSNAP precision on your shape, perhaps a midpoint on the top side as seen in Figure 6.11.
- AutoCAD will say: `Specify next point:`

Step 3
Move your mouse at a 45° angle away from the first point and click again at a reasonable distance away. You've created the *leader line and arrowhead* as seen here:

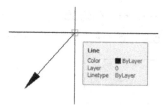

- AutoCAD will say: `Specify next point or [Annotation/Format/Undo] <Annotation>:`

Step 4
Turn Ortho back on and draw a short line to the right, clicking when done; this is your *horizontal landing* as seen here:

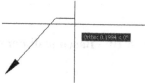

- AutoCAD will say: `Specify next point or [Annotation/Format/Undo] <Annotation>:`

Step 5

You're done, so press Enter.

- AutoCAD will say: `Enter first line of annotation text or <options>:`

Type something in, pressing Enter if you wish to do another line, and Enter again if you are done as seen here:

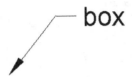

Multileader:

This method is relatively new and is meant to add more flexibility and usefulness (and inadvertently some complexity) to the leader command. You will need to bring up the multileader toolbar, seen here in Figure 6.12.

Figure 6.12 – Multileader toolbar

There is something new to multileader that you didn't have with the basic one. You can set up a multileader style, so all leaders will have the exact look you want. In our sample case, we will give our multileader a bubble in which to add text. Press the very last button on the right (mleader with a paint brush) and the following dialog box will appear (Figure 6.13).

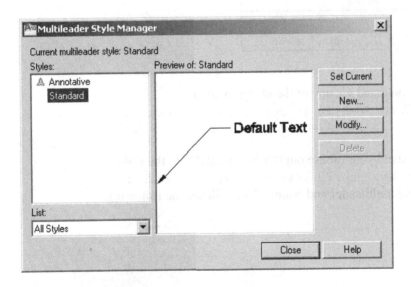

Figure 6.13 – Multileader Style Manager

Press New… and give the multileader a name, such as *Sample Style* and press Continue. You will be taken to the Modify Multileader Style dialog box (Figure 6.14). Examine the three tabs, Leader Format, Leader Structure and Content, carefully. Most of the options are reasonably self-explanatory, and you will see some of them again later on. Under the Content tab, select Multileader type: as Block and Source block: as Detail Callout. Finally press OK, Set Current and Close.

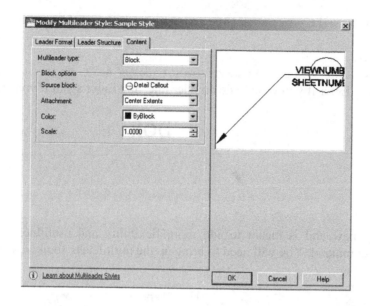

Figure 6.14 – Modify Multileader Style

Now let's try out our new multileader style.

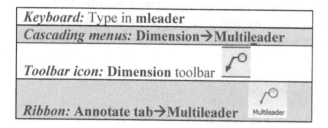

Step 1

Start the multileader command via any of the above methods.

- AutoCAD will say: Specify leader arrowhead location or [leader Landing first/Content first/Options] <Options>:

Step 2

Click anywhere and extend your mouse out (Ortho off) at 45° to the right.

- AutoCAD will say: Specify leader landing location:

Click again to place the multileader and bubble. You will see the following:

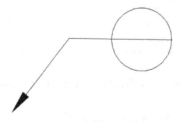

Step 3

- AutoCAD will say: Enter view number <VIEWNUMBER>:

Enter in some value.

- AutoCAD will say: Enter sheet number <SHEETNUMBER>:

Enter in another value.

Once you do this, you will see the final result (Figure 6.15); the appearance of which can of course be fine-tuned.

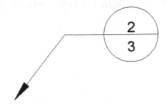

Figure 6.15 – Multileader with Block

Now that you have one leader on the screen, it's easy to experiment with other interesting options. Move across the toolbar, starting first with the Add Leader (Figure 6.16), and then get rid of it via Remove Leader. Then after adding it back in try the Align Multileaders (Figure 6.17), which aligns the leaders with one that you pick as the primary. Finally Collect Multileader (Figure 6.18) combines them into one string.

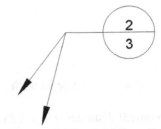

Figure 6.16 – Add Leader

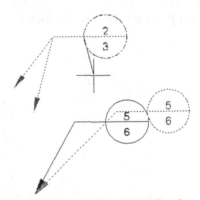

Figure 6.17 – Align Multileaders

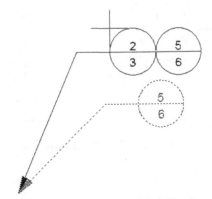

Figure 6.18 – Collect Multileaders

It was mentioned that there are some secondary dimensions, meaning they are not used as often. We'll briefly mention them here, and you may want to go over them on your own in detail, especially if you feel one or more of them may be extremely useful to you.

- Arc Length
- Ordinate
- Jogged

Arc Length, as you may have guessed, measures the length of an arc. Simply select the toolbar icon and click the arc you want to measure (Figure 6.19).

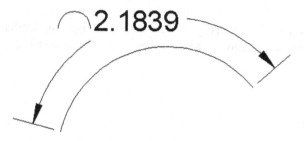

Figure 6.19 – Arc Length

Ordinate dimensions measure horizontal or vertical distances from a datum point (0,0 in this case). They are used in manufacturing to prevent accumulation of errors that can occur using continuous measurements. If you draw the shapes shown below and start up the ordinate dimension, you can then click major points along the way in either horizontal or vertical directions to get precise values at that location from the origin.

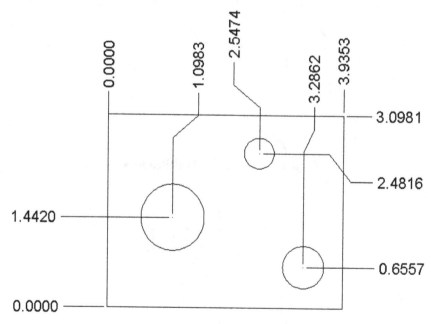

Figure 6.20 – Ordinate

Jogged dimensions are useful when the radius or diameter dimension's center is off the sheet of paper and cannot be shown directly, so a "break line" of sorts is used. You simply indicate select the arc or circle and then the center location override and a jogged dimension appears (Figure 6.21).

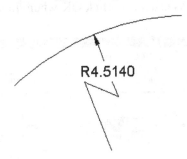

Figure 6.21 – Jogged

There are other dimension options we haven't yet explored! We will come back to some of them in Level 2. For now we must move on and learn how to do editing.

Sec 6.3 - Editing Dimensions

You should now be quite familiar with the types of dimensions available and how to apply them to basic shapes. The next step is to be able to edit them, or change their values if needed. This is a very simple and short topic.

You've already learned that you can edit text and mtext by double clicking it. The text in the dimensions is essentially editable mtext, so you would be tempted to double click the text to edit it. Not so fast though. If you do that, then the Properties dialog box pops up. You've seen it in Chapter 3. Here's what it looks like again when the dimension text is double-clicked:

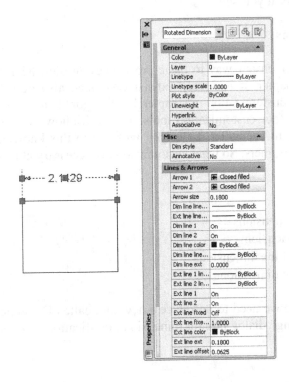

Figure 6.22 – Dimension Properties

While this is useful later on in advanced topics it's not going to let you edit the dimensions. However, in Chapter 4 we covered **ddedit**, which it mentioned was also useful for editing. Go ahead and create a horizontal dimension again on your square, type in **ddedit** and press Enter. The mtext editing box will pop up and you can now type in any value you need as shown below. Edit as usual and click OK when done.

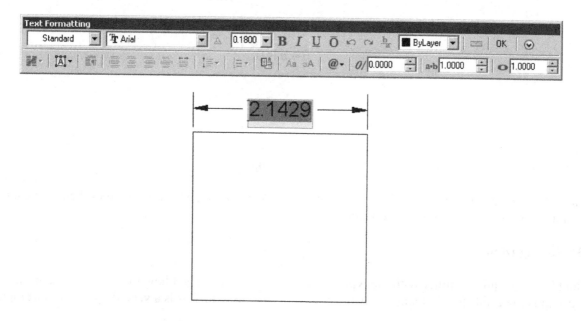

Figure 6.23 – Mtext edit of dimensions

Here's a useful tip. If you want to reset your forced dimension value back to its natural value, and you forgot what it was, just type in < >. These two "alligator teeth" brackets will restore the natural value. Type them in instead of the forced value and click OK. Try it yourself.

Sec 6.4 - Customizing dimensions

We have one more topic to cover on our way to basic proficiency in dimensioning. It is customization, and as mentioned at the start of the chapter, it is a very extensive subject, also necessitating it being split into the fundamentals here and advanced customization in Chapter 13. The goal here is to show you what experience has shown to be the most important four customization tools. This will allow you to be productive in most of the drafting situations you are likely to encounter. Later you will add to this knowledge by learning tools used by CAD administrators and senior designers. So what customization is necessary at this level?

We need to learn how to:

- Change the *units* of the dimensions
- Change the *font* of the dimensions
- Change the *arrowheads* of the dimensions
- Change the *fit* (size) of the dimensions

To be able to change any of these we need to introduce a new tool called Dynamic Style dialog box or **dimstyle** for short. It is a "one-stop-shopping" dialog box for dimension modification that you will eventually need to learn well.

> **Ddim**

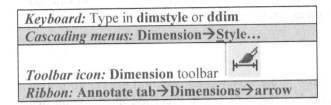

Keyboard: Type in **dimstyle** or **ddim**	
Cascading menus: **Dimension→Style…**	
Toolbar icon: **Dimension** toolbar	
Ribbon: **Annotate tab→Dimensions→arrow**	

Use any of the above methods to bring up the Dimension Style Manager, as seen in Figure 6.24.

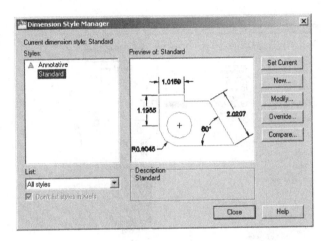

Figure 6.24 – Dimension Style Manager

In Chapter 13 we will create a brand-new dimension style and heavily modify it. For now let's just go ahead and change the current one to reflect the changes suggested in the previous list. Go ahead and press Modify…from the menu on the right. You will see the following Modify Dimension Style: Standard dialog box (Figure 6.25).

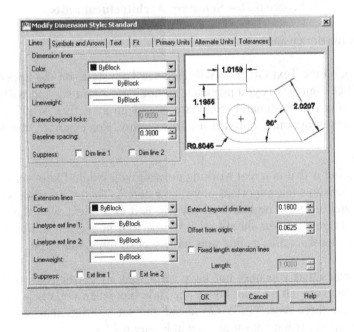

Figure 6.25 – Modify Dimension (Lines tab)

The dialog box may or may not open to the Lines tab as seen in Figure 6.25, but regardless take a look at the available tabs at the top. As noted before, we will cover every command and option under those tabs in Chapter 13 and return to them again when we learn Paper Space. Let's focus right now only on what we need to change.

Step 1 - Change the *units* of the dimensions

Pick the Primary Units tab and simply select the drop-down menu at the very top left where it says Unit format. The default value is Decimal; change it to Architectural and select the appropriate Precision in the menu just below it. That's all we need from that tab for now. Notice how the preview window on the upper right reflects your choice of units as seen in Figure 6.26.

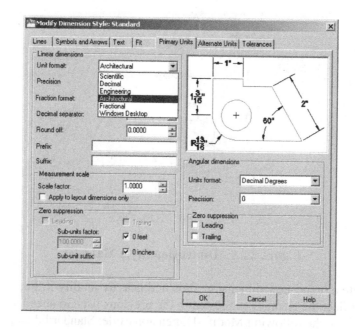

Figure 6.26 – Selecting Architectural units

Step 2 - Change the *font* of the dimensions

To do this you need to click on the Text tab. You will then see Text style: on the upper left. If you just opened a new AutoCAD file to practice dimensions you probably do not have a font set and will only get Standard as your choice if you click the down arrow. Fortunately Standard is now an attractive Arial font, not the ugly Simplex it used to be, so changing fonts is optional, but still useful to know. For example, the RomanS font is a popular alternative.

Something else to keep in mind is that in a real working drawing you would likely set your text style (using **style**) before you work on any dimensions (an important point; take note!). Then you would just select the font from the list. The idea here is to match up your regular font used in text and mtext with the font used in dimensioning (another important point!). In general you should stick to one font in a given drawing and just vary the size as needed. The titleblock is exempt from this, as that may have special fonts, logos, etc.

AutoCAD of course anticipates that you may not have set a font style when you first set up dimensions, and allows you to do this from the Text tab by bringing up the style box when you click the button just to the right of the Text style menu (with the three dots, "…"). Go ahead and set a different font if you wish – RomanS .25" perhaps and you will see this new setting appear as seen in Figure 6.27.

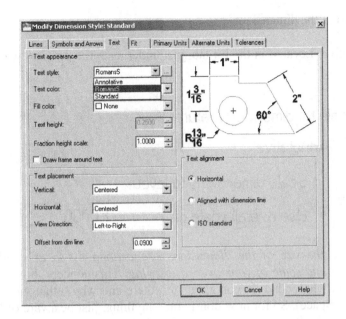

Figure 6.27 – Setting new font

Step 3 - Change the *arrowheads* of the dimensions

The default for all dimensioning is the standard arrowhead. You can easily change that to any other type including the Architectural tick, popular with Architects. You can even create a custom arrowhead (not something we will try to do here).

Click over to the Symbols and Arrows tab. In the upper left corner under the Arrowheads section you will see First:. That is the first arrowhead type, and if you change it AutoCAD assumes you want the second one to be the same and changes it as well. Simply click the down arrow and select the Architectural tick. Leave everything else the same. The result is shown in Figure 6.28.

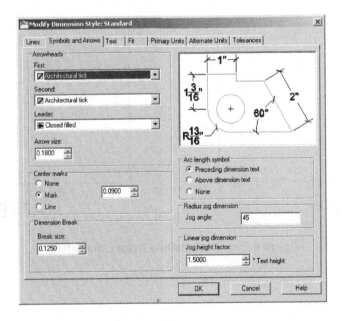

Figure 6.28 – Changing the arrowheads

Step 4 - Change the fit (size) of the dimensions

This function is quite deceptive. Located in the Fit tab, just underneath the preview window it features only a few lines and looks like this:

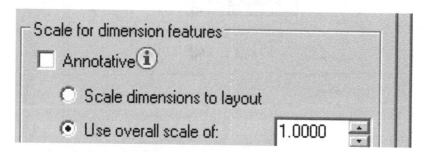

The idea here is to *increase the size of the dimensions proportionately so everything scales up evenly and smoothly in step with the size and scale of your overall drawing.* This concept, however, opens up a number of questions (chief among them: what value do you enter into there and why?) that we aren't ready to discuss yet, but will in Chapter 18 (Paper Space). So for now don't enter anything, just be aware of how to do it when needed. When you get to dimensioning the floor plan, just enter 15 in that space.

This is it for now; just click OK and AutoCAD will return you to the Dimension Style Manager where you can just press Close. Try creating a dimension and see what it looks like now. Below is a typical horizontal dimension before the changes we just went through (except for the font left as Arial) and another one afterwards. The difference is obvious. For now, this is all you need to make professional looking dimensions. Review and memorize everything you learned and go on to the floor plan below, adding dimensions to complete the drawing.

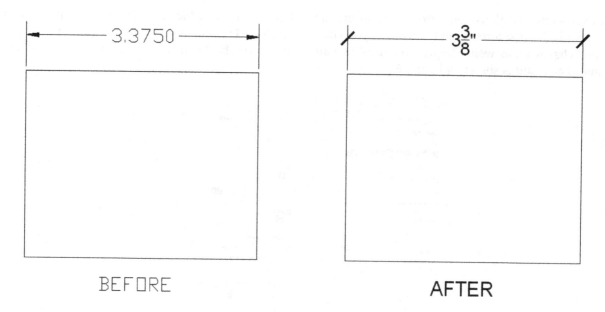

Figure 6.29 – Dimensions before and after

In Class Drawing Project – Adding Dimensions to Floor Plan Layout

Let's now apply what you learned to the floor plan. Open the file and freeze all the layers except the walls and windows. If you really want to get fancy, put the wall solid hatch on its own (visible) layer and freeze the regular floor hatch layer so the carpeting and floors don't show. Next create a new layer for the dimensions, A-Dim. Finally set up the dimensions: Use Architectural units, Arial 6" font, leave the arrowheads as they are and change the fit to 15. Dimension the floor plan any way you want; what is shown below is a guide, but is not the only way.

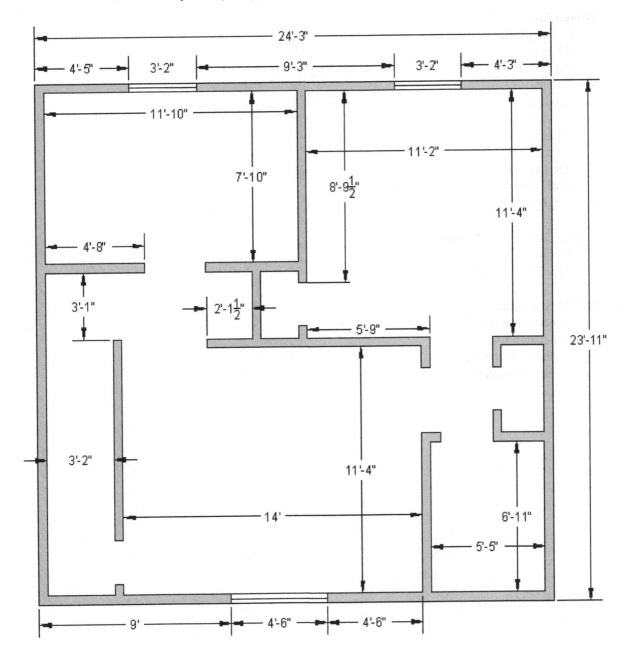

Chapter 6
Summary

You should understand and know how to use the following concepts and/or commands before moving on to Chapter 7:

- **Dimensions**
  - Linear (horizontal and vertical)
  - Aligned
  - Diameter
  - Radius
  - Angular
  - Continuous
  - Baseline
  - Leader and Multileader
  - Arc Length
  - Ordinate
  - Jogged

- **Editing Dimension Values**
  - ddedit
  - The effect of: < >

- **DDIM command**
  - Dimension units
  - Dimension font
  - Dimension arrowheads
  - Dimension overall size

Chapter 6

Review Questions

Answer the following based on what you learned in Chapter 6.

1. List the 13 types of **dimensions** discussed.

2. What command is best for **editing** dimension values?

3. What is the difference between **forced and natural** dimensions?

4. How do you **restore** a natural dimension value when you forgot what it was?

5. List the four items that needed to be **customized** in our dimensions.

6. What is the command to bring up the **Dimension Style Manager**?

7. What type of **arrowhead** do architects usually prefer?

Chapter 6
Exercises

Exercise #1 – In a new file draw and dimension the following mechanical part.
(Difficulty level: Easy/Moderate. Time to completion: 20-25 minutes)

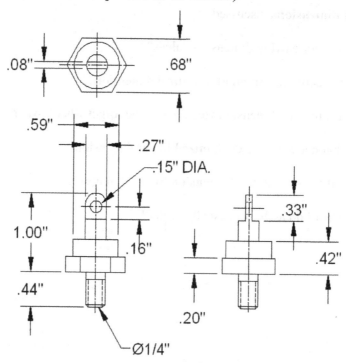

Exercise #2 – In a new file draw and dimension the following mechanical part. Any dimensions not given can be estimated.
(Difficulty level: Moderate. Time to completion: 30-35 minutes)

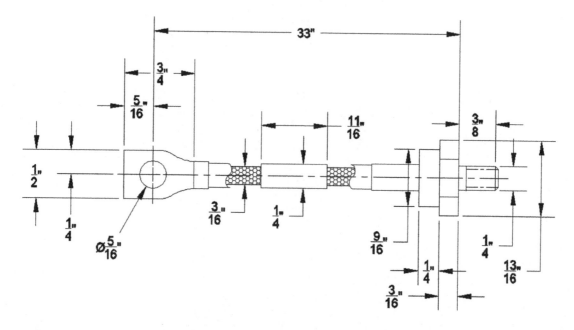

CHAPTER  7

Blocks, Wblocks, Dynamic Blocks and Purge

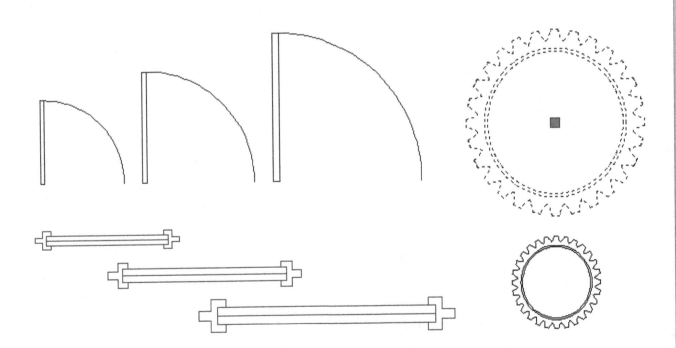

Chapter 7

Learning Objectives

In this chapter we will introduce the concept of Blocks and discuss the following:

- Creating Blocks
- Retain, Convert, Delete options
- Redefining Blocks
- Inserting Blocks
- Purge
- Creating Wblocks
- Creating Dynamic Blocks

At the end of this chapter you will be able to create blocks, wblocks and dynamic blocks as well as insert symbol libraries and purge your drawing.

Estimated time for completion of chapter: 1-2 hours

Sec 7.1 - Introduction to Blocks

This chapter is about a very useful AutoCAD concept that will greatly simplify your workload. Groups, Blocks, Wblocks and Dynamic Blocks are all members of the same family. They exist around the idea that objects can and should be grouped together if possible, to allow for easier handling and storage for future re-use. We will not look at groups, as they are rarely used, but instead will focus on blocks and wblocks, with an additional discussion of their more sophisticated cousins, dynamic blocks.

Let's formally define a block. It is a collection of objects that are grouped and held together under some identifying name. These objects can then be copied, moved, erased and just about anything else, all as one unit. Think of blocks as electronic "glue" that binds the objects they contain to each other.

The benefit of this should be apparent. If you have an office table with eight chairs around it, and you need to move them all around many times while optimizing a room layout, it will be a lot easier to click a block once than attempt to select all nine pieces using window or crossing or one at a time. Imagine an engineering example involving a simple gear with all its intricately designed teeth. There may be dozens if not hundreds of lines and arcs that make up the gear. It would be not just wise, but mandatory to form them all into a single block.

There is however another very important reason. That is the concept of *re-use and symbol libraries*. Blocks can be saved and re-used for future drawings that may need the same exact object. That way you never have to draw the same thing twice; an AutoCAD "Golden Rule". A collection of these objects catalogued in some orderly fashion becomes a symbol library, an indispensable part of just about any designer's toolset. Likely candidates for these libraries include door, window and furniture symbols, bolts, standard parts and fasteners; really just about anything that can be used in more than one design or drawing.

➢ Difference between Blocks and Wblocks

Hopefully you're sold on the importance of blocks, and we can go through the specifics. The differences between blocks and wblocks are blurred; they're alike in many ways. The key difference is that blocks are *internal* to the current file you're working in, and wblocks are *external*. This means that blocks are saved inside whatever file is open and you are working on. As such, their main use is for grouping things together and temporary re-use in that particular AutoCAD drawing. While they can certainly be moved between files (Clipboard Copy/Paste is a great tool), this is not really what they are for.

Wblocks (write blocks) are intended for creating symbol libraries. Wblocks are saved externally, anywhere you want to put them, and become independent stand-alone AutoCAD drawings, which you can open and work on if needed. Both types of blocks can of course be easily inserted back into drawings and we will cover that procedure in detail.

➢ Creating a Block

To create a block you first need something to make a block out of. Once you draw something, only three things need to be done after that. You need to:

- Select the objects
- Give the block a name
- Select a base point

Let's try this out. Draw a chair as shown in Figure 7.1. It is made up of three rectangles and three lines. Use accuracy tools!

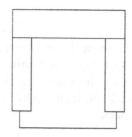

Figure 7.1 – Basic lounge chair

We need to now group them all together into one object.

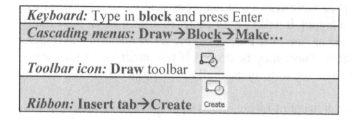

Keyboard: Type in **block** and press Enter
Cascading menus: **Draw→Block→Make...**

Toolbar icon: **Draw** toolbar

Ribbon: **Insert tab→Create**   Create

Start the block creation command via any of the above methods. The Block Definition dialog box will appear as seen here in Figure 7.2.

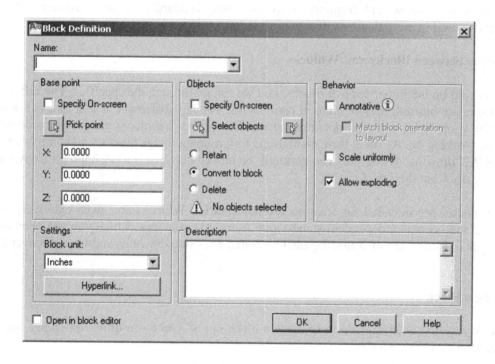

Figure 7.2 – Block Definition

Step 1 - Press the Select objects button at the top of the middle column. Select the entire chair and press Enter.
Step 2 - Enter a name for this block in the top left field under Name.
Step 3 - Pick a base point with the Pick point button at the upper left. Select a corner or midpoint somewhere on the object. The significance of this will become more apparent when we insert the object back into the file.

There are of course a few other items in the dialog box under Settings, Behavior and Description, but they are not usually needed (though feel free to enter a description). Annotative will be covered in advanced chapters. The one setting of importance, however, is the "Retain", "Convert to Block", "Delete" choices under the Select objects button. They are explained below and will appear with almost similar wording in wblock as well.

Retain
This means that the block will be created and stored in memory, but the original object (the chair you are looking at) will remain in separate pieces. Useful if you wish to make several blocks of the chair, one after the other, each slightly different yet based off the same original.

Convert to Block
This option is what you are likely to use the most. It will create the block and store it in memory and also make a block out of the original object. Used when you're sure one block is all you'll need to make out of this one object, and you have a need for it right away.

Delete
Here the block is created and stored in memory, but the original is deleted from the drawing. This is useful when you're making a block for future use but have no need for it now.

The common theme with all three of these choices is that the block gets made either way and saved to memory. The only real choice is what to do with the original.

Finally you're ready to click OK. The Block Definition dialog box should look like this (Figure 7.3). Notice the small preview window in the upper middle of the dialog box.

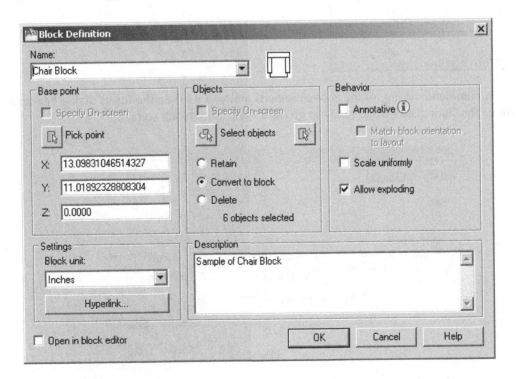

Figure 7.3 – Block Definition, completed

The chair is now a block, which can be proved by clicking it once with the mouse. You will notice that only one blue grip point will appear (wherever it is you picked the base point). If you now erase it, the entire chair will disappear. Go ahead and erase it; we will need to do this to introduce two new commands before we discuss wblock.

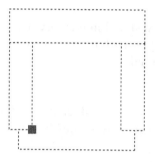

Figure 7.4 – Chair as a block

Sec 7.2 - Insert

As mentioned before, the process of making the block adds it into the memory. Then you can insert it back into the drawing when needed.

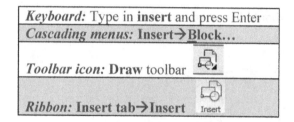

Start up the insert command via any of the above methods and you will see the Insert dialog box appear as seen in Figure 7.5. Notice the chair cued up and ready to go (as it is the only block so far).

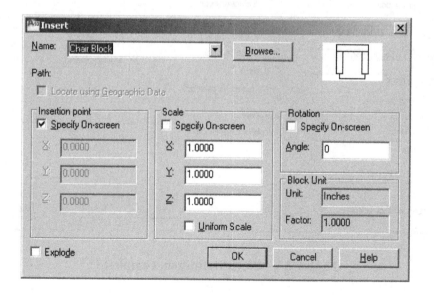

Figure 7.5 – Insert

Look over the options such as insertion point, scale, rotation and block unit, but at this point you don't really need to do anything extra. Go ahead and click OK, and the chair will appear attached at the insertion point to the mouse. The point of attachment was specified when you made the block. Click anywhere on the screen and the block will become part of the drawing. This process can be repeated as often as necessary as long as the block remains in memory.

There is something important to remember concerning redefining blocks. If a block is saved under a certain name, and another block is created and saved under that same name, the first block will be redefined, a sort of over-writing process. Sometimes this is bad as you are destroying the old block. Sometimes however this is desirable such as when you are looking to update the appearance of the block. When there are many blocks in the drawing under the same name, they will all be updated in this manner.

Sec 7.3 - Purge

Between introduction of the block and wblock concepts is a good time to take an intermission of sorts and discuss a command called Purge. Purge is a very useful tool to "clean up" your drawing of unwanted and often unseen geometry and data. It is especially useful when it comes to blocks and wblocks, hence its inclusion here.

Conceptually, here is the idea. A complex AutoCAD drawing often grows in size and becomes weighed down, similar to an iceberg floating in the ocean. Much like the (mostly hidden under the surface) mountain of ice, what you see on-screen may only be a small fraction of what there actually is present in the file.

The reason why becomes apparent when you think about the drafting process for a moment. If you create 50 layers for anticipated future use, but only end up needing 40 of them, the other 10 are uselessly hanging out in your layer dialog box. Same thing with unused fonts or linetypes (remember loading ALL of them, but only using a few). Layers, linetypes and fonts really don't take up much space, but with blocks and wblocks it's a whole different story.

Wblocks are miniature drawings all to themselves, and while many are just simple doors or window symbols, others can be quite large (such as entire furniture sets or appliances in an architectural layout). Worse yet, these blocks rarely come alone; they bring friends, lots of them. The author has seen and worked with complex multi-story sprawling building layouts that feature not dozens but *hundreds* of blocks. They all take up room in a file; ballooning it up to a rather huge size.

If all of these blocks are needed, then fine, this is a necessary evil. But if not - let's say a type of chair was deleted from the specs - they need to be removed. Erasing them all is easy enough, but are you done? Are they all gone for good? Unfortunately, no! They may not be visible but they are still there weighing down your file, slowing down computer performance and just about everything else. This is where purge comes into play.

Purge will get rid of items that are not actually used in the drawing file, and permanently delete them. You can use the command to selectively purge items or purge all at once. The two main points to remember are:

- Purge will not get rid of objects that are visible and present in the drawing – it is NOT an erase command.

- Purge is permanent (unless you immediately undo it), so be careful of what you purge. If you anticipate eventually needing something, don't purge it out.

Let's give the command a try.

Keyboard: Type in **purge** and press Enter	
Cascading menus: File→Drawing Utilities→Purge…	
Toolbar icon:…**none**	
Ribbon: **none**	

First of all, erase the chair from your screen! Then start up the purge command via one of the two methods and the following dialog box will appear (Figure 7.6).

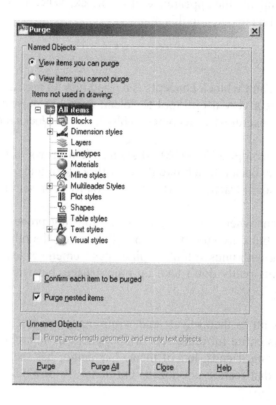

Figure 7.6 – Purge

As you can see from looking at the list of items in the dialog box, there are quite a few items eligible for purging. Some were already mentioned like Layers and Linetypes, while others you may not have heard of like Materials and Shapes. Either way, what is important is whether or not the item has a plus next to it. If it does, you can click the plus sign and expand the folder to see what specific items are in those general categories. In our case you will see the chair block under the blocks category. You will also see Annotative under some of the other categories. We will look at Annotative in later chapters, so for now just ignore it and we will purge Annotative alongside the chair block.

The actual process of purging is easy. You can either select each item one at a time (usually only done when you want to selectively purge), or you can click "Purge All" at the bottom. Be sure the "Purge nested items" box is checked and the one above it is not as seen in Figure 7.6. When the items are all purged there will be no more plus signs and the buttons will grey out as well. Press "Close" after you're done and that is it.

Experienced users purge often and keep little useless junk in the file unless absolutely necessary. Purging is certainly something that should be done prior to saving the file for the day, emailing it or archiving it for storage and record keeping. Though be careful and don't get rid of needed items. Always think before purging!

Sec 7.4 - Wblocks

Pretty much *everything* said about blocks applies to wblocks as well, and as a result this section will be short, as you have seen this before. The central idea to keep in mind is that a wblock is external, and as such you will need to tell AutoCAD where to put the wblock as you are making it. Let's try this out. Clear your screen and make sure all items are purged out from the previous exercise. Now go ahead and draw another furniture object, perhaps a sofa this time.

The only way to start up this command is via typing **wblock** and pressing Enter. Once you do that, the following dialog box appears (Figure 7.7).

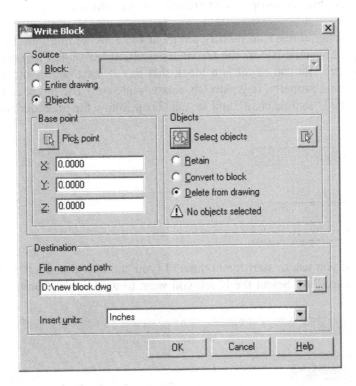

Figure 7.7 – Write Block

For the screen shot above, a specific path has not been selected, but you can do this by clicking the button with the three dots "…" to the right of the "File name and path:" and browse around until you find where it is you want the file to go. Be sure to also change the name of the file as well at the end of the string so it doesn't just say "new block". Then select the sofa and pick one of the three options to retain, convert or delete the original. When done press OK and the wblock will be created and dropped off where you indicated in the path.

> **Inserting Wblocks**

Go ahead and insert your sofa back in using the **insert** command. The procedure is essentially the same, except that now you naturally have to browse for the file. The Browse button is just to the right of the name field. Look around for your block and when you find it, click on it and press OK. It will insert just as with the previous block command.

Sec 7.5 - Dynamic Blocks

Dynamic Blocks were introduced relatively recently to AutoCAD (in Release 2006). Like many new features, they are not necessarily a radically new concept. Rather, they are an evolution of the regular block or wblock. The idea here is to be able to modify a block in response to design conditions. A simple example is of a door. If you insert a block of a 3' door, rotated at 45°, and later realize that you need a 3' - 6" door, rotated to a closed position, you can change that in place without redefining the block or inserting another one.

So what we are doing essentially is adding a level of intelligence and automation to the standard block concept. This is especially useful when it comes to scaling, stretching and otherwise modifying the sizes, as that is cumbersome to do with regular blocks (as opposed to rotating them, which is easy).

To accomplish all this, the designer needs to modify a regular block by adding Parameters and Actions. So there are three things making up a Dynamic Block: the *Geometry* itself, at least one *Parameter* and at least one *Action* associated with that parameter. The end result is a block that has at least one or more special grip points, each defined with a certain modifying property, (e.g., stretch, rotate, scale etc.) The block can then be modified in place as needed. Create a block called "sample chair" and save it. Then follow the procedures outlined.

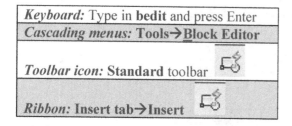

Keyboard: Type in **bedit** and press Enter
Cascading menus: **Tools→Block Editor**
Toolbar icon: **Standard** toolbar
Ribbon: **Insert tab→Insert**

Regardless of which method you use to start the command (you can also just double click the block itself), you will see this dialog box (Figure 7.8). Select the block you want to work with, which in this case is the only one available.

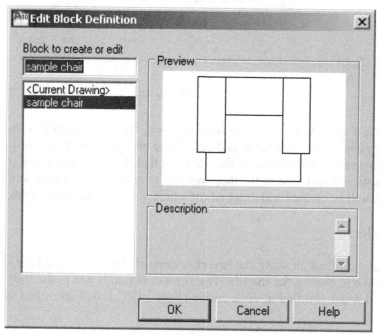

Figure 7.8 – Edit Block Definition

After you select the "sample chair" you will be taken to the main editing screen called a Block Editor (recognizable by its yellow background), shown in Figure 7.9. This is where the settings are applied.

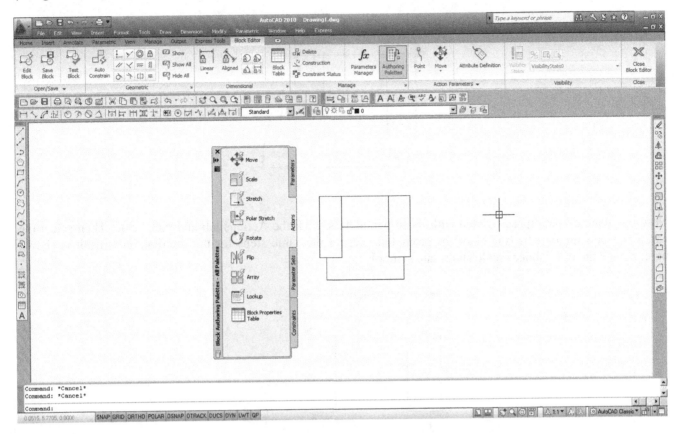

Figure 7.9 – **Dynamic Block Edit mode**

A Block Authoring Palette will also appear. Sets of parameters and actions are available from this palette, and the block can be set up to have any of those. When done, Close Block Editor is pressed (top right of screen) to return back to the regular drawing space. So what can we do here? Well, let's add two *actions* and *parameters* to the chair that will give it some scaling and rotational abilities.

Open the Parameter tab on the palette and pick the linear parameter. That after all is the way "scale" works, by linearly scaling up an object. Pick first the upper left corner start point, then the upper right corner end point of the chair (sort of like adding a horizontal dimension). The result is shown here.

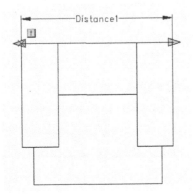

While you're there, add a rotation as well by running through the required base point, radius and angle prompts. The result is shown here.

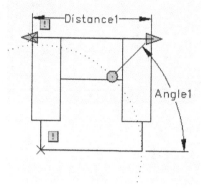

Now you can add actions associated with those parameters. Pick the Actions tab and pick scale. Then select the parameter and the objects (the chair) as prompted. Repeat the same steps for the rotation. The result is shown here. Notice the new (faded) scale and rotate symbols.

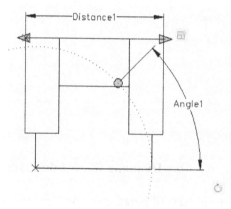

Close the Block Editor (button at upper right) and you will be prompted to save your work as seen here.

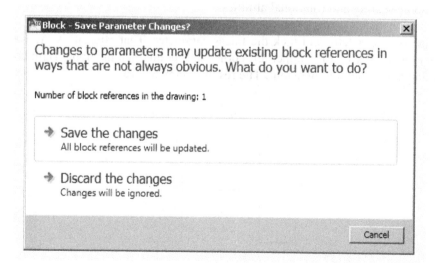

Finally you are back where you started. Click the new block and you will notice the dynamic grips as seen here.

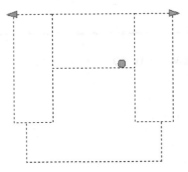

You can now use these to change the size and rotation of the block as seen here with the scaling.

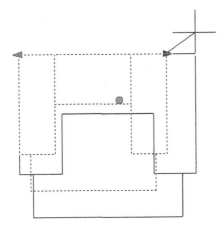

There are of course a few more tricks to these dynamic blocks, chief among them being are the new parametric tools, but we will address that in the advanced level.

So how do dynamic blocks fit into the big picture? Are they as useful as they seem? In the author's opinion the results are mixed. The idea is certainly a good one, but the effort involved to add meaningful and extensive dynamic intelligence to a large group of different blocks is considerable. Existing libraries of blocks have sufficed for most users, who have diligently built them up over the years, and experience has sometimes shown that there has been a reluctance to redefine these libraries. A lack of time and money to tinker with AutoCAD (as opposed to using AutoCAD for billable drafting work), has always been a problem. In a sense it may be a shame as dynamic blocks, by allowing multiple variations on a single block, may actually reduce the size of libraries. A door may just be a door, adjustable to any size on the go, instead of 20 doors for every conceivable situation. It's a useful, if underused, idea.

Chapter 7
Summary

You should understand and know how to use the following concepts and/or commands before moving on to Chapter 8:

- **Block**
  o Select objects
  o Name the block
  o Select base point

- **Wblock**
  o Select objects
  o Select path and enter block name
  o Select base point

- **Options**
  o Retain
  o Convert to Block
  o Delete

- **Insert command**

- **Purge command**

- **Dynamic Blocks**
  o Parameter
  o Actions

Chapter 7
Review Questions

Answer the following based on what you learned in Chapter 7.

1) What are **blocks** and why are they useful?

2) What are the differences between **blocks** and **wblocks**?

3) What are the essential steps in making a **block**? A **wblock**?

4) How do you bring the block or wblock back in to the drawing?

5) Explain the purpose of **Purge**.

6) Explain the basic idea behind **Dynamic Blocks**. What are the steps involved?

Chapter 7

Exercises

Exercise #1 – In a new file, create the following architectural table, chair, keyboard, phone and screen, and make a block out of the individual pieces as well as the overall "assembly." Nested blocks such as these are quite common in AutoCAD design. Insert the block back into the file. Dimensions not given (most of them) can be approximated, and naming blocks is up to you, just make them descriptive.
(Difficulty level: Easy. Time to completion: 15-20 minutes)

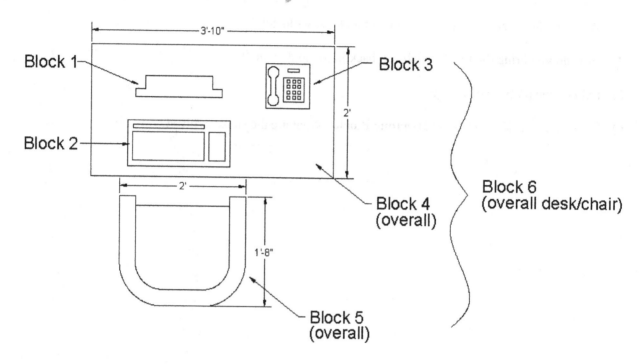

Exercise #2 – In a new file, create the following set of doors and make a wblock out of each, dropping them off in a folder named "Doors". Insert each back into the file. Repeat the process using Dynamic Blocks and the scale action, linear parameter.
(Difficulty level: Easy. Time to completion: 5-10 minutes)

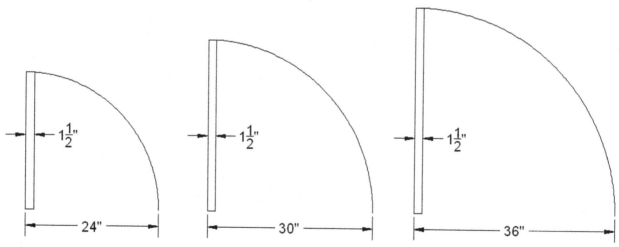

CHAPTER  8

Polar and Rectangular Arrays

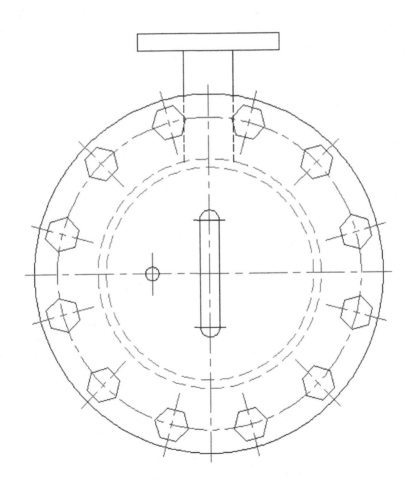

Chapter 8

Learning Objectives

In this chapter we will introduce the concept of an Array and discuss the following:

- Creating a Polar Array
- Object, Center, Quantity, Degrees
- Creating a Rectangular Array
- Object, Rows and Columns, Distances

At the end of this chapter you will be able to create polar and rectangular arrays. You will then start a new in-class mechanical project.

Estimated time for completion of chapter (Lesson + Project): 3 hours

Arrays are a very useful tool in AutoCAD to create patterns of objects. These patterns can be of a circular or linear type, hence the "Polar Array" and "Rectangular Array" discussions below. In a strict sense this tool is not something entirely new, but merely automates what you can do (albeit tediously) by regular commands learned in Chapter 1, and is a huge time-saver.

Sec 8.1 - Polar Array

A polar array is a collection of objects around some common point arranged in a circle (or part thereof). An example in architecture is a group of chairs arranged around a conference table. The idea is to draw the table, then one chair, and after positioning the chair exactly where it will be in relation to the table, use polar array to copy it around the table to create, let's say, 10 of them. The array command will copy and rotate them into position and space them out evenly – a task that would have taken a while to do one chair at a time.

In mechanical engineering an example may be teeth of a sprocket gear. You draw the circle representing the wheel of the gear, then put one tooth properly drawn, detailed and centered on top, then array to get the full gear. Below are two illustrations of what was just discussed. There are of course many other examples.

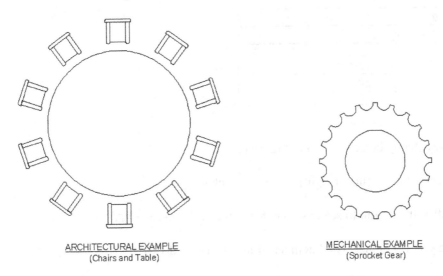

ARCHITECTURAL EXAMPLE
(Chairs and Table)

MECHANICAL EXAMPLE
(Sprocket Gear)

Figure 8.1 – Example of arrays

➢ **Steps in creating a Polar Array**

First of all we need to create something to array. So go ahead and draw a circle of any size to represent the table above, and then create a convincing-looking chair. You may want to make a block out of it after you're done for convenience. Then position it at the top of the table (Figure 8.2). Be sure to line it up perfectly on the vertical axis (or else the array will be skewed off center), and also move it back a bit from the table with Ortho on. So now what does AutoCAD need to know?

- It needs to know *what object* to array (that would be the chair, NOT the table)
- It needs to know the *center point* of the array (center of the table)
- It needs to know *how many* objects to create (lets stick to 10 or so)
- It needs to know if the pattern will go all the way around (that is, 360° or less)

It's good to learn these steps and realize what you need to do before even starting up the array command; that way you have a clear idea of what buttons to push and what data to enter once you have the dialog box in front of you. We used this approach when learning hatch, block and wblock, and will continue to do so.

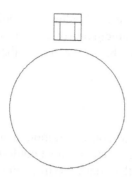

Figure 8.2 – Chair and table - initial setup

Step 1
Once you have the table/chair drawn, follow the steps shown to create the polar array. Start up the array command via any of the methods shown.

Keyboard: Type in **array** and press Enter
Cascading menus: Modify→Array...
Toolbar icon: **Modify** toolbar ⊞⊞
Ribbon: **Home tab→Modify** ⊞⊞

Step 2
You will see the array dialog box as shown in Figure 8.3.
Step 3
Use the "Select objects" button (upper right) to select the chair.
Step 4
Just below to the left is the button to select the center of the table (use OSNAPs).
Step 5
Below that is where you fill in the "Total number of items:" (let's do 10).
Step 6
Finally, below that is "Angle to fill" (we'll go with the full 360°).

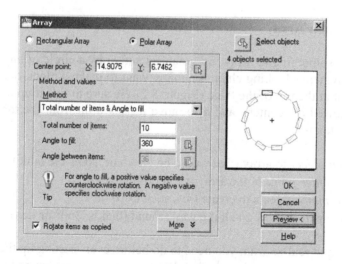

Figure 8.3 – Array dialog box with Steps 3 thru 6 completed

Notice how one of the items we did NOT specify is the angle between the objects. That's something that AutoCAD will figure out for you and space the objects accordingly.

Step 7
Be sure the "Rotate items as copied" button is checked.
Step 8
Important: before you press OK, press "Preview<", which will allow you to see the results of what you did without committing to the final array. If you want to modify some feature of the array, then press Esc and return to the main dialog box, and if not, then press Enter, and you're done. You should see an array similar to the example in Figure 8.1.

Note that the entire array isn't a block, it's just a collection of objects, and to redo the whole thing you will need to erase all the copies, leaving just the table and the original chair and repeat everything again. That's why the preview button is so important. Don't be so quick to press OK right away!

Sec 8.2 - Rectangular Array

Here the idea is basically the same. We are looking to replicate objects, but this time in rows (horizontal) and columns (vertical). An example, as shown in the illustration below, may be columns of a warehouse. If they are spread out evenly, then this is the perfect tool for this. You simply draw one I-Beam with furrowing, and array it up and across the appropriate area.

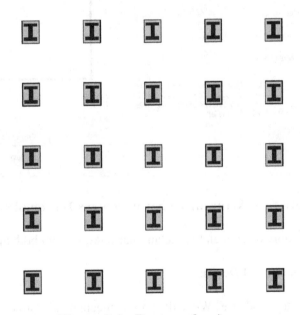

Figure 8.4 – Rectangular Array

Steps in creating a Rectangular Array

The first thing is to draw a convincing looking I-beam (everything to scale this time) surrounded by furrowing, as above. Let's make the furrowing rectangle 20" wide by 26" tall with the I-beam inside. What does AutoCAD need to know?

- It needs to know *what object* to array (that would be the column)
- It needs to know *how many rows* and columns to create
- It needs to know the *distance* between the rows and the columns

Step 1

Start up the array command just as in the previous polar array exercise. The array dialog box will appear as before. Be sure to pick the other choice, "Rectangular Array" this time (Figure 8.5).

Step 2

Use the "Select objects" button (upper right) to select the column.

Step 3

Just below you can enter the number of rows and columns. Let's do ten "Rows" and ten "Columns".

Step 4

Below that you can enter the offset distance for the rows and columns. That is the distance between two similar points on two sample objects. So for example you can say it will be 10 ft center-line to center-line of two columns. That is why it was important to draw them to scale so you know how much to offset. Let's enter 72 for the "Row offset:" and 72 for the "Column offset:". Leave the "Angle of array:" as is.

Step 5

Finally preview the array (it may be partially off the screen) and press OK if all is well.

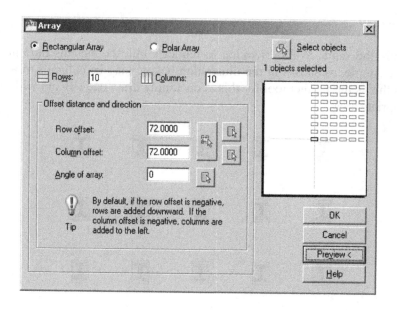

Figure 8.5 – Array dialog box with Steps 1-4 completed

In general, polar arrays are used more often than the rectangular ones, though both are important to know.

In-Class Drawing Project – Mechanical Device

Let's now put what we learned to good use. We will draw a mechanical gadget (a suspension strut of sorts), created years ago for my classes. Though what it is exactly is still debatable, the gadget proved to be an excellent drawing example, and features quite a bit of what we covered in this and preceding chapters as well as a few minor nuances of AutoCAD drafting, which once learned will raise your skill levels considerably. The trick is to make your drawing look exactly like the example. If you approximate and cut corners you will miss out on some of the important concepts that will give a more professional look to your drawings.

A full page view of the completed device is found towards the end of the chapter. Take a good look at it and decide on a strategy to draw it. Over the course of the next few pages we will develop a step by step process to actually do it. Follow the steps carefully; though don't worry too much about the thick border, that's a polyline, to be covered in Level 2. The title block is just a simple collection of lines and text.

Step 1
For starters open a new file, give it a name (Save As…) and set up your layers. We need at a minimum:

M-Part, Color: *Green* (for most of the parts of the device)
M-Text, Color: *Cyan*
M-Dims, Color: *Cyan* (for the dimensions)
M-Hidden, Color: *Yellow*, Linetype: *Hidden* (for all hidden lines)
M-Center, Color: *Red*, Linetype: *Center* (for the main center lines)
M-Hatch, Color: *9* (a Grey for cross-section cutaways)

You may also want additional layers for the spring and for the title block. You can decide on the exact layers and their colors. The above is just a guide.

Step 2
Set up all the necessary items for smooth drafting (if desired) such as the text style (Arial, .2"), the dim style, units (either architectural or mechanical is fine), and get rid of the UCS icon if you do not like it.

Step 3
To begin the drawing itself, we need to start with the hex bolts. Go ahead and draw one of them according to the information below (be VERY careful…all sizes are diameters NOT radiuses; you must tell AutoCAD this by pressing **d** for diameter while making a circle). This has tripped up many of my students in the past. Also don't forget: Make layer M-Part current! If the bolt is anything but green in color, you've missed this step. The dimensions are only for your reference for now – no need to draw them.

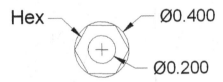

Step 4
Now make a block out of the above, calling it Bolt - Top View.

Step 5
Complete the rest of the top view by drawing the remaining circles and positioning the bolt at the top as shown. Use accuracy (OSNAPs), though the bolt can be any distance from the top of the circle. Be careful to use diameters as indicated.

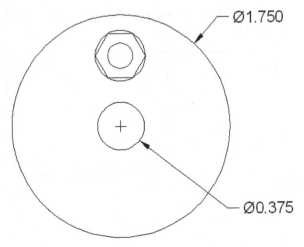

Step 6
Array the bolt around the top view four times as shown next.

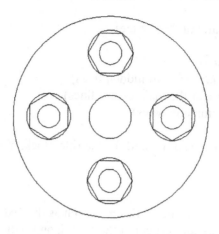

Step 7
You now have to project the main body of the part based on this top view. The best way is by drawing two horizontal guidelines of any length starting from the top and bottom quadrants of the big circle, then an arbitrary vertical line to mark the beginning of the part on the right as shown below.

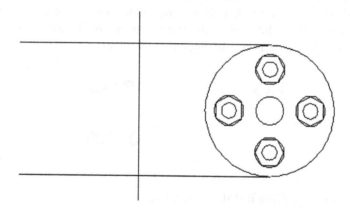

Step 8
Trim or fillet the main vertical guideline and continue offsetting based on the given geometry to "sketch out" the basic outline of the shape as seen below. Do not yet add dimensions; they are for construction only.

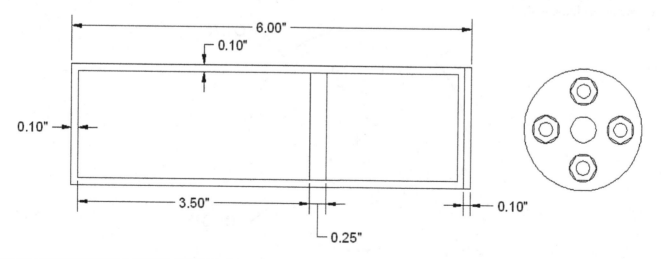

Step **9**

Continue adding the basic geometry to the side view, as well as hidden and center lines as show below. Set LTSCALE to .4.

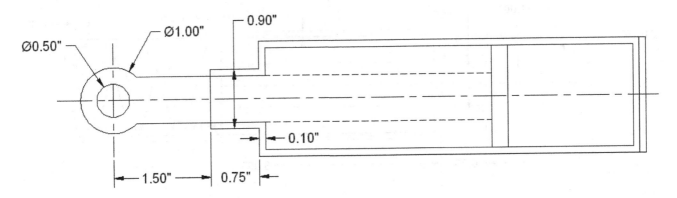

Step 10

Add the bolt projections as seen here (depth into the side view can be approximated). Finally, add hatch patterns, using the Angle option to reverse the hatch directions, which is a common technique in mechanical design to differentiate the parts from each other in cross-section.

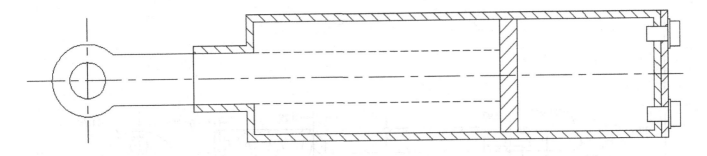

Step 11

Now, how do you do the spring? Well, this is where the Circle/Ttr comes in. Draw one circle at the upper left of the area where it's supposed to go with the correct diameter (see main diagram). Then one way of proceeding is to draw a straight arbitrary line down from the center of that circle, rotate the line 15° about that center point, then offset .1 in each direction, erase the original line, extend the two new ones to the bottom (if necessary) and repeat the process.

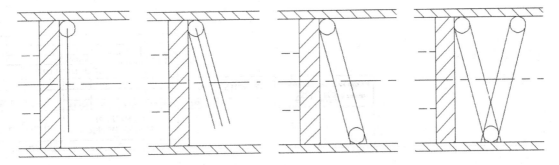

Once a set is done, you can then mirror the rest over and trim carefully to create the impression of a coiled spring.

Step 12

Finally add dimensions and a title block by simple use of rectangles as seen below in the final screen shots. The title block and its contents are typical of a mechanical engineering drawing.

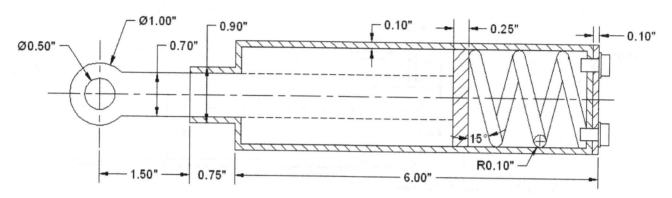

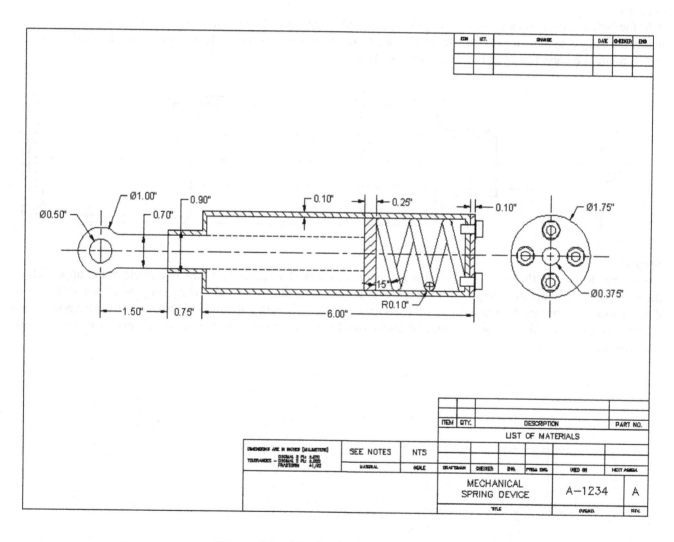

Figure 8.6 – Mechanical Project - final

Drawing the star:

As promised, the technique of drawing the star from Chapter 2 is as follows. Create the pentagon, then, noticing that it is a 5-point geometric shape, draw lines from endpoint to endpoint connecting those vertices. Then erase the pentagon and use trim to erase the inner lines, as shown in the progression below.

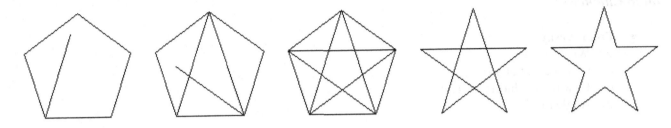

The reason students do this simple exercise is to underscore a very important point. AutoCAD is used across a vast array of industries, each with its own symbols and unique objects. It would be impossible for the software to include all of them in its database. Instead AutoCAD gives you basic tools and leaves it up to your creativity and skill to create the shape you need. So when at a loss as to how to draw something you imagined in your mind, stop and think about the basics that you've learned and how you can put them together into something new.

Chapter 8

Summary

You should understand and know how to use the following concepts and/or commands before moving on to Chapter 9:

- **Polar Array**
  - What to array
  - Center point of array
  - How many objects to create
  - Angle to fill

- **Rectangular Array**
  - What to array
  - How many rows
  - How many columns
  - What distance between rows
  - What distance between columns

Chapter 8

Review Questions

Answer the following based on what you learned in Chapter 8.

1) List the two **types** of arrays available to you.

2) What are the four steps needed to create a **polar** array?

3) What are the three steps needed to create a **rectangular** array?

Chapter 8

Exercises

Exercise #1 – In a new file, draw the image found on the cover page of Chapter 8, reproduced again below with dimensions. You should be able to draft everything shown, including the dimensions.
(Difficulty level: Easy/Moderate. Time to completion: 10-15 minutes)

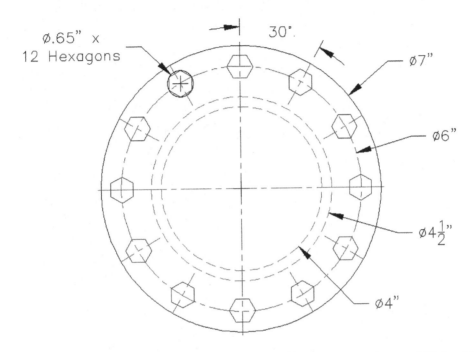

Exercise #2 – In a new file, draw the following rectangular array. The chairs are 20" by 20" and the offset between them is 50". The array is also at a 45° angle.
(Difficulty level: Easy. Time to completion: 5 minutes)

CHAPTER  9

Isometric Drawing

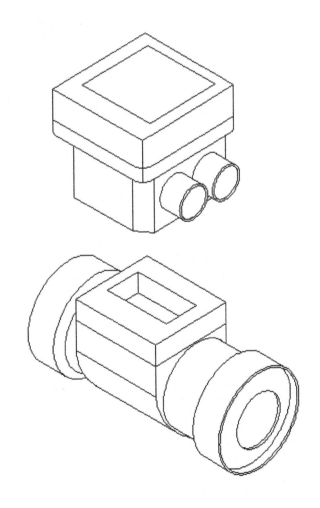

Chapter 9

Learning Objectives

In this chapter we will introduce Isometric Perspective and how to use it.

- What is isometric perspective?
- When to use it
- When not to use it
- Setting isometric perspective
- Changing planes (F5)
- Ellipses in isometric
- Text in isometric

In the course of the chapter you will create several isometric designs including a computer desk and mechanical devices.

Estimated time for completion of chapter: 1 hour

Sec 9.1 – Introduction to Isometric Perspective

By strict definition Isometric Perspective means representing a three-dimensional object in two dimensions. As applied to AutoCAD it means we are drawing a 3D object without resorting to the complexities of 3D space, and sort of "faking it" in 2D, by angling horizontal lines 30 degrees (the accepted Isometric drafting standard, though there is also Axonometric which is 45 degrees). This creates the illusion of 3D and is good enough to get an idea across; indeed, that's how it's done with paper and pencil. In our case we will sometimes leave the horizontal line flat (as with the box below) and sometimes angle all of them 30 degrees (as with the computer table exercise at the end of the chapter).

➤ **Why use isometric instead of 3D?**

There are a few reasons. One is that it's much easier to learn isometric drawing as opposed to developing proficiency in real 3D. So for someone who only needs a quick 3D sketch once in a while just to accent an otherwise excellent 2D presentation, this perfectly fits the bill. Another reason is that 3D takes up more time and computing resources, something that isn't always available, and there's also some difficulty in combining 2D and 3D drawings on one sheet. Many students also don't go on to take AutoCAD 3D right away (nor should they) and isometric fills that temporary void quite well, so it's taught in Level 1. For those that do go on, isometric serves as an excellent introduction to 3D and should be reviewed before starting the 3D instruction.

➤ **When NOT to use isometric**

Isometric is inappropriate when precision 3D drawings are required for design, testing and manufacture. Remember, isometric perspective is only for visualization. These are NOT true 3D objects and have no real depth, only the illusion of depth. As a result, true measurements and the use of the offset command are not possible in the traditional sense.

Sec 9.2 - Basic Technique

There is nothing inherently special about drawing in isometric. Draw a straight (Ortho) horizontal line, then rotate it 30 degrees counter-clockwise. Then attach a perfectly straight line to one of the tips using ENDpoint. Copy the original horizontal line from where it meets the vertical to the other end (using ENDpoint as well). You're on your way to drawing an isometric box. It's the same as taking a T-square in hand drafting and putting a 30° - 60° - 90° triangle on it as illustrated below:

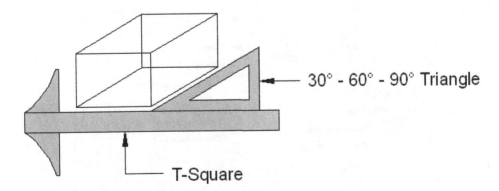

That wasn't hard, but it was tedious. There is of course a way to have AutoCAD preset the cross-hairs so you're always drawing in isometric. To explain it we need to introduce the idea of planes, so here's a short intro to the plane concept.

In isometric you are always drawing in one of three available planes: Top, Right or Left (see Figure 9.1). Because you are using a 2D pointing device (a mouse) which exists only in a 2D world, you need to be able to easily move from one plane to the other to be to draw on it. To toggle between planes press the F5 key. Your cross-hairs will line up accordingly, and now you just draw using the line command. Here's one final rule: you almost always have Ortho on while in isometric mode, otherwise the lines will not be straight and rotating them 30 degrees will be meaningless. Let's put it all together and draw an isometric box. After you are able to do this, more advanced isometric drawings will be surprisingly easy.

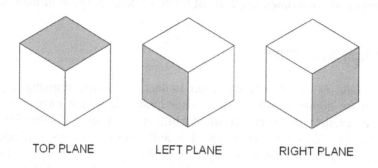

TOP PLANE   LEFT PLANE   RIGHT PLANE

Figure 9.1 – Isometric Planes

Step 1
Open a blank file.
Step 2
From the cascading menu select: **Tools→ Drafting Settings…** and a dialog box will appear as seen in Figure 9.2.
Step 3
Pick the "Snap and Grid" tab, go to "Snap type" (lower left), pick "Isometric snap" and finally press OK. Here's a screenshot of the Drafting Settings dialog box with the Isometric snap selected (Figure 9.2).

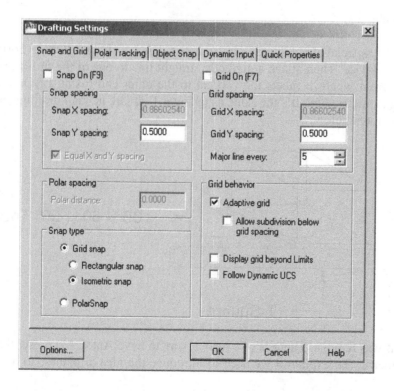

Figure 9.2 – Drafting Settings, Snap type - Isometric snap

Step 4
Make sure Ortho is activated and the cross-hairs are set at 100% for easier visualization of planes. Notice also how the cross-hairs are positioned now that you are in isometric.

Step 5
Draw straight lines (as they should be with Ortho) of any size and continue drawing until you have the front face of a box.

Step 6
Very Important: For the final line, don't try to connect it to the first line; you'll never get it exactly right. Instead overshoot and click after passing it by, as shown next.

Here's a screenshot of Step 5 and 6:

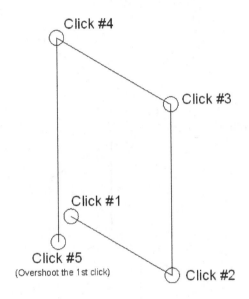

Step 7
Now use the fillet command to fillet the line between Clicks 4 & 5 and Clicks 1 & 2.
Here's the result:

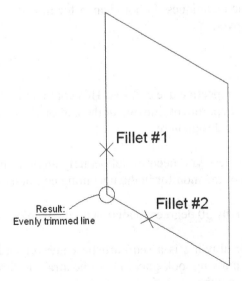

Step 8
Now press F5. The cross-hairs will shift to another plane and you can draw another surface.

Step 9
You can press F5 again for the final surface or also use the copy command. Be sure to use the ENDpoint OSNAP to cleanly connect lines.

Here is what you get after a few lines:

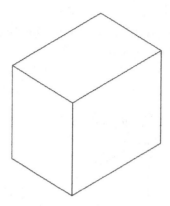

And the final result after a few more lines:

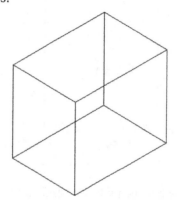

What makes isometric easy is that the techniques do not change for more complex objects. There may be more lines but it's the same thing over and over.

Here are some additional pointers:

- To draw circles in isometric perspective use ellipses. This command was introduced in Chapter 2 and was used to make the bathtub in the apartment drawing at the end of Chapter 4. We will say a few more words about using ellipses in isometric design here.

- Always stay in Ortho mode unless you need to deliberately angle a line at some odd angle. You see an example of this with the back of the monitor in the upcoming computer desk drawing.

- Text can be angled up or down by 30 degrees to align with the isoplanes. We will discuss this shortly.

- Have fun with this. Isometric drawing is a semi-artistic endeavor and not as rigid as drafting. Eyeball distances but try to make the drawing look good. Don't be afraid to throw in some hatch patterns, as was done on the next drawing. Remember though, if you need more precision you'll have to stick around for the real 3D.

Sec 9.3 - Ellipses in Isometric

A circle angled away from you is an ellipse, and these shapes find wide application in isometric drawing. You will need ellipses for the base of the monitor in the upcoming computer desk drawing as well as the power button and extensive end-of-chapter exercises.

There is no sure fire way to create the perfect ellipse; it will always require a bit of adjusting to get right and a bit of a creative eye to do so. Reproducing what was first shown in Chapter 2 we have the following:

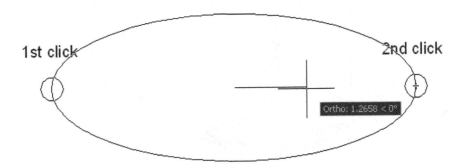

Once the ellipse is positioned and is the right size, you can rotate it until it looks correct and in the same plane as the rest of the isometric geometry. Here is an example of some ellipses in an isometric box.

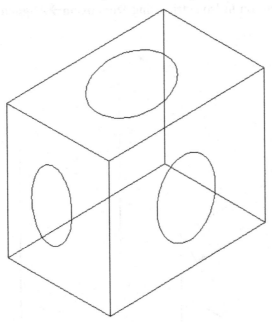

Sec 9.4 - Text and Dimensions in Isometric

Text can easily be aligned with isoplanes by rotating it plus or minus 30° (or other values such as 90°), depending on the plane as shown in Figure 9.3.

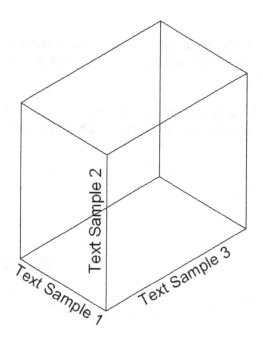

Figure 9.3 – Text in isometric

Although not ideal, you can dimension in Isometric using **Dimension→Aligned** and vertical dimensions as seen here in Figure 9.4.

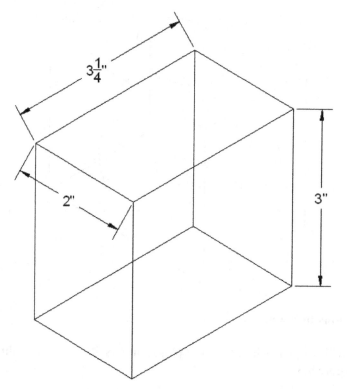

Figure 9.4 – Dimensions in isometric

Chapter 9

Summary

You should understand and know how to use the following concepts and/or commands before moving on to Chapter 10.

- **Isometric**
  - What is it?
  - When to use it?
  - When not to use it?
  - Isometric settings

- **Planes**
  - What are the three planes?
  - Switching between them (F5)

- **Ellipses in isometric**

- **Text and dimensions in isometric**

Chapter 9

Review Questions

Answer the following based on what you learned in Chapter 9.

1) List some instances when isometric is **appropriate**.

2) List some instances where **3D** would be better than isometric.

3) Name the three **planes** that exist in isometric.

4) In what dialog box is **isometric snap** found?

5) What **F-key** switches the cursor from plane to plane?

6) What is almost always on when drawing in isometric?

7) What is the equivalent of a circle in isometric?

Chapter 9
Exercises

Exercise #1 – Draw the following computer desk. You will need your basic isometric knowledge and construction techniques developed in this chapter. You will also make use of the ellipse command and hatch the desk itself. All sizes are approximate and may be estimated.
(Difficulty level: Easy/Moderate. Time to completion: 20-30 minutes)

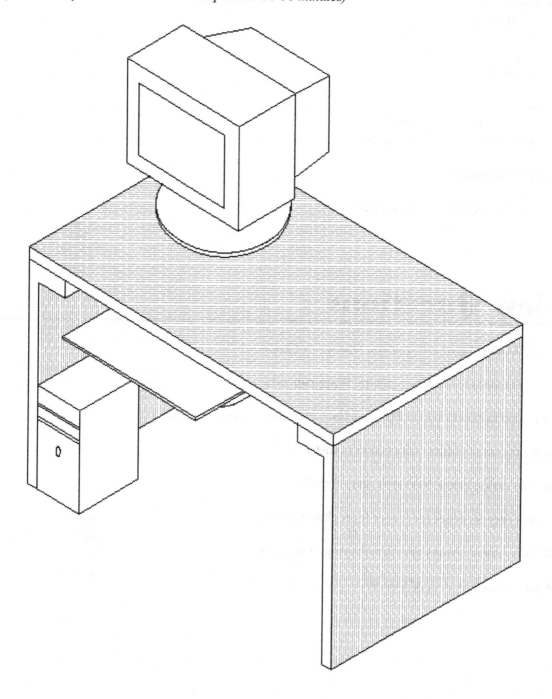

Exercise #2 – Draw the following mechanical device. You will need your basic isometric knowledge and construction techniques developed in this chapter. You will also make extensive use of the ellipse command. All sizes are approximate and may be estimated.

(Difficulty level: Moderate. Time to completion: 30 minutes)

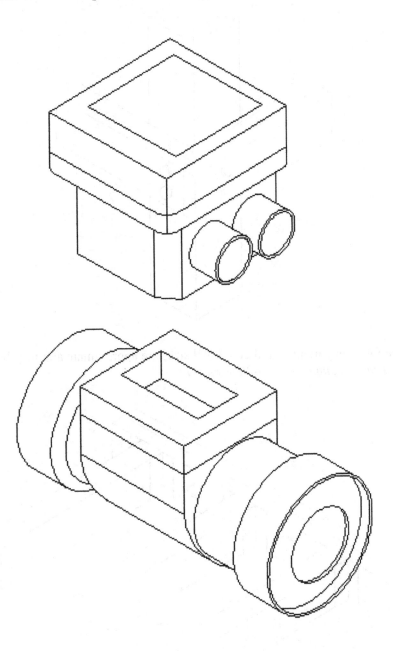

Exercise #3 – Draw the following architectural detail. All sizes are approximate and may be estimated.
(Difficulty level: Easy. Time to completion: 10-15 minutes)

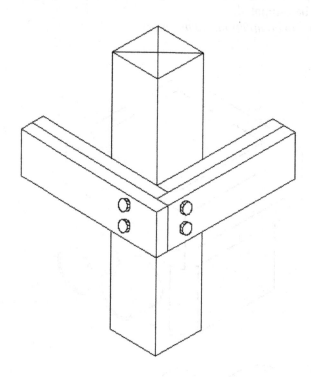

Exercise #4 – Draw the following architectural detail. All sizes are approximate and may be estimated.
(Difficulty level: Easy. Time to completion: 15-20 minutes)

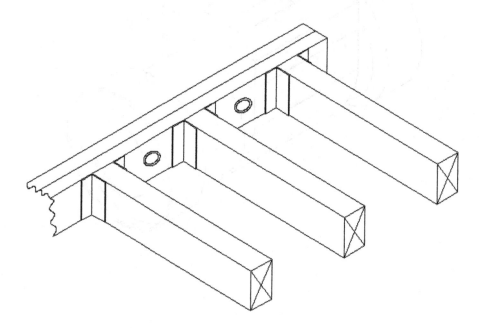

CHAPTER  10

Basic Printing and Output

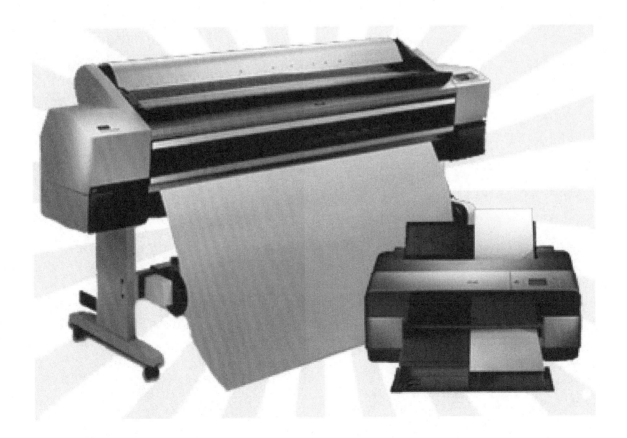

Chapter 10

Learning Objectives

In this chapter we will introduce the Plot dialog box and the Page Setup Manager, and go through all the associated options to produce a print or a plot, including the following:

- What *printer* or *plotter* to use
- What *paper size* to use
- What *area* to plot
- What *scale* to plot at
- What *pen settings* to use
- What *orientation* to use
- What *offset* (if any) to use
- The page setup manager

Upon the completion of this chapter you will be able to recognize and set up all the features needed to print or plot any type of drawing from any AutoCAD workstation.

Estimated time for completion of chapter: 1 hour

Sec 10.1 – Introduction to Printing and Plotting

In almost every situation the on-screen AutoCAD design is plotted or printed and indeed the fundamental "product" of most design work is paper output. This output is then used to manufacture or build the design and it also may be needed for sales or archiving purposes.

Because designers and their clients want something tangible to look at and use, all design work in AutoCAD must be created with the ultimate goal of displaying it on a piece of paper, not just on-screen. This fact influences the evolution of a drawing from the outset, and a skilled AutoCAD designer will select paper size, output scales, and line weights early in the process so the inevitable output will proceed smoothly. We will eventually talk about all of these and more.

Professional-looking output in AutoCAD is no accident, and comes with its own set of unique challenges. Though you may be able to walk up to an expertly set up drawing and just select **File→Plot...→OK**, such situations will only result from a thorough understanding of what's involved. More importantly, if the requirements call for a different output of the same drawing (e.g.: for an informal check of a small section of the drawing versus a final full-size submission to the client), you must know what to change and by how much.

AutoCAD allows for a wide array of output, and the devil is in the minute details. Output spans this chapter as well as part of Chapter 18 (Paper Space) and Chapter 20 (Advanced Output and Pen Settings). Our goal right now will be only to familiarize you with the essentials and what AutoCAD needs to know from you in order to print or plot a drawing.

Take note as you go through this that, although some specifics (such as type of printer or plotter) may be unique to your particular office, home or school, most of what you need to know will stay constant from situation to situation and from one release of AutoCAD to another (even if the location of the buttons and menus change); thus, it is important that you memorize and understand the essential concepts.

Sec 10.2 - The Essentials

You will need to tell AutoCAD the following information when setting up a drawing for printing or plotting:

- What *printer* or *plotter* to use
- What *paper size* to use
- What *area* to plot
- What *scale* to plot at
- What *pen settings* to use
- What *orientation* to use
- What *offset* (if any) to use

There are of course some other subtleties, but we will focus on these essential ones for now. Note that we will not look at the actual Plot dialog box until later, as you truly need to understand what buttons you are pushing, and why, before you actually go and do it.

➢ **What Printer or Plotter to Use**

This is where you select what output device you will be using. In most architecture offices, engineering companies and schools you will have a plotter (color or black and white) for C, D and E-sized drawings (paper sizes discussed next), and a LaserJet printer (color or black and white), for A and B-sized drawings. Your typical AutoCAD station will usually be networked to access both, as requirements may change daily. You will need to select which device will be getting the drawing (AutoCAD doesn't care – it will send a floor plan of a sports stadium to a postcard if you let it). This choice is of course tied in with the scale and the ultimate destination of the printed drawing (check plot or to a client). Much more on this later.

➢ **What Paper Size to Use**

You would think that paper sizing is a cut-and-dried, straightforward matter (somewhere a professional printing company manager is laughing), but this is certainly not the case. Paper comes in a bewildering array of international sizes and standards in both inches and metric (millimeter) sizing. Some common standards include ANSI, ISO JIS and Arch. For purposes of this discussion we need to simplify matters and present all paper as existing in the following basic sizes as viewed in Landscape mode, meaning the first (larger) value is the horizontal size:

- A SIZE – 11" × 8.5" sheet of paper (letter).
- B SIZE – 17" × 11" sheet of paper (ledger)
- C SIZE – 22" × 17" sheet of paper.
- D SIZE – 36" × 24" sheet of paper.
- E SIZE – 48" × 36" sheet of paper.

Generally speaking, these are the maximum paper dimensions. Some standards list essentially the same sizes but with the border accounted for, so a D-sized sheet will be 34" × 22" and so on. Note that legal (14" × 8.5") sized paper, while it certainly can be used, is not common in engineering or architecture.

As a side note, we can now define precisely the difference between printing and plotting.

Printing:
Usually refers to output of drawings on A or B-sized paper, which is easily done on most office LaserJet printers. Most home printers can only do A size, but this of course varies with the size (and cost) of the printer.

Plotting:
Usually refers to output of drawings on C, D, or E-sized paper which has to be done on plotters, as a LaserJet generally can't accept paper of this size.

Memorize the paper sizes presented above, as it is pretty much the industry standard as far as AutoCAD design is concerned. You will hear these sizes asked for and referred to come printing time, and if there is any deviation from the above, rest assured it won't be anything radical, as there is only so much flexibility in what printers and plotters can accept (at least among the models that small companies and schools can afford).

➢ **What Area to Plot**

What "area to plot" can be taken quite literally – as in "what do you want to see on the printed paper?" The essential choices boil down to really two. Do you want to see the entire drawing (usually) or parts thereof (sometimes). Your choices in that regard include the following.

Extents

This option prints everything visible in the AutoCAD file. However, this means everything visible to AutoCAD (and not necessarily to you!), so if you have a floor plan with some stray lines that you didn't notice located some distance away, AutoCAD will shrink the important floor plan to make room for those unimportant stray lines. This is exactly the same as typing in **Zoom→Extents** (from Chapter 1). So it is a very good idea to zoom to extents prior to printing. Remember this acronym: WYSIWYG – What You See Is What You Get. If you see it on the screen it will print unless prior arrangements are made (Freeze the layer, etc.) or the geometry is on the Defpoints layer. For a carefully drawn layout, Extents is the only choice when you want to see the entire drawing.

Window

As you may have guessed, this is for those situations where you want to see just a piece of the drawing. You simply select that choice and click the window button that will appear to the right of it and draw a selection window around what you want to see.

Display and *Limits*

The remaining two choices are of little use in most situations. "Limits" plots the entire drawing from 0,0 to the preset limits (something most users don't set anyway), and "Display" plots the view in the current viewport in the Model tab or the current paper space view in a Layout tab (something we haven't covered yet). One can generally use Window in those situations, but Display is fine as well if you want only the current viewport printed.

➤ **What Scale to Plot**

Of all options in Plot, this one causes the most confusion to a beginner AutoCAD user for the simple reason that scale is closely tied in with a very important topic: Paper Space. The essential idea here is that while yes, you can assign a scale from the Plot dialog box, in reality this isn't usually done for most drawings except for simple and small mechanical pieces such as bolts and brackets. What is done instead is that a title block is set up in Paper Space and the design (in Model Space) is scaled by means of Viewports (much more on this in Chapter 18), so then in the Plot dialog box you can select 1:1 (one-to-one) and a properly set up drawing will print to scale.

What if you don't care about scale and just want a print to look over? Here the pendulum swings in the other direction and things become very easy. Just check off the "Fit to Paper" box above the scale and the drawing will fit upon whatever paper you're using, with no regard to scale whatsoever. While this technique is NOT for final prints of to-scale architectural layouts, it is used very often for non scaled work such as electrical wiring schematics, network design, conceptual layouts, and of course check plots.

➤ **What Pen Settings to Use**

This is a topic that has an entire chapter all to itself (Chapter 20), and we will not go into detail here. Suffice it to say, if there is one thing that will instantly scream "amateur," it is lack of Pen Settings. The concept is simple, critically important in hand drafting, yet sometimes overlooked in the computerized world of AutoCAD. It deals with the fact that linework in a drawing is not all the same size. Primary features (such as walls on a floor plan) should appear darker on a print, while secondary features (furniture, for example) should appear slightly lighter. This was the reason for the existence of 2H, 4H and other pencil leads (for those old enough to remember), and the reason today for the existence of pen settings.

With AutoCAD, different colors will represent the different pencil leads, and a strict standard needs to be followed as to what drawing features will have what color assigned to them. In this manner, drawings come out with different shades of linework, something that is noticed by the eye, and as a result the drawings are much more readable and professional looking. These settings are created, customized, adjusted and saved under the "ctb" extension, for all to use according to company standards set by management.

For now you will select "monochrome.ctb" as your ctb file, and the entire set of customization tools will be presented in Chapter 20 as promised.

> ### What Orientation to Use

This is simply a question of whether you want the drawing to be oriented Portrait or Landscape; in other words do you want the longer side of the sheet of paper oriented up/down or across. This should be familiar to just about every computer user that has printed anything from a word processor or other software.

One twist here is that you can check off a box to plot upside down. A useful application for this option, as noticed one day by the author, appears when you are printing many sets of similar looking (but different) drawings. If you want to know which drawing is coming out of the plotter without waiting until it plots completely, send it in backwards. That way the title block (with the identifying drawing number) comes out first and that question is instantly resolved.

> ### What Offset to Use

This refers to whether or not you need to shift your drawing around to account for a border and stapling area on the left (or elsewhere). If you check off "Center the plot" it will be automatically centered on the paper, and in most cases this is what you would do.

> ### Miscellaneous

There are a number of other features in Plot. For example you can plot to a file (*.plt) and take it with you to another computer; useful for large volume drawing duplication at stores that don't have AutoCAD, only printers, such as Kinko's®. You can also turn on a plot stamp so information such as date and time of the printing can appear in the border. There are also some plotting features unique to 3D and solid modeling. We will explore some of these topics in Chapter 20. Finally there is a preview window and a "Number of copies" to print selection, the function of those being self-explanatory.

Sec 10.3 - The Plot Dialog Box

Let's now take a close look at the Plot Dialog Box and identify the locations of all the features described earlier. Open up the floor plan from the earlier chapters and bring up plot via any of the following methods.

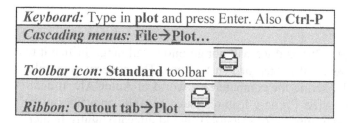

Keyboard: Type in **plot** and press Enter. Also **Ctrl-P**
Cascading menus: **File→Plot...**
Toolbar icon: **Standard** toolbar
Ribbon: **Outout tab→Plot**

The Plot – Model dialog box will appear as seen in Figure 10.1. If it is not full size, press the small arrow at the bottom right to expand it fully.

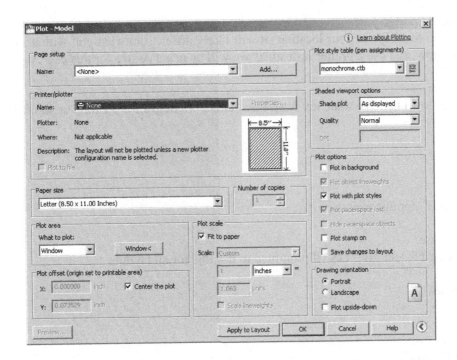

Figure 10.1 – Plot - Model

Here are the features we discussed in Figure 10.2.

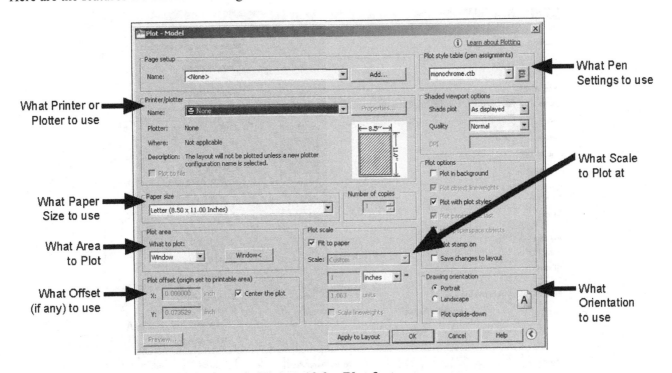

Figure 10.2 – Plot features

Go through all of the settings, referencing the explanations in the previous discussion until you completely understand what each one does. Set the final settings as shown next.

What printer or plotter to use:
Set whatever printer is available at your location (it will likely differ from what you will see here). If you have a plotter, ignore it for now, and use a printer instead, as we will be using A-size for this exercise.

What paper size to use:
Select from the drop-down list. We want any A-size available; "Letter (8.5 × 11 in.)" or similar is fine.

What area to plot:
Select "Extents".

What scale to plot at:
Check off "Fit to paper" for this exercise

What pen settings to use:
Select "monochrome.ctb", and say "Yes" if prompted to use this for all layouts.

What orientation to use:
Select "Portrait".

What offset (if any) to use:
Check off "Center the plot".

Do NOT press OK yet! This is what the Plot box should look like:

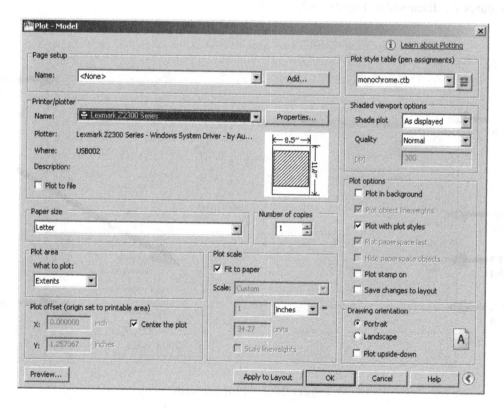

Figure 10.3 – Plot features selected

➢ **Preview**

Much like the situation with creating an array or hatch, a preview button is available. With printing/plotting drawings it is crucially important to first preview the drawing before sending to plot. Paper and ink cost money, and everyone regardless of skill level will make mistakes in setting up printing. This really should become a habit, as errors can be costly, especially on large print jobs. On an industry-wide scale, this simple preview button literally saves tons of paper and gallons of ink. The author has caught not only printing set-up errors but even linework drawing errors during the preview process!

Go ahead and press "Preview…" (lower left corner) and the following preview screen will appear (Figure 10.4). You can zoom and pan around using standard mouse techniques, and when you are ready to send the drawing to the printer (or return back to plot for adjustments) right click and select Exit or Plot, respectively.

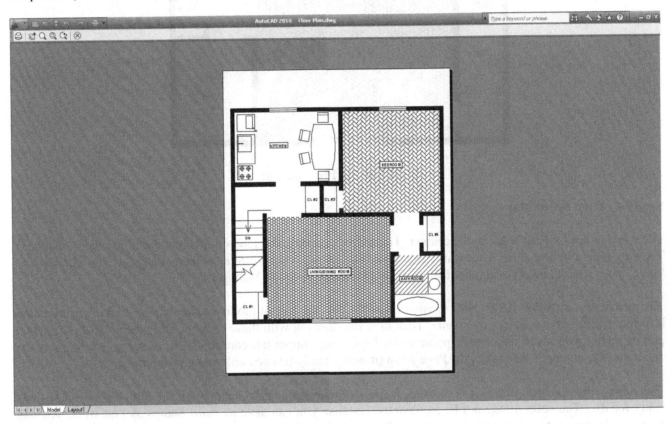

Figure 10.4 – Plot Preview

Go ahead and plot. If everything was set up correctly and your computer is connected to the printer, the floor plan should print as seen in Figure 10.5.

If something doesn't look right anywhere in the process, review the entire set of steps one by one, returning to the preview screen to check up on the effect of the changes.

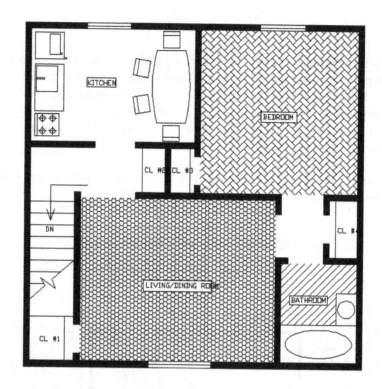

Figure 10.5 – Final output

Sec 10.4 - Page Setup Manager

There is one last topic to learn in this chapter. The settings on the print job you just sent are not saved in memory. Try it by typing in plot again. Unless it's been saved somehow, you have to reset them all again. The technique to "save" your settings involves another feature of Plot called Page Setup Manager.

The idea here is as follows. You need to enter all of the data for a print job in another similar looking print dialog box and save it under a descriptive name. Then save the drawing with these settings, and when you do the actual plot, the name you saved under will appear under Page setup (upper left corner). Then, no more entering settings one item after another – just select the Page setup (it may already be cued up), preview and print.

This has two advantages. The first is obvious: it speeds things up a lot by keeping all the drawing's print settings ready for use. The second may not be as obvious. Because you can save as many setups as you want, you can have one drawing set up for multiple printers and/or multiple settings.

For example, your office may need the final output sent in color to a D-sized plotter at a certain scale, but in-house checks are done on 17" × 11" sheets, in black and white and no scale. You can easily set up two distinct setups and name them "Client_Layout" and "Check_Layout". The author has on occasion used up to a dozen setups on one drawing, each with its own purpose: presentation, archiving, optimized for color, check plots, etc, etc.

A question usually arises when this topic is presented. Does this need to be done every time a new drawing is created from a blank file? The answer is yes, but it's something that only needs to be done once and will save you much time down the road. Obviously, using templates will negate the need to do this at all, as all settings will carry over.

The Page Setup Manager is found using the cascading drop-down menu: **File→Page Setup Manager...** or the Output tab on the Ribbon and is shown in Figure 10.6.

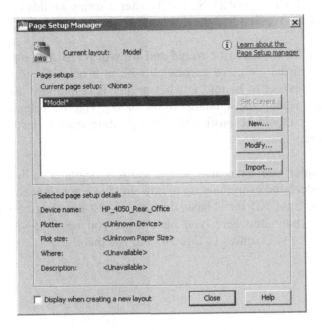

Figure 10.6 – Page Setup Manager

Press the New...button and the New Page Setup will appear as shown below. Enter the name of the new setup and press OK.

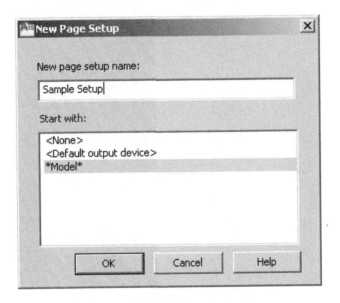

Figure 10.7 – New Page Setup

You will then be taken to the now familiar looking Page Setup dialog box. Upon completion of the settings, a preview (yes, still needed), and a press of the OK button, you will be redirected back to the Page Setup Manager ready for another setup if needed. If not press Close.

To see the benefit of this, go to plot and select the name of the setup from the Page setup in the upper left corner and all your settings will appear. Plot the drawing and save it. You will never have to enter settings again (unless something changes like arrival of a new printer). So finally after learning all this new material, you're now able to reduce printing to a pressing of the Preview and the OK buttons - our goal all along.

As stated before, there is still a lot we haven't covered and we will revisit this topic again in Chapter 18 and 20. Printing is something you will usually receive assistance with at a new job, training center or consulting gig. However, the most you should expect to be told is what printer is commonly used (up to 10 connected to one computer is not unheard of) and what is the standard ctb file in use. The rest is up to you, and with the knowledge you gained in this chapter, you should be proficient in all printing matters just shy of the level of AutoCAD management and administration.

As of completion of this chapter you have also passed from a beginner to intermediate level of AutoCAD knowledge. Hopefully, you understood everything up to now and have not skipped any exercises or reading. If so, then good job! If all you need AutoCAD for is basic drafting or just knowing enough to oversee other full time AutoCAD designers, then you may stop here, you have what you need. If your work requires expert level knowledge, you will need to go on to Chapters 11 thru 20. See you there!

Chapter 10

Summary

You should understand and know how to use the following concepts and/or commands before moving on to Level 2 and Chapter 11.

- **What Printer or Plotter to use**
  - Differences between plotting and printing

- **What Paper Size to use**
  - A-Size
  - B-Size
  - C-Size
  - D-Size

- **What Area to Plot**
  - Extents
  - Window
  - Display
  - Limits

- **What Scale to plot at**
  - 1:1
  - Fit to Paper
  - Other Scale

- **What Pen Settings to use**
  - ctb files

- **What Orientation to use**
  - Portrait/Landscape

- **What Offset (if any) to use**
  - Center Plot
  - X ort Y Offset

- **Preview**

- **Page Setup Manager**

Chapter 10

Review Questions

Answer the following based on what you learned in Chapter 10.

1) List the seven essential sets of information needed for printing/plotting:

2) List the five paper sizes and their corresponding measurements in inches.

3) What's the difference between printing and plotting?

4) List the four plot area selection options.

5) What scale setting ignores scale completely?

6) What is the extension of the Pen Settings file?

7) Explain the importance of Preview before printing/plotting.

8) Why is Page Setup Manager useful?

Chapter 10

Exercises

Exercise #1 – Open the mechanical device drawing from earlier chapters and practice printing with a variety of scales and other settings.
(Difficulty level: Easy. Time to completion: 10 minutes)

- Select printer.
- Select paper size – stay with A-size.
- Select Extents for area. Also try again with Window; observe the results in Preview.
- Try a variety of scales such as 1:1, 1:2 and so on; observe the results in Preview.
- Pick the monochrome ctb file.
- Try both Landscape and Portrait orientation.
- Center the plot.

Spotlight On: Architecture

AutoCAD is the undisputed CAD software of choice among most architects, and indeed the software finds its widest application in architecture and related fields. Therefore, it makes sense to feature architecture in our first "Spotlight On:" feature.

Architecture is the art and science of designing and constructing buildings and other physical structures for human shelter or use. It's a far reaching field that encompasses not only construction and layout of structures but also some civil engineering, landscape design, interior design and project management.

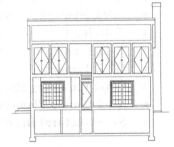

Architects are often described as a cross between engineers and artists. They not only have to have extensive knowledge of statics, strength of materials and construction methods and codes, but also artistic flair and well developed aesthetics. Although architects often work with specialists to complete projects, many are well versed in electrical, plumbing, HVAC, landscape and lighting design. Add to that the necessary ability to manage projects and deal with clients as well as financial and legal matters, and one can see that architecture is a demanding and multifaceted profession.

It is also not an easy profession to get started in. In the United States a student needs to complete a Bachelors of Architecture followed by a Masters of Architecture. Depending on individual programs and how they are structured, this can take anywhere from 5 to 7 years. The student then has to gain work experience, often while earning a relatively low salary, before being eligible to sit for the Architect Registration Examination (ARE), which will license them in the state of their residence. While earnings for partners and senior architects can be substantial, it does take quite a while to get established and build up a reputation and client base.

Virtually all architects today have switched to computer aided drafting, and a group of CAD drafters is a common site in most offices. Drafting is time consuming, so most senior architects focus on bringing in new business and creating top level concepts and designs, leaving details and drafting to junior architects and CAD designers (which is sometimes one and the same person). This works out well, as junior architects have an opportunity to learn by doing actual useful work while "learning the ropes" of the business. Acquiring cutting-edge AutoCAD skills is therefore a great way to get your foot in the door of the profession.

According to the 2006 - 2007 Occupation Outlook Handbook published by the US Department of Labor, the median salary of architects in the United States was $62,960 with the middle 50% earning between $46,690 and $79,770. For 10 years' experience, the base compensation level increases significantly to an average range of $62,608 - $79,919; that range reaches $72,678 - $96,928 for architects with 15 years' experience.

Spotlight On: Architecture

Senior architects and partners typically have earnings that exceed $100K annually. It is not unusual for an officer or equity partner to earn a base salary of $235,000, with a bonus of $200,000. Due to the major stake in ownership that equity partners may have, they can earn incomes approaching, and occasionally surpassing, seven figures.

So how do architects use AutoCAD, and what can you expect? A typical medium-sized project, such as a residential home, will feature a set of plans that may include:

- **A-1 Site Plan** (The property surrounding the structure)
- **A-2 Demo Plan** (What needs to be demolished to make room for the new design, if applicable)
- **A-3 Construction Plan** (The actual design, which may be multiple floors)
- **A-4 RCP Plan** (The Reflected Ceiling Plan, which may include lighting)
- **A-5 Electrical Plan** (The outlets, switches, panels and lighting, if not on RCP or separate plan)
- **A-6 Kitchen Elevations** (Shows cabinets and appliances)
- **A-7 Bathroom Elevations** (Shows bath, shower, sink and other features)
- **A-8 Window and Door Schedule** (Lists all doors/windows, and sizing, quantity, etc.)
- **A-9 Cellar** (Shows basement or cellar plan if needed)
- **A-10 Roof** (Shows construction of roof beams, rafters, braces, joists and other elements)
- **A-11 Wall Sections** (Shows the various walls used, and their internal structure)
- **GN-1 General Notes** (Describes the overall job and any special instructions to the contractor)
- **GN-2 Plumbing Riser Diagram** (Describes flow of pipes and plumbing features)
- **S-1 Framing Plan** (Describes the construction of the house frame)
- **S-2 Structural Details** (Additional details of design, may be multiple pages)

Extensive use of AIA layering, or an in-house system based on it, is employed as seen here:

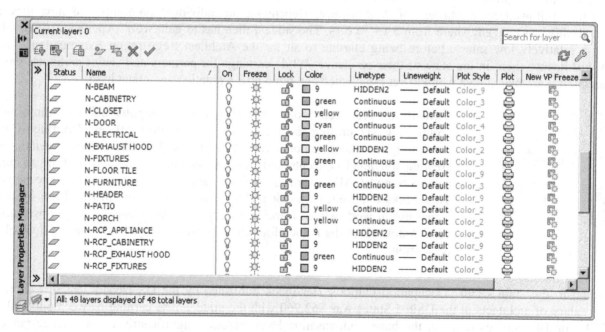

You should also expect to see extensive use of Level 2 concepts such as:

- **Xrefs** (The main plan is Xref'd to the electrical and RCP plans)
- **Paper Space** (Each of the plans listed above is a separate Layout tab)
- **Attributes** (Found in the titleblocks, door and window schedules, and other elements)

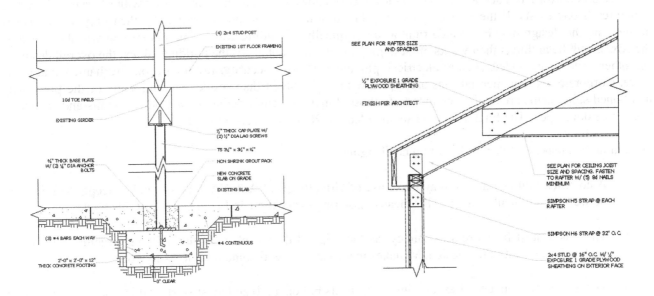

Architectural Details

Residential Architectural Floor Plan

Images courtesy of RM Architect, PC
www.rma-nyc.com

Architectural drafting is a demanding and skill-intensive endeavor. Many jobs are started with a survey of existing structures if one exists. If the project calls for a brand-new building, not a renovation, then only a site plan is needed and the design may be created right away. Typically a building core is laid out (which will then become the Xref), and the architect then works on designing the internal details of the building. When the overall design is set, other systems are added, such as electrical, plumbing, HVAC, security, fire protection, mechanical and much more. Corporate and commercial jobs are typically more involved than residential, and some of the previously mentioned subsystems are applicable mostly to those. Finally extensive construction notes, details and schedules need to be developed. The final document set may feature 20 to 40 sheets or more.

Several factors make architectural drafting challenging:

- A complex building design may make use of virtually all of AutoCAD's tools and concepts; you need to be well versed in all of them. This means mastering every topic in Level 1 and 2 of this book.

- There is a need for extreme accuracy, as the original design plans of the building structure will be the anchor for the rest of the design. Mistakes made early on will come back to haunt you.

- There is rarely a luxury of excess time. Architects bid on a job and must work within a set budget. A slow draftsman that is not proficient (or efficient) will take too long to create the drawings and diminish profits. Working fast and accurately is a must. Rarely do you learn the basics and are allowed to mature "on the job" anymore; it is just too expensive for a small architecture office.

In spite of the above challenges, or maybe because of them, architectural drafting is enjoyable and rewarding, especially if you are a new graduate and are picking up knowledge in your profession along the way. Even if you just do non design drafting, the pay is good (salaries vary from $35,000 - $55,000), and you will acquire some frightening AutoCAD skills very quickly in this fast-paced environment. Most of today's AutoCAD experts (including the author) can trace their roots to using AutoCAD in architecture; it's the cutting edge of computer-aided drafting.

Level 2

Chapters 11 - 20

Introduction to Level 2

Now that you have completed Level 1 – Beginner to Intermediate, you should have the basic skills necessary to work on a drawing of moderate complexity with a certain amount of speed and accuracy. Level 2 is the second part of your journey in becoming proficient in AutoCAD. The goal at the end of this level is for you to know most of the practical information about AutoCAD 2010 in two dimensions. This will allow you to not only work on very complex drawings, but also set up and manage them. You will then, with some additional work experience, have the skills of a CAD manager or administrator, a necessity in many engineering or architectural firms, even if your job is not just AutoCAD.

Starting with the next chapter, this textbook will not feature any of the basic introductions concerning AutoCAD and what it is. It is assumed you have already read through all the introductory information and are familiar with all the material in Chapters 1–10. There will be no further explanations or review of earlier material, so if you need to review anything from Level 1, do so before proceeding. With this book, we will dive into the new material right away and not look back as there is much more to cover.

Level 2 consists of three distinct parts. Chapters 11 through 13 are all about advanced features of basic concepts: lines, layers and dimensions. What we are doing is adding significant depth to these previously introduced Level 1 topics. Chapters 14 through 16 cover a broad set of topics that introduce a wide variety of advanced features such as interacting with other software, the Design Center, Express tools, the CUI and many shorter topics. All this leads up to the "core" of Level 2, namely External References, Paper Space, Attributes and Advanced Output and Pen Settings, that make up Chapters 17 through 20. These are very important topics, especially External References and Paper Space, as those are critical to most architectural designs, and you have to master them for your AutoCAD knowledge to be taken seriously. Indeed, many architectural designs and layouts depend on both to function.

The pace of the course will pick up somewhat as well, and more responsibility will be placed on the student to follow up and do their own research into additional AutoCAD functions. We will also introduce a drawing project that spans the entire level from Chapters 11 to 20. It is a more in-depth floor plan that, while still greatly simplified, does attempt to replicate a real-life architectural layout. There will also be fewer generic drawings to do, in favor of more topic-specific exercises at the end of each chapter. One constant will remain: you will still need to practice to ensure success. Good luck in your studies of AutoCAD and Computer Aided Design!

CHAPTER  11

Advanced Linework

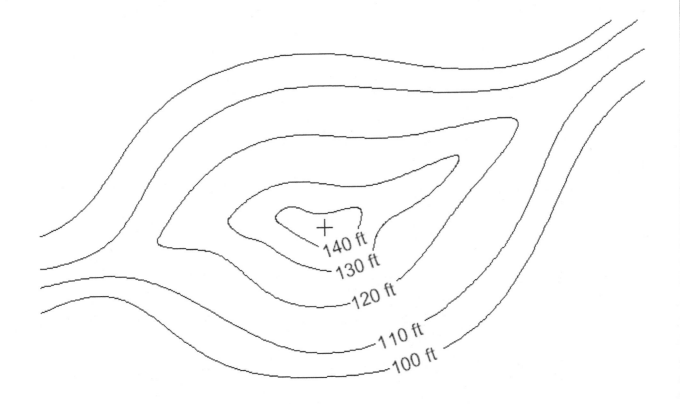

Chapter 11

Learning Objectives

In this chapter we will revisit the line concept and expand upon it by discussing the following:

- Polyline and Pedit
- Xline
- Ray
- Spline
- Mline, MLstyle and MLedit
- Sketch

By the end of the chapter you will have learned how to use and apply these new types of lines, and be able to incorporate them into future design work.

Estimated time for completion of chapter: 2-3 hours

Sec 11.1 - Introduction to Advanced Linework

We begin Level 2 where we started Level 1, with the simple line command. At the time this was enough to start with, and no mention was made of the line command's six other relatives. Experience has shown this is sure way to scare off brand new students. Now that you are more advanced and have hopefully built up skills and confidence, we can come back and go in-depth.

As it turns out, while the line command is useful for most drawing situations, it is inadequate or inconvenient for a small but important number of other applications. Here we will look at other variants of the line including the **Pline** (Polyline), the **Xline** (Construction Line), its close relative the **Ray**, a special case of the polyline called a **Spline**, an **Mline** (Multiple Line) and the most unusual one of all, the **Sketch** command.

Sec 11.2 - Pline (Polyline)

A pline is a regular line with some unique properties:

- Property 1 - A pline's segments are joined together.
- Property 2 - A pline can have a thickness.

The above two properties add tremendous importance and versatility to the old line command, which only created un-joined segments that maintained a constant zero thickness. Here's an illustration of the first property. The upper "W" is drawn with four line segments. As you can see they are all separate, the outer two only have been clicked on, and the grips reflect that. The "W" below it is drawn using the pline command and is all one unit as seen by clicking it once.

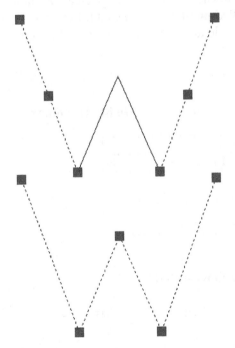

Figure 11.1 – Line vs. Pline (Property 1)

One example of applying this feature may include drawing lengthy runs of cabling or piping. Then if the run needs to be deleted, only one click is necessary to remove all the segments. Let's try to draw a pline.

Keyboard: Type in **pline** and press Enter.
Cascading menus: **Draw→Polyline**
Toolbar icon: **Draw** toolbar
Ribbon: **Home tab→Polyline**

Step 1
Start the pline command via any of the above methods.
Step 2
- AutoCAD will say: `Specify start point:`

Step 3
Left click anywhere on the screen.
- AutoCAD will say: `Current line-width is 0.0000`
  `Specify next point or [Arc/Halfwidth/Length/Undo/Width]:`

Step 4
Ignoring the lengthy menus for now, click around until you get the same "W" shape seen earlier.
- AutoCAD will say: `Specify next point or [Arc/Close/Halfwidth/Length/Undo/Width]:`
When done press Enter or Esc. Then click your new pline to see how it is all one unit, not separate lines.

➤ **Pedit**

Now what if you wanted to apply a thickness to the pline, or create a pline out of a bunch of lines. For this you need pline's companion command called **pedit** or polyline edit. Let's try this out on the pline you already created and give it thickness. You will need to add another toolbar to our collection called Modify II (Figure 11.2).

Figure 11.2 – **Modify II toolbar**

Keyboard: Type in **pedit** and press Enter.
Cascading menus: **Modify→Object→Polyline**
Toolbar icon: **Modify II** toolbar
Ribbon: **Home tab→Modify panel→Edit Polyline...**

Step 1
Start the pedit command via any of the above methods.
Step 2
- AutoCAD will say: `Select polyline or [Multiple]:`

Step 3
Click any of the "W" polyline segments
- AutoCAD will say: `Enter an option [Close/Join/Width/Edit`
  `vertex/Fit/Spline/Decurve/Ltype`

Step 4
Press **w** for `Width`, and press Enter.
- AutoCAD will say: `Specify new width for all segments:`

Type in a small value, say .5 and press Enter. The pline will adopt the new width for all segments and is shown in Figure 11.3. You can press Esc to completely exit out of the pedit command.

Figure 11.3 – Pline with thickness (Property 2)

Draw the "W" once again out of ordinary lines. The procedure to create a pline out of lines is as follows:

Step 1
Start the pedit command via any of the previous methods.
Step 2
- AutoCAD will say: `Select polyline or [Multiple]:`
Step 3
Click any of the "W" line segments
- AutoCAD will say: `Do you want to turn it into one? <Y>`
Press Enter to accept "yes".
Step 4
- AutoCAD will say: `Enter an option [Close/Join/Width/Edit`
  `vertex/Fit/Spline/Decurve/Ltype gen/Undo]:`

Select **j** for `Join`, and select all relevant lines, pressing Enter when done. The collection of separate lines is now a polyline and you may press **w** for `Width` as before. Press Esc to completely exit out of the pedit command.

Another notable option in the pedit command is `Spline`, and if you press **s** while in the pedit command, your W will look like this:

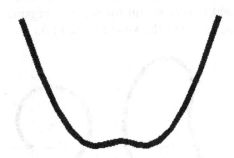

Figure 11.4 – Pline with the Spline option

To undo this you can press **d** for `Decurve` (another pedit option). We will study spline as a separate command later in the chapter.

> ➤ **Exploding a pline**

Plines can be easily exploded using the explode command. The effect is two-fold:

- The pline loses its thickness
- The pline is broken into individual lines

To put it back together requires the pedit command, so explode and pedit are inverses of each other. Be careful though - don't explode plines unless absolutely necessary, as there may be many individual lines to account for once exploded.

> ➤ **Additional pline options**

There are a few more options worth mentioning under the pline command. First of all you can start out drawing a pline with a preset width (or half-width). Here's the command sequence for that, presented in a shortened "actual" format as seen from the command line by the user.

```
Command: pline (typed or via any of the previous methods)
Specify start point:
Current line-width is 0.0000
Specify next point or [Arc/Halfwidth/Length/Undo/Width]: w
Specify starting width <0.0000>: .25
Specify ending width <0.2500>: .25
Specify next point or [Arc/Halfwidth/Length/Undo/Width]:
Specify next point or [Arc/Close/Halfwidth/Length/Undo/Width]:
```

As you can see, the W option was chosen and a width of .25 was entered for both start and end. If those lengths are not similar, you will get something akin to Fig 11.5, a zero start width followed by a .75 end width.

Figure 11.5 – Pline with varying width

Another option is **l** for Length. This is similar to direct distance entry from Level 1. Finally there is **a** for the Arc option. This creates arcs on the fly, and is very useful for drawing organic looking shapes for everything from landscape design to furniture to cabling/wiring as shown in Figure 11.6.

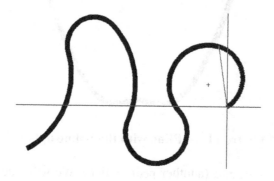

Figure 11.6 – Pline with Arc option

Sec 11.3 - Xline (Construction Line)

From the relatively involved pline we swing over to the very simple xline. This is a continuous line that has no beginning or end. The zoom command will have no effect on it. The "construction" part of the name refers back to the days of hand drafting where you would draw, in light pencil, a guideline that stretched from one end of the paper to the other, and was used as a reference starting point for a string of elevations (or something else) to be drawn one after another.

To use it, make sure Ortho or OSNAP aren't activated, and do the following:

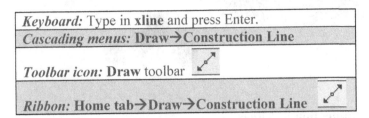

Keyboard: Type in **xline** and press Enter.
Cascading menus: **Draw→Construction Line**
Toolbar icon: **Draw** toolbar
Ribbon: **Home tab→Draw→Construction Line**

Step 1
Start the xline command via any of the above methods.
Step 2
- AutoCAD will say: `Specify a point or [Hor/Ver/Ang/Bisect/Offset]:`
Step 3
Click anywhere and an xline will appear.
- AutoCAD will say: `Specify through point:`
Step 4
Click randomly anywhere along the screen and multiple xlines will appear. Here is one result of clicking in a circular manner (Figure 11.7).

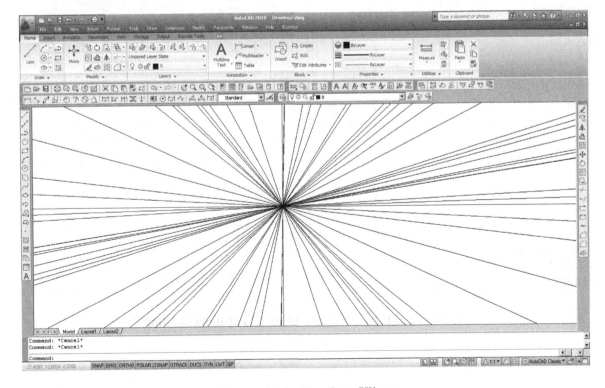

Figure 11.7 – Random Xlines

Figure 11.7 was perhaps not the most useful of creations, but landscape designers may find some use for the result of placing a circle at the intersection of those lines and trimming. Then by some creative scaling, rotating and copying you will get this reasonable representation of trees or shrubs (Figure 11.8).

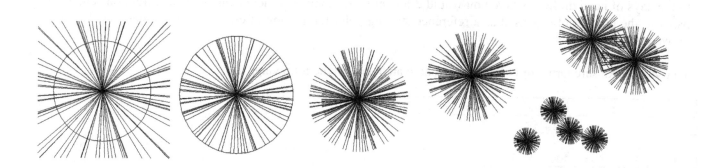

Figure 11.8 – An application of Xlines

As seen in the menu below, you can also create perfectly horizontal, vertical or angular xlines. The same can be accomplished by having Ortho on, although the options below will allow you to create multiple xlines easier. Clear your screen, start up xline again and when you see the following menu, try out the various options.

```
Specify a point or [Hor/Ver/Ang/Bisect/Offset]:
```

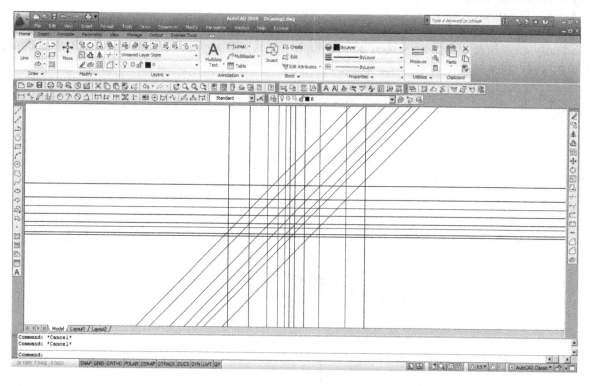

Figure 11.9 – Hor/Ver/Ang (set to 45°) options in Xline

Sec 11.4 - Ray

A ray is simply an xline, but only infinite in one direction. It is easy to use and quite similar to xline, except there are no options for horizontal, vertical or angular; to get a perfectly straight ray you just have to turn on Ortho.

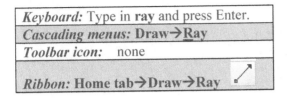

Keyboard: Type in **ray** and press Enter.
Cascading menus: **Draw→Ray**
Toolbar icon: none
Ribbon: **Home tab→Draw→Ray**

Step 1
Start the ray command via any of the above methods.
Step 2
- AutoCAD will say: `Specify start point:`

Step 3
Click anywhere and a ray will be created (Figure 11.10).

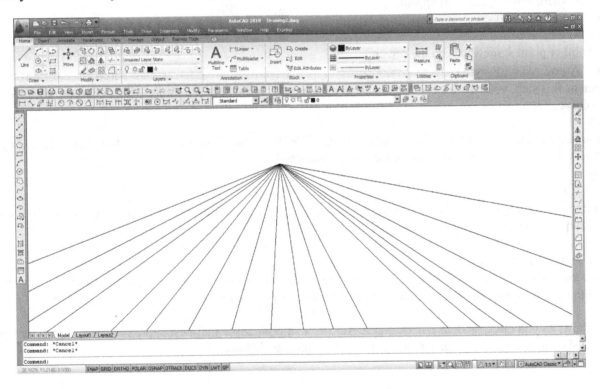

Figure 11.10 – Ray command

Sec 11.5 - Spline

A spline is a unique and valuable command. Its origins are in the mathematical field of numerical analysis, and it is called a piecewise polynomial parametric curve (also referred to as a NURBS curve: Non-Uniform Rational B-Spline). While knowing the details isn't necessary for using this command, what's important is that a spline is ideal for approximating complex shapes through curve fitting. What you will do is create a path made out of point clicks, and AutoCAD will interpolate a "best fit" curve through those points.

The result is a smooth, organic shape useful for a great number of applications, from contour gradient lines in civil engineering to landscape design to electrical wiring and many other shapes and objects. The term spline, by the way, comes from the flexible spline devices used by shipbuilders and draftsmen to draw smooth shapes in pre-computer days. To draw a spline do the following.

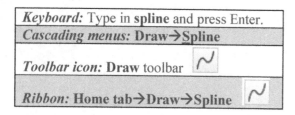

Keyboard: Type in **spline** and press Enter.
Cascading menus: **Draw→Spline**
Toolbar icon: **Draw** toolbar
Ribbon: **Home tab→Draw→Spline**

Step 1
Start up the spline command via any of the above methods.
Step 2
• AutoCAD will say: Specify first point or [Object]:
Step 3
Click anywhere and a spline will appear.
• AutoCAD will say: Specify next point:
Step 4
Click another point somewhere on the screen.
• AutoCAD will say: Specify next point or [Close/Fit tolerance] <start tangent>:
As soon as this step, you can select **c** for Close to close up the spline. Instead though, let's continue with the spline curve.
Step 5
Click another point somewhere on the screen. AutoCAD will say the same thing as in Step 4. At this point press Enter, as we're now just looking to finish the spline command.
Step 6
• AutoCAD will say: Specify start tangent:
This refers to the angle of incidence of the start point. You will notice the cursor jump to it and you are able to control the angle by movement of the mouse. Press Enter.
Step 7
• AutoCAD will say: Specify end tangent:
This refers to the angle of incidence of the end point. You will notice the cursor jump to it and you are able to control the angle by movement of the mouse. Press Enter one final time.

Notice how, as you click random points, AutoCAD fits a smooth curve that best approximates the path. A typical spline is shown below with the grip points activated via mouse clicking.

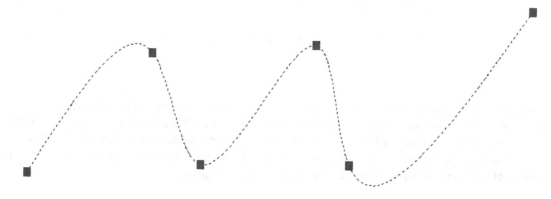

Figure 11.11 – Spline command

Through the examination of the spline in detail, you should now have a good idea of how to create and control it. The points of the spline do NOT have to be random. You can set up a framework for the shape and use OSNAPs to precisely place where you want the spline to go, as shown in Figure 11.12.

> ➤ **Offsetting the spline**

A spline can be offset using the standard offset command, though be careful in areas where the spline takes a sharp turn. If the offset distance is too great and the spline can't make a smooth turn after the offset, it will break down into separate lines. The offset spline will also have many more grip/control points than the original spline. A 3" spline offset is shown in Figure 11.12.

> ➤ **Grips and spline**

As you may have noticed, everywhere you click while creating the spline, a grip point will appear. You can manipulate the position of these grip points in the standard manner, with the spline following suit. This makes the spline a very flexible tool, adjustable to the exact demands of the designer. The original spline's grip points are also shown in Figure 11.12.

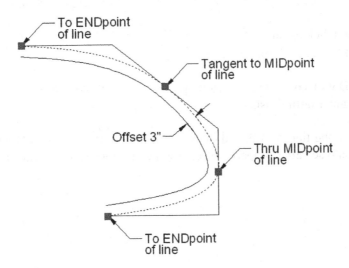

Figure 11.12 – Spline with framework, offset and grips

Examples of spline use are numerous. Already mentioned were contour lines (shown below), electrical wiring, piping, network cables, landscape design (golf courses, curved driveways and walkways), interior design (Jacuzzis, swimming pools, furniture), and much more.

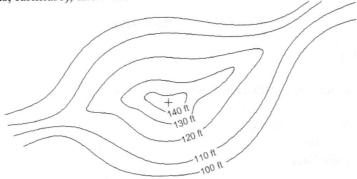

Figure 11.13 – Contour gradient Splines

Sec 11.6 - Mline (Multiline)

A multiline is a very useful tool, and does pretty much what you would expect from the name; it is used to draw multiple lines all at once. This, in theory, should save you time when drawing everything from architectural walls to multi-lane highways. As we will see soon, the basic mline is easy to draw, but not too useful. The trick will be to set it up for useful applications using additional tools.

Keyboard: Type in **mline** and press Enter.	
Cascading menus: **Draw→Multiline**	
Toolbar icon: none	
Ribbon: none	

Step 1
Start up the mline command via any of the above methods (no toolbar or Ribbon available).
Step 2
- AutoCAD will say: `Current settings: Justification = Top, Scale = 1.00, Style = STANDARD`
  `Specify start point or [Justification/Scale/STyle]:`

Step 3
A generic mline will appear. Click again.
- AutoCAD will say: `Specify next point or [Undo]:`

Step 4
Another click and AutoCAD will say `Specify next point or [Close/Undo]:`
You can continue in this manner until finished.

The default distance between the lines is 1.0, and of course if the mline is too small or too large you can zoom in or out. Here's what you should see after properly executing the previous steps (Figure 11.14).

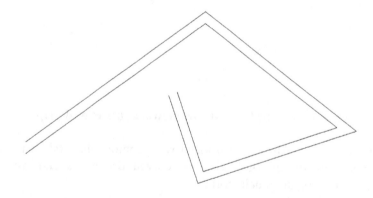

Figure 11.14 – Basic mline

➢ **Modifying the mline**

While simple, the above wasn't too useful either. Specifically you may want to change one or more of the following properties.
- Number of lines making up the mline
- Distances between those lines
- Linetypes making up the lines
- Colors of the lines
- Fills, end caps and other details

All the previous and more can be input and modified, thereby making the mline more useful and sophisticated. The key command for setting up mlines is **mlstyle**.

> ➢ **MLStyle (Multiline Style)**

The multiline style command is a dialog box that allows you to set all of the above parameters. Type in **mlstyle** and press Enter or select **Format→Multiline**…from the cascading menu. The following will appear (Figure 11.15).

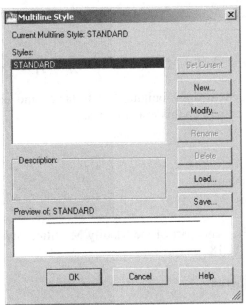

Figure 11.15 – Mline Style

Let's create a new style. Press New, give it a name (no spaces allowed in the name) and press Continue. You will then see the main dialog box as seen in Figure 11.16.

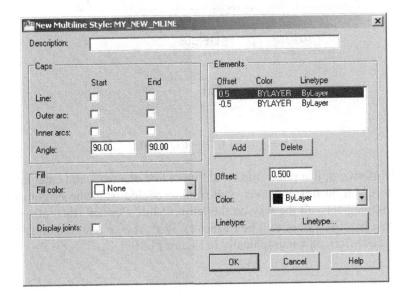

Figure 11.16 – Mline Style – main box

We can now go through the essential changes to the mline. Our goal is to create a typical wall used in construction (see Figure 11.17), and shown often in cross-section in architectural elevations. For simplicity, it will be 6" wide and will feature a 1" drywall panel on either side, with insulation in between, and room for a wooden stud (though not shown here).

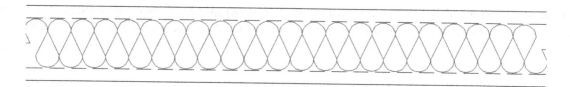

Figure 11.17 – Mline modified for a typical wall

This mline, once set up, can be used to draw building walls faster and easier than offsetting and changing properties one line at a time. The process to do this is as follows.

Step 1

- Number of lines making up the mline
- Distances between those lines

Both of these can be set using the Elements part of the Modify Multiline Style dialog box using the **Add** button and the **Offset** field shown in Figure 11.18.

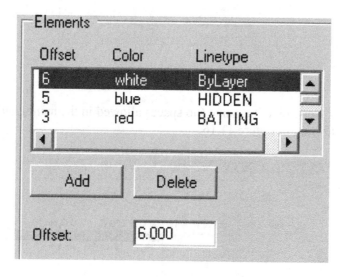

Figure 11.18 – Elements

The process is very straightforward; just press Add to add more line elements and enter an offset distance for each highlighted element. Before you do this, however, you need to decide what the origin of the offset will be. The two best choices are positioning zero all the way on the leftmost line element (then all values will be positive) or in the middle element (then half the values will be negative). With choice one, the outer walls will be at 0 and 6, and with choice two at -3 and +3 as shown graphically in Figure 11.19.

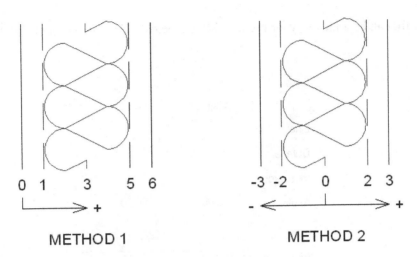

Wall is 6" wide with 1" drywall

METHOD 1   METHOD 2

Figure 11.19 – Offset origin

We'll choose method 1 and set the walls to 0,1,3,5 and 6 inch positions (there are no feet units in mline; you use decimal inches throughout).

Step 2

- Linetypes making up the lines
- Colors of the lines

Both of these can be set using the **Color:** drop-down menu, and the **Linetype...** button as shown below. Simply highlight the line element you wish to change and change it. Remember that you may not have the linetypes loaded yet, so you may need to do this (refer back to Chapter 3 for a reminder, if needed).

In the Figure 11.17 example, the outer wall color was selected to be black (white if using a black background), and the inner walls were blue with a hidden linetype. The center line is a special linetype, often used for insulation, called BATTING. Feel free to choose your own colors and linetypes; these are just for illustration, but may be a good starting point.

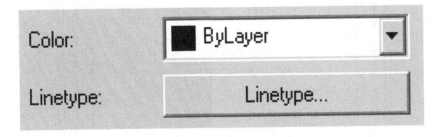

Figure 11.20 – Color and Linetypes

Step 3

- Fills, end caps and other details

End caps and even a fill can be added to your mline as shown below. They are not added in this example, but feel free to experiment.

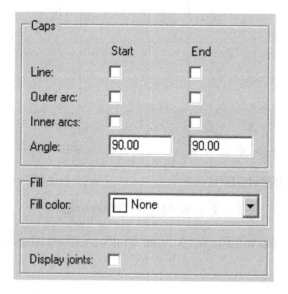

Figure 11.21 – Caps and Fills

After your new mline has been created, go ahead and click OK and you will be taken back to the dialog box of Figure 11.16, with your new mline previewing in the window at the bottom. The other choice should be STANDARD, the default mline you created by just typing in mline and seen in Figure 11.14. Set your own mline as the current one (Set Current button), and click OK. Now go ahead and type in **mline** and draw part of a rectangle as seen below.

Figure 11.22 – The new Mline

The linetypes may not look right in this example. This is due to a low linetype scale being set. If you recall from Chapter 3, the way to fix this is to type in **ltscale** and enter a new (more appropriate value). The value used was 4.8. The result is shown in Figure 11.24.

Figure 11.23 – The new Mline with an ltscale adjustment

> **MLEdit (Multiline Edit)**

Multiline edit is a dialog box that deals with the practical applications of mline, such as how to trim intersecting mlines, and how to deal with joints, cuts and vertexes. To try out some of the features draw two intersecting mlines as shown in Figure 11.24.

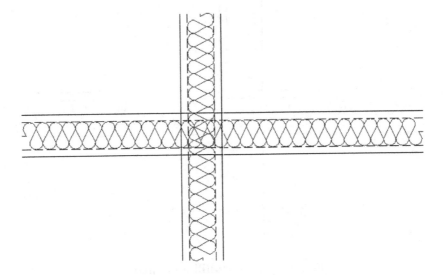

Figure 11.24 – Intersecting Mlines

Now go ahead and type in **mledit** and press Enter (or **Modify→Object→Multiline** from the cascading menus) and the following dialog box will pop up (Figure 11.25).

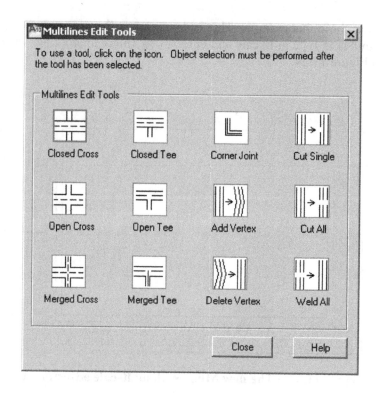

Figure 11.25 – Multilines Edit Tools

Select Closed Cross (upper left). The dialog box will disappear and you will be asked to select the first mline (click on it), then the second mline (click that too). After the second click the following (Figure 11.26) will appear. The mline's intersection has been modified.

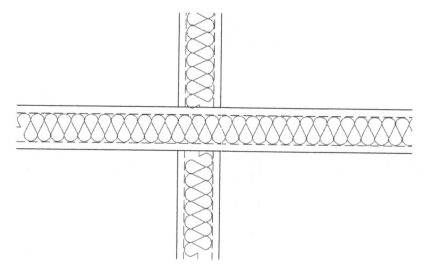

Figure 11.26 – Multilines edited

In the same manner try out all the features of mledit at your own pace; they are all relatively straightforward and self-explanatory. A cut will cut the line; a vertex will add a "crink" in the mline. Close, open, merge, will deal with intersections and so on.

> **Other mline properties**

Grips - All mlines will have grip points at either top or bottom of the overall line (this is set using Justification when first creating the mline). Feel free to use these grips to stretch the line around as needed.

Exploding the mline - Mlines can be exploded, which will separate them in to the individual component lines. This may not necessarily be a bad thing. An architectural floor plan may be quite complex and you may not be successful in getting an mline to do exactly what you want all the time. If the mline is exploded, then you can use fillet, trim and extend to coax it into the shape you want. Much time has already been saved by using an mline in the first place, so exploding it may be the right solution to finish off the floor plan linework.

Trim/Extend - Mlines can be trimmed and extended without exploding them (filleting, however, is not possible). However there is an option that pops up as seen below:

```
Enter mline junction option [Closed/Open/Merged] <Open>:
```

The Closed, Open, Merged options have to do with the intersections, so you will need to pick one option to continue the trim or extend. Go through them all as practice.

Modification of mlines – Once created, a new mline style cannot be modified once you start using it (but can be beforehand). If you wish to change an mline after you have a few drawn, you need to create a new style that incorporates your desired change.

Sec 11.7 - Sketch

This is perhaps one of the most unusual tools in AutoCAD. It allows you to draw objects in freehand. This may seem counterproductive and unnecessary; after all, you spent so much time learning precision and accuracy, but sketch actually finds a wide array of applications, as we'll soon see.

To truly sketch in freehand on a computer you need bitmap graphics (such as what Photoshop® does). Then the maximum resolution is determined by the pixels. In almost all cases the pixels are so small that it truly seems like you're drawing smooth (and continuous) freehand lines and shapes. You would have to zoom in pretty close up to see what is called "pixilation" or the breakdown of smooth lines into the little squares (pixels) that make up the drawn geometry.

This is all fine and good, and would be useful for AutoCAD if not for one problem. AutoCAD is a "vector program," meaning it doesn't use pixels. Instead, all the geometry is represented by mathematical formulas, which completely describe the shape, size and location of just about everything seen on the screen. This approach is necessary for many reasons, not the least of which is to insure the integrity of the geometry when zoomed in close. So how to sketch in freehand? The solution is simple; instead of pixels use very short lines, which look very much continuous if they are small enough and you are not too close to them.

This is where the **record increment** setting comes in. We will need to set it to a very small value (use 0.01). Let's try this out. Type in **sketch** and press Enter. AutoCAD will ask you for a record increment – enter the .01 and press Enter. Then click anywhere on the screen and sketch away. To lift the pen click once, then click again to put it down again. The drawn lines will be green, then to set them press Enter, and they will turn white (or some other color depending on current layer). Be careful not to press Esc. as everything will disappear. Here is what we get after a few strokes of the mouse (Figure 11.27).

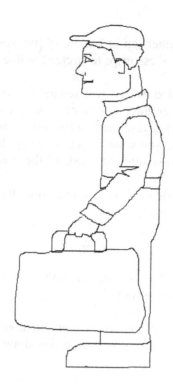

Figure 11.27 – Basic sketch

> **Applications of sketch**

Hopefully, whatever you tried to draw was better than the odd-looking little man with a briefcase. It isn't that easy to draw something well with a mouse; occasional erasing and fixing up is needed to get a good sketch. Sometimes a graphic design tablet and a pen-like stylus may be used. Zoom in closely on whatever you created and you will notice the figure is made up of the small lines we talked about. In this close-up of the hand and briefcase handle, you can clearly see the individual line segments. Be careful making a dense drawing. All these lines will increase the size of the file very quickly!

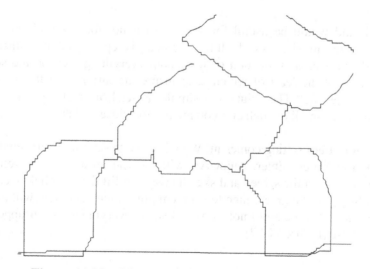

Figure 11.28 – Line increments of sketch command

Sketch has many applications such as:

- Architectural elevations featuring sketched people for comparative sizing
- Sketched people for traffic sign design and civil engineering studies
- Trees, shrubs and other landscape design objects
- Realistic drawings of small cables and wiring
- Anytime an organic non-rigid shape is needed

A website mentioned in the appendix (www.cben.net) is a great source of free sketches, both in plan and in elevation. Here are a few sketch examples from that website.

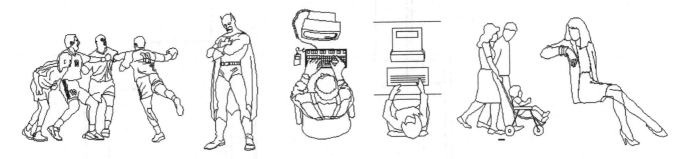

Figure 11.29 – Creative use of the sketch command

Level 2 Drawing Project (1 of 10) – Architectural Floor Plan

The following drawing project will be started now and completed, chapter by chapter, at the end of Level 2. It will be an architectural layout, similar to the apartment you did in Level 1, but more in-depth. The project will take advantage of, and have you put to use, a lot of what you learned in each of the chapters (though maybe not right away). The house design is turned on its end to better fit the "Portrait" orientation of this textbook, but by the time we get to Paper Space, you will need to turn it 90° counter-clockwise as that is how the house would best fit the "Landscape" orientation of a typical set of architectural prints.

Step 1
Open a new file and save it as House_Plan.dwg.
Step 2
Set up units as Architectural, and create the following layers. Use any color you see fit.

- A-Deck
- A-Doors
- A-Hatch
- A-Walls-Ext
- A-Walls-Int
- A-Windows

Step 3
Draw the floor plan shown on the following page. The dimensions are only for your reference, as is the wall hatch, which is added for clarity only. Use the proper layers and consistent colors.

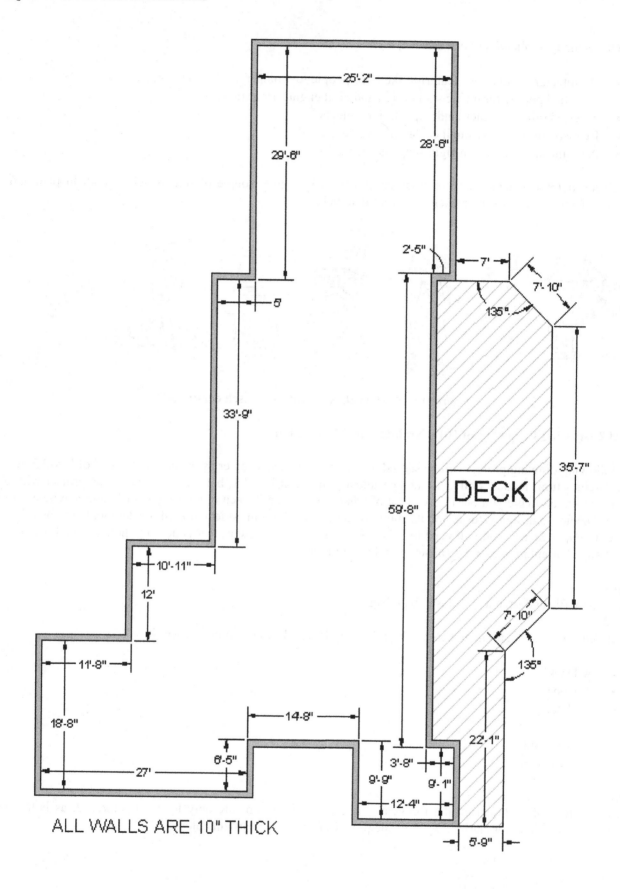

ALL WALLS ARE 10" THICK

Chapter 11

Summary

You should understand and know how to use the following concepts and/or commands before moving on to Chapter 12.

- **Pline (Polyline)**
  - Creating a pline
  - Pedit
  - Setting constant width (thickness)
  - Setting variable width (thickness)
  - Changing lines into plines
  - Adding spline curves and de-curving
  - Other pline options (length, arc, offset)
  - Exploding a pline

- **Xline (Construction line)**
  - Creating a random xline
  - Creating a Hor/Vert/Ang xline

- **Ray**
  - Creating a random ray
  - Creating a Horizontal/Vertical ray (w/Ortho)

- **Spline**
  - Creating a random spline
  - Offsetting the spline
  - Grips and modifying the spline

- **Mline (Multiline)**
  - Creating a generic mline
  - MLStyle
  - MLEdit
  - Other mline properties

- **Sketch**
  - How sketch works
  - Setting a record increment
  - Applications of sketch

Chapter 11
Review Questions

Answer the following based on what you learned in Chapter 11.

1) What are the two major properties a pline has and a line does not?

2) What is the key command to add thickness to a pline?

3) What is the key command to create a pline out of regular lines?

4) What option gives the pline curvature? What option undoes the curvature?

5) What two major things happen when you explode a pline?

6) What other options for pline were mentioned?

7) Define an xline and list a unique property that it has.

8) How is a ray the same and how is it different from an xline?

9) Describe what a spline is and what are some of its advantages?

10) List some uses for a spline in a drawing situation?

11) What is an mline? List some uses.

12) What are some of the items likely to be changed using mlstyle?

13) What are some of the uses of mledit?

14) Describe the sketch command? What is a record increment? What size is best?

Chapter 11

Exercises

Exercise #1 – Draw a pline with 10 segments, each 2 units long and at right angles to each other. Give the pline a thickness of .2 units. spline the pline. The progression is shown below. Upon completion, decurve the pline and explode it.
(Difficulty level: Easy, Time to completion: 1-2 minutes)

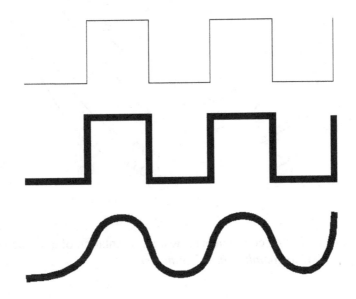

Exercise #2 – Using xline, circle, trim, copy and rotate, draw a tree symbol as seen in Figure 11.7, reproduced below. Repeat the same procedure using the ray command.
(Difficulty level: Easy, Time to completion: 1-2 minutes)

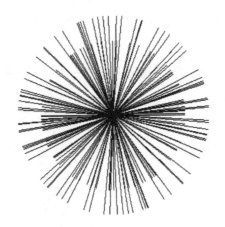

Exercise #3 – Using spline, break and text commands, draw contour lines as seen in Figure 11.13, reproduced again below.
(Difficulty level: Easy, Time to completion: 5-7 minutes)

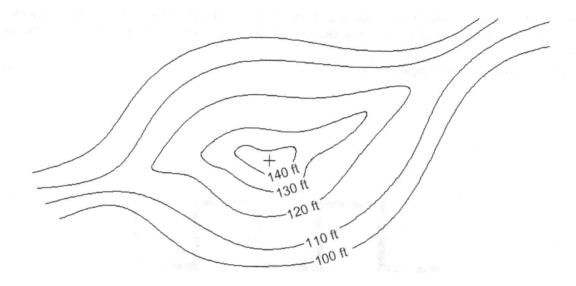

Exercise #4 – Using mline, line and hatch commands draw a representation of a divided highway as seen below.
(Difficulty level: Intermediate, Time to completion: 5-7 minutes)

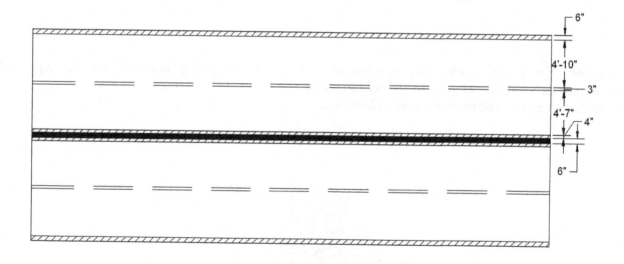

LEVEL 2

CHAPTER 12

Advanced Layers

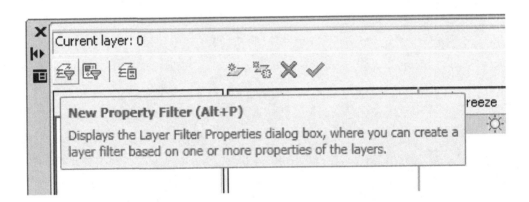

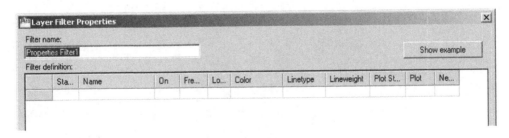

Chapter 12
Learning Objectives

In this chapter we will introduce advanced layers and discuss the following:

- The need for automation
- Script Files
- Layer State Manager
- Layer Filters

By the end of the chapter you will have learned how to use and apply these new types of layer tools, and be able to incorporate them into future design work where a large number of layers are present in the file.

Estimated time for completion of chapter: 1-2 hours

Sec 12.1 - Introduction to Advanced Layers

After the basic commands in Level 1, we introduced the Layers concept. This was just a basic primer, showing you how to create them and set a variety of items, i.e., color, linetype, etc. Also, several methods of control were covered, such as freezing, locking, etc. Here we will take a look at additional powerful features that allow for unprecedented control over your layers. They are called Script Files, the Layer State Manager, and Layer Filtering.

Imagine the following scenario. You work at a large Architecture/Engineering consulting firm that specializes in complete commercial building and site design.

Among your staff there are a number of:

- **Architects and Interior Designers** – to layout the overall exterior and interior of the building.
- **Civil Engineers** – to assist in structural design and site plans.
- **Mechanical Engineers** – to design the HVAC and mechanical systems.
- **Fire Protection/Security Engineers** – to design the electrical, sprinkler and security systems.
- **Landscape Designers** – to integrate landscape design into the site plan.
- **Additional Design Professionals** – as needed for specialized subsystems.

All of these people will contribute to the complete design in their own way. Generally all of their work will be based on (or end up in) one main building design file. This common file will then contain not only the architectural layout of the building but also the data from all of these other disciplines. Each discipline will have its own set of layers (usually following the AIA convention, or another in-house layer naming method). Therefore architects will have their own set of layers starting with an A- (ex: A-Demo-Wall, A-Bay-Window, etc.). Civil engineers will have their own (C-Struct-Column, etc.), electrical their own (E-Light-Switch, etc.) and so on.

You may already see where this is going as far as layers are concerned. As the project progresses, the layers grow in number. Toward the end, when everyone has added their respective design, the file may contain as many as 500 layers (something the author has seen more than once in actual use!). There needs to be a way to impose some order and control over them, so only the needed layers are visible to each discipline. This can certainly be done by manually freezing and thawing layers but is very time consuming. If you guessed that some sort of simple automation is available to you for this purpose then you are correct.

This in essence is advanced layer management, a process where the layers, once set, are saved under a descriptive name such as Arch_Layout, or Elect_Layout, and then recalled when necessary by simply clicking the name. AutoCAD will then freeze and thaw the layers automatically exactly the way you initially told it to. This is what Script Files, the Layer State Manager and Layer Filtering do. These tools allow sorting, classifying and filtering of layers according to a variety of properties and are indispensable when working on a large project.

Sec 12.2 - Script Files

A script (in most general terms) is simply a text file with one command on each line. Its purpose is to automate a sequence of commands when executed. It is perhaps the simplest type of programming you can possibly do. All you need is a text editor (such as Notepad). You then list the commands you want the script to go through – one per line. Then you save it under a name and a .scr extension (ex: MyScript.scr), close the file and then run it from AutoCAD's command line by typing in **script** and pressing Enter (or alternatively via the Ribbon's Manage tab→Run Script). The script dialog box pops up, the script is double clicked (executed), it does its thing, and if you got the command sequence correct, a short task is automatically performed.

Scripts are generally not used as much now for purposes of layer control as they once were, but we will briefly cover them because knowing how they perform will give you insight into the other remaining tools. It's also a good skill to have overall. You can automate a variety of AutoCAD routines. It may be worth your time to read up on further script file details in the AutoCAD help files (F1), as we will not go too in depth here.

Here is a very simple example of a script file. It freezes all layers after first thawing them and setting "0" as the current layer. As scripts don't work with dialog boxes, everything has to be command line driven such as this version of the layer command (generally many commands that are preceded by a "-" will go to the command line version). After the -layer is a set of answers to prompts: t is "thaw", * means "all", s is "set" and f is "freeze". This exactly answers everything AutoCAD asks for if you were to do this by hand at the command line. The last two keystrokes (→) are Enter and Enter.

-layer
t
*
s
0
f
*
→
→

This reads as Layer, Thaw, All, Set, 0, Freeze, All, Enter, Enter. Try this out with an AutoCAD file that has a bunch of layers. In a real application you would of course have to manually type in the layers deemed not relevant to the particular state (and therefore need to be frozen) in place of the last "*" and before the two "Enters". The steps prior to that (t, *, s, 0) are necessary in case a layer that needs to be frozen is set to current (you can't freeze the current layer) so the zero layer is set as current.

This was a lengthy process, and is the main reason why script files have gradually fallen out of favor and replaced by the Layer State Manager for layer control, as shown next. Keep in mind, however, that the Layer State Manager (LSM) works in pretty much the same way, except you select layers and do everything graphically through the Layers dialog box. Now that you have read through the underlying theory above, we can rapidly go through the functions of the LSM.

Sec 12.3 - Layer State Manager (LSM)

As described previously, the LSM is essentially a graphical version of script files. Since you should now have a basic understanding of what you need to do, let's just set up a collection of layers and go through the mechanics of how to set up several layer states. While our example will be simple – we will only set up three layer states, an Arch_Layout, Mech_Layout and Elect_Layout - the theory can be easily extended to a large number of layers and states.

Open up a new AutoCAD file and set up the following layers:

A-Walls-New
A-Walls-Demo
A-Window-Bay

E-Wiring-Main
E-Switch-Main
E-Outlet-Main

M-Diffuser
M-Return_Air
M-Panel

Assign color to these layers as shown in Figure 12.1. What we want to do now is create a layer state called Arch_Layout. This means only the A- layers are needed with the Electrical and Mechanical layers frozen. So go ahead and freeze the six layers named E- and M- also as shown in Figure 12.1.

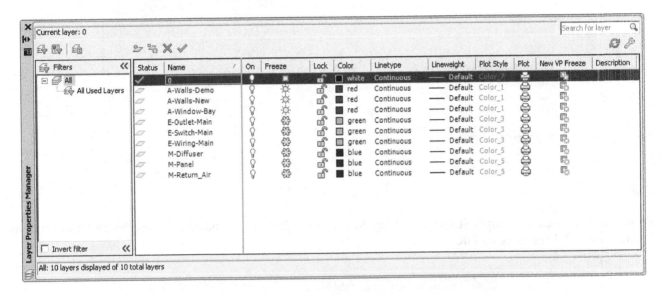

Figure 12.1 – Layers for LSM

Now that the layers have been frozen, we need to save this "state" under the appropriate name. Click the Layer State Manager button in the upper left of the Layer dialog box as seen in Figure 12.2.

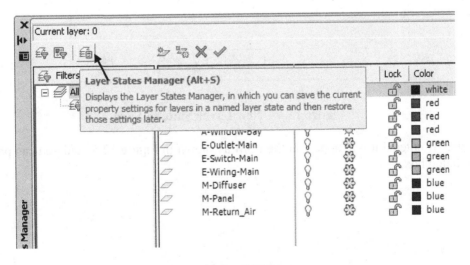

Figure 12.2 – LSM icon

The LSM will then appear. It can be expanded (if it isn't already) by pressing the arrow on the lower right.

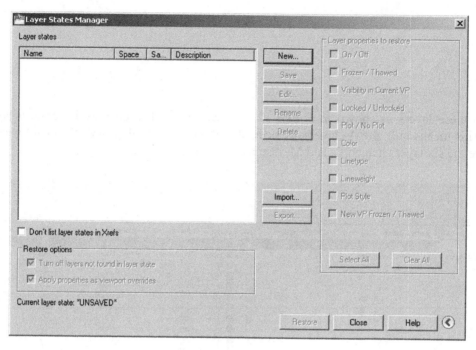

Figure 12.3 – LSM

We will now create a new layer state by clicking **New…** and filling in the name and description (if desired) as shown in Fig. 12.4 and clicking OK.

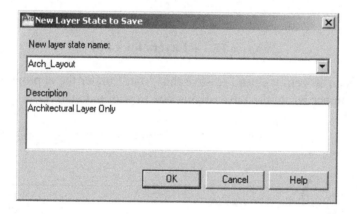

Figure 12.4 – New Layer State to Save

Once the layer state is entered it will appear in the LSM as shown in Figure 12.5, and you can press **Close** to exit out of the LSM dialog box.

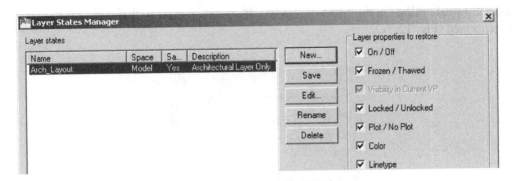

Figure 12.5 – LSM with new layer state

In the same manner, go back to the Layers dialog box and create the other two states: Elect_Layout (only A- and E- layers visible) by turning off the unneeded layers and activating the LSM, and Mech_Layout (only A- and M- layers visible). The result is shown in Figure 12.6.

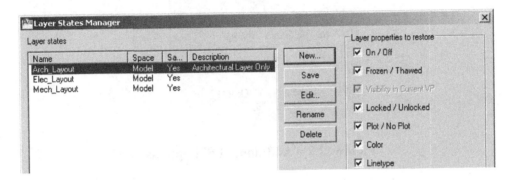

Figure 12.6 – LSM with remaining layer states

To activate these states when needed, you simply go to the Layers dialog box, click the LSM icon and when it appears double-click the state name or highlight it with one click and press **Restore**. Instantly the correct layers will be frozen or thawed (you can see this happen in real-time in the Layers dialog box as you click one state or another).

The purposes of the buttons found to the right of the main LSM window are as follows:

- **New** - Displays the New Layer State to Save dialog box, which we have already used.
- **Save** - Saves the selected named layer state.
- **Edit** - Displays the Edit Layer State dialog box, where you can modify the layer state.
- **Rename** - Allows editing of the layer state name.
- **Delete** - Removes the selected layer state.
- **Import** – Layer states (*.las extension) can be imported from another file by means of this option, though this is not done often in practice.
- **Export** – Layer states (*.las extension) can be exported to another file by means of this option, though this is also rarely done in practice.

There are some additional options in the LSM dialog box if you click the small arrow in the extreme bottom right as previously mentioned (Fig. 12.7).

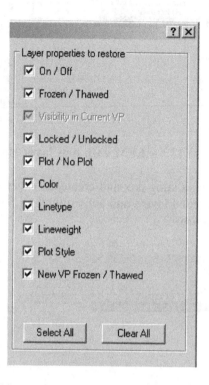

Figure 12.7 – Additional LSM options

Layer properties to restore is simply a listing of every property a layer typically has and you can check off if you'd like that particular property restored. If they are all checked, as is more typical in practice, all the layer's properties will be affected.

Sec 12.4 - Layer Filtering

Layer Filtering is a tool that allows you to filter out (or selectively choose) layers that fall under a certain description. You can create filters of your own choosing and assign them names. For example, if for some reason you only wanted the red layers to show or ones that are assigned a hidden linetype, this can be done. Layer names can also be filtered. If you want only those layers that start with an A- to show, this can be arranged, though in this case you're essentially doing what was described previously with the LSM.

The key to layer filtering is to set up filter definitions. These can include full names or partial names (using wildcard "*" character) and property definitions (color, linetype, etc.). The entire left side of the layer dialog box since AutoCAD 2005 has been dedicated to filters. When used properly these are very effective tools. We will highlight some of the more important features essential to using them in this chapter. The reader is encouraged, as with many topics mentioned in Level 2, to explore them further on their own.

There are two types of filters available as icons on the top left of the Layers dialog box as shown in Figure 12.8.

New Property Filter

New Group Filter

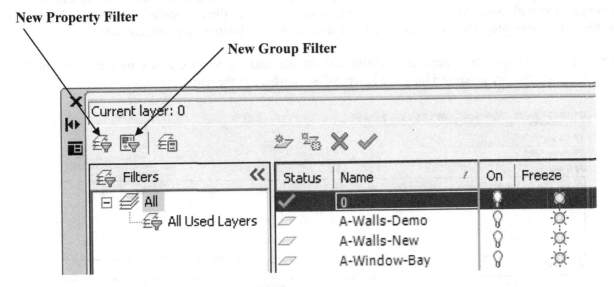

Figure 12.8 – Layer Filters

The **New Property Filter** is what you will use to set up the definitions. If you press that icon the following will appear (Figure 12.9), with the previously entered layers visible.

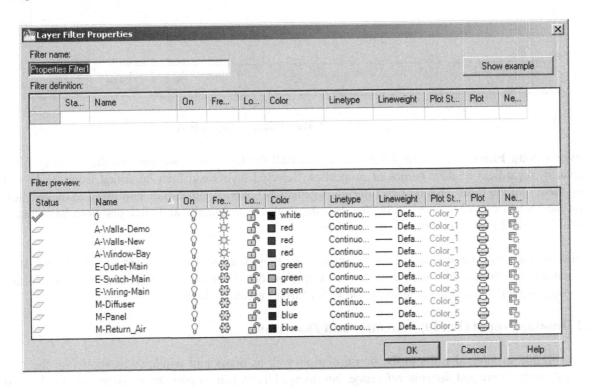

Figure 12.9 – Layer Filters Dialog Box

Here you will first enter the name of your filter in the text field in the upper left. Then it is just a matter of careful selection of the properties or names (or both) of the layers you would like to see. You may use the wild character "*" in conjunction with typed parts of the layer name to indicate to the filter to include anything before or after the specific letters. Any property can also be used as a filter; color and linetype are common ones.

Go through the filter at your own pace and experiment with the settings. Shown, as a simple example, in Figure 12.10 are the layers that are green and have the letter "w" somewhere in the name.

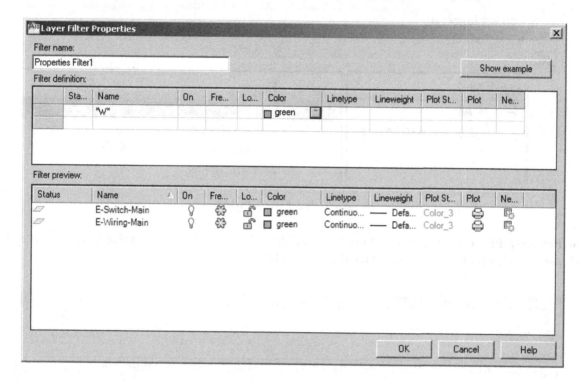

Figure 12.10 – Layer Filters Dialog Box

The **Layer Group Filter** is a general filter that includes all the layers that are put into the Property filter when you define it, regardless of their names or properties. Selected layers can then be added from the layer list by dragging them into the filter. This type of filtering is generally not used, but can be employed to make a new filter based on the layers of another filter. To begin, click the icon, create a name, and click and drag layer names from the right side of the Layers box into the Group Filter.

Remember an important point with filters. By themselves they don't do much except sort the layers. It is up to you to then do something useful with this, such as freeze or lock them as needed. Therefore think of filtering as simply a way to get the layers in some sort of order for further action on them.

Level 2 Drawing Project (2 of 10) – Architectural Floor Plan

Here we will add in internal walls to the house plan as well as doors and windows as shown on the following page. The dims are there just for your reference, but the wall hatch is now permanent, so be sure to also add layer **A-Wall-Hatch** as well! All internal walls are 6" thick, and if you don't see a dimension, just assume an appropriate value. The floor plan is of course greatly simplified from the original design.

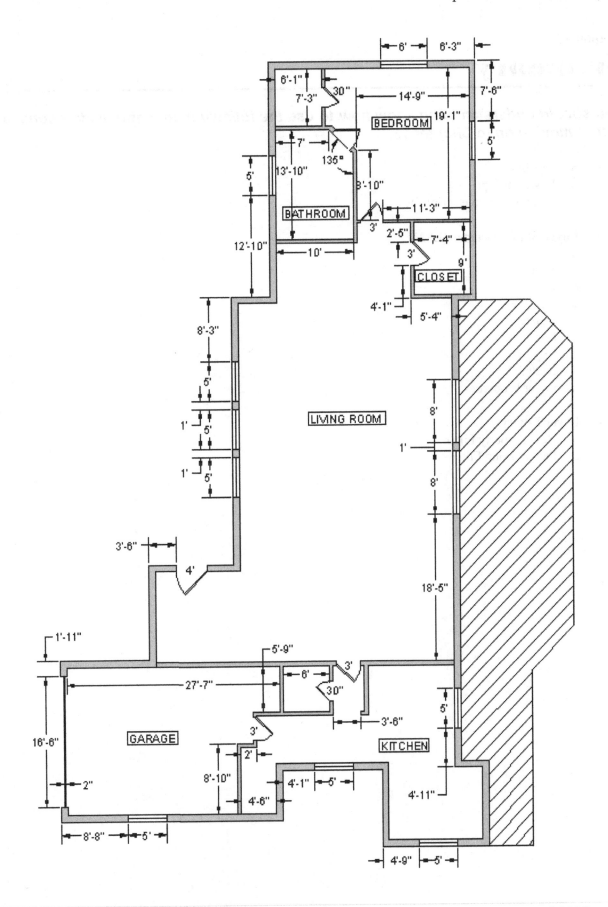

Chapter 12

Summary

You should understand and know how to use the following concepts and/or commands before moving on to Chapter 13.

- **Script Files**
  - Basic concept
  - Writing a simple script file

- **Layer State Manager (LSM)**
  - New
  - Save
  - Edit
  - Rename
  - Delete
  - Import
  - Export
  - Restore
  - Layer Properties

- **Layer Filters**
  - New Property Filter
  - Layer Group Filter

Chapter 12
Review Questions

Answer the following based on what you learned in Chapter 12.

1) Describe what is meant by Layer Management. Why is it needed?

2) Describe the basic premise of a script.

3) What can a Script File do for Layer Management?

4) Describe what a Layer State Manager will do.

5) What are the fundamental steps in using a Layer State Manager?

6) What is Layer Filtering? What are the two types?

Chapter 12
Exercises

Exercise #1 – Create six layers as following:

A-Walls_1
A-Walls_2
A-Walls_3
A-Walls_4
A-Walls_5
A-Walls_6.

Color all odd layers Green and all even layers Red. Write a script file to thaw all layers, set layer 0 as current. Then freeze all and thaw just the even layers.
(Difficulty level: Easy, Time to completion: 10 minutes)

Exercise #2 – Using the same six layers, set up layer states "Odd_Layer" and "Even_Layer" and run them.
(Difficulty level: Easy, Time to completion: 3 minutes)

Exercise #3 – Using the same six layers, create filters to only sort and capture the Green layers.
(Difficulty level: Easy, Time to completion: 2 minutes)

CHAPTER

Advanced Dimensions

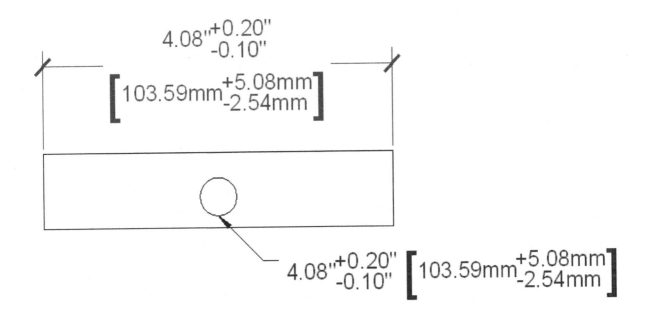

Chapter 13

Learning Objectives

In this chapter we will introduce advanced dimensions and discuss the following:

- The Lines tab
- The Symbols and Arrows tab
- Text tab
- Fit tab
- Primary Units tab
- Alternate Units tab
- Tolerance tab
- Geometric Constraints
- Dimensional Constraints

By the end of the chapter you will have learned additional advanced features of the Dimension Style Manager dialog box and be capable of setting up complex dimensions. We will also introduce two new AutoCAD 2010 features: Geometric and Dimensional Constraints.

Estimated time for completion of chapter: 2-3 hours

Sec 13.1 - Introduction to Advanced Dimensions

This chapter is meant to complete your knowledge of AutoCAD's dimensioning features. Specifically we will be looking at the New Dimension Style dialog box and Parametrics. In Level 1 we focused primarily on defining the types of available dimensions and how to properly dimension geometry. Little was mentioned of this dialog box except for the four essential features deemed most important: Arrowheads, Units, Fit and Text Style (not used as much in AutoCAD 2010, as the better looking Ariel is the new default font). Here we will go through the entire Dimension Style dialog box and discuss other options and features, some more than others according to their usefulness.

The reason you need this knowledge is because AutoCAD allows for a tremendous amount of power, flexibility and variation with its dimensioning. As a regular user, you may not come to appreciate what's available because much of it may already be "set up" for you. As a CAD manager and advanced user (for whom Level 2 is geared for), you may be charged with doing this "setting up" and need to know what to do. We will conclude the chapter with a new feature just introduced in AutoCAD 2010 – Parametric Dimensions and the idea of Constraints. These are routinely found in advanced 3D software, but just now are appearing in AutoCAD.

Sec 13.2 – Dimension Style Manager

To start off, let's take a closer look at the Dimension Style Manager. You've seen it in Chapter 6, so as a reminder here's how we got it up on our screen.

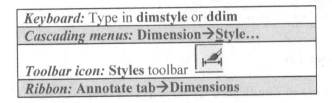

Create a new style (by pressing the New… button) and naming it Sample Style (Figure 13.1). Click Continue when done. You will then see the familiar Dim Style Mgr. Box (Figure 13.2). We will now take a much closer look at the seven tabs titled Lines, Symbols and Arrows, Text, Fit, Primary Units, Alternate Units and Tolerances.

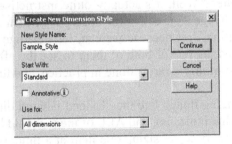

Figure 13.1 – Create New Dimension Style

An important point to remember here is that our goal is NOT to go through every possible button and menu in extreme detail, but rather give a basic overview of them, followed by a narrow focus on (and exact description of) the most useful features ONLY, and how they fit in with the design process and how they benefit you as an AutoCAD user.

Let's take a look at the first tab all the way on the left called **Lines**. Throughout the process, carefully observe the preview window at the upper right to see the results of changing or adjusting whatever it is we discuss.

➢ **Lines Tab**

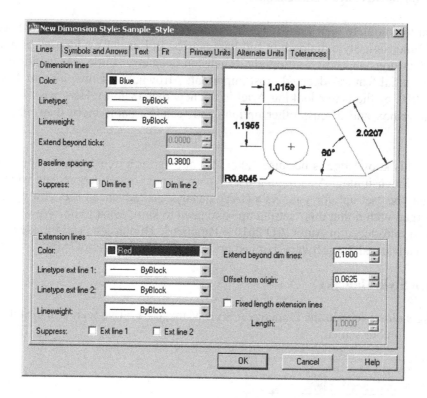

Figure 13.2 – Lines Tab

This tab is the only one where you will likely not change anything of significance. There is a good reason for that. It focuses on dimension lines and extension lines and various sizing, linetype, color and suppression options. However, as far as sizing is concerned, all these features can (and should) be adjusted all at once using the Fit tab's overall scale feature (to be discussed shortly).

So what can we learn from this tab then? Well, it's actually quite instructive to see what exactly dimension and extension lines are and how you can change their colors (or linetypes). Much of what is shown here falls under the "just in case you may need it" definition. Just because we can, let's change the dimension lines to blue and the extension lines to red. Also try suppressing one or both of them to see the effect. We won't change the linetypes; it should be clear what that does. Most of this is rarely done in practice, and the dimensions are left as they are.

It's important to understand that changing things like the "Extend beyond dim lines" value (bottom right), really shouldn't be touched (though know what it means), and ALL of them should be changed up or down all at once as discussed later under the Fit tab. If you try to set sizing one item at a time, you will end up with lopsided, oddly shaped dimensions – there are just too many items to change in correct proportion to guarantee you'll get it correct, and too much effort for something unnecessary.

Except for the suggested color change, leave everything as it is – ByBlock. Your NDS dialog box should look like Figure 13.2.

➢ **Symbols and Arrows Tab**

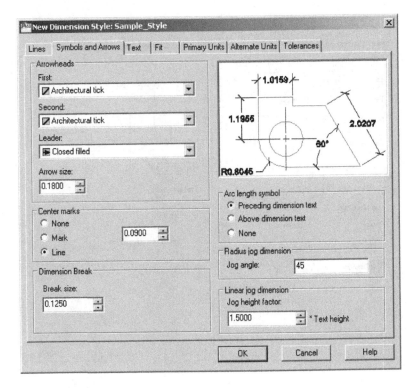

Figure 13.3 – Symbols and Arrows Tab

Under this tab you have already practiced modifying the arrowheads in Chapter 6 to architectural ticks. Go ahead and do this again, making a note of the other available ones. The leader is generally left as an arrow since its main purpose is to point at something.

Moving below the arrowheads we have the "Center marks." This has to do with measurements of a circle and what you want to see in the center of one when a diameter or radius dimension is added. A "Mark" or "Line" is typically selected; let's go with "Line". Notice how the preview window reflects both this and the arrowhead changes.

On the right we have some relatively minor options. "Arc length symbol" concerns the placement of the symbol and the "Radius jog dimension" has to do with the angle of the jog "wiggle." Leave them, as well as the "Linear jog dimension," as default all around.

What is most important in the Symbols and Arrows tab is how to set the arrowheads to other types by selecting them from the drop-down menu. Your NDS dialog box should look like Figure 13.3.

➢ **Text Tab**

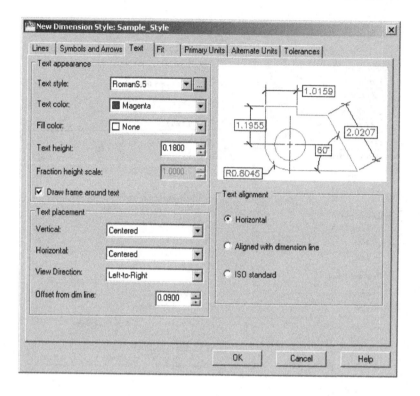

Figure 13.4 – Text Tab

Under this tab you have already practiced changing the text under "Text style:" (upper left). The goal was to make the dimension text the same as the rest of the text in the drawing (always the same font, and usually the same size too). In Chapter 6, when this was done, the text you selected on the floor plan was Arial 6". Here we will make one from scratch by pressing the button to the right of the Text style field (with the three dots) and setting a font style called RomanS.25 (as outlined in Chapter 4 when we discussed the Style box). Review this if necessary. Moving down the "Text appearance" category we find:

- **Text color** – This is usually not set separately, but we'll do it here, setting Magenta.
- **Fill color** – This creates a color highlight around the text. Try it out. To add a frame around the text there is a check box for this further on down, which we will check off.
- **Text Height** – Not adjusted here! Use the previously mentioned Style dialog box.
- **Fraction height scale** – Applies to the size of fractions. Those will appear only when we set architectural units (coming soon).

Finally we have the two remaining categories "Text placement" on the bottom left and "Text alignment" on the right. The default values for Text placement (mainly the first two: Centered and Centered) are correct for most situations, but do try them all out to see the effect in the preview window. Offset from dim line is not to be changed.

With "Text alignment" the common default is Horizontal because people generally don't want to tilt their heads to the side every time they check a dimension on a printout. Do check the other two options (aligned and ISO) to see what they do though. Your NDS dialog box should look like Figure 13.4.

> **Fit Tab**

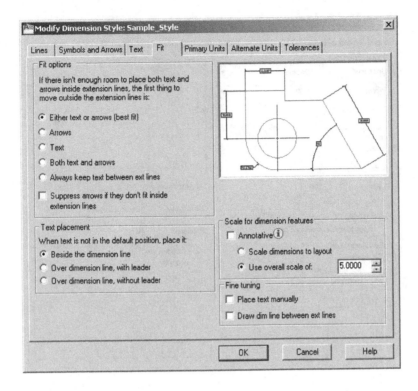

Figure 13.5 – Fit Tab

Under this tab you have already practiced changing the "Use overall scale of:" text field (on the right, underneath the preview window) to a value of 15 in Chapter 6. We really did not discuss why this value was chosen (we'll leave this for the Paper Space discussion in Chapter 18), but this field is by far the most important one in this tab. What it is for is to simply boost up the overall dimension size evenly and uniformly, as opposed to manually adjusting every possible value (under the Lines tab – and a few other places). We will see in later chapters how this value is tied-in with the overall printing scale and viewports. The annotative option will also be discussed. Change the value in the field to 5.0000 for now, as that corresponds well to a ½" height text we selected earlier.

The rest of the categories are quite straightforward. "Fit options" simply address what to stick outside of the extension lines if there isn't enough room. Best fit is usually the safe choice, as it lets AutoCAD decide how to best proceed. If you force a specific action, it may not be appropriate for every situation, so let AutoCAD handle it.

"Text placement" has some minor adjustment options, as stated in the choices (leave as default), as does "Fine tuning", where you should leave both boxes unchecked. Placing the text manually isn't necessary in almost every case, and drawing a dim line between extension lines effectively "crosses out" the dimension value; a distracting feature.

What is most important under the Fit tab is adjusting the "Use overall scale of" under the "Scale for dimension features" category. As mentioned, use 5.0000 as a value for now. Your NDS dialog box should look like Figure 13.5.

➢ **Primary Units Tab**

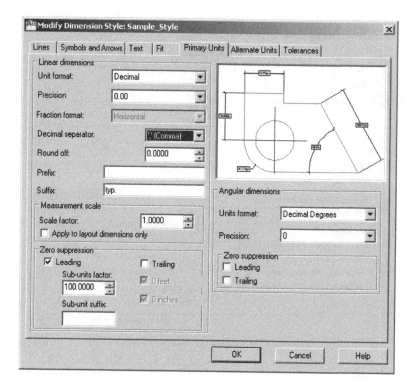

Figure 13.6 – Primary Units Tab

Under this tab you have already practiced changing the "Unit format" to architectural in Chapter 6. Generally the dimension units are the same as the drawing units. Here we'll leave the units as decimal, not architectural. Note in both cases the precision available to you. This closely parallels what you saw in the Units dialog box from Chapter 2. Proceeding further down the "Linear dimensions" category we have:

- *Fraction format* – Various ways to represent fractions with architectural dimensions; grayed-out when decimal is selected. The choices are Horizontal, Diagonal, and Not Stacked. Try them out.
- *Decimal Separator* – Various ways to represent decimal numbers; grayed-out when architectural is selected. The choices are Period, Comma and Space, and simply represent different ways to show the same thing. Europeans, for example, sometimes use a space or decimal, not comma (ex: 5 000, not 5,000).
- *Round off* – Rounds off dimension values; not necessary in common usage.
- *Prefix/Suffix* – A very useful feature allowing you to add text, symbols or values before and/or after the dimension values. We will add the abbreviation "typ." in the suffix as an example.
- *Measurement scale* – Scales dimension values; not necessary in common usage.
- *Zero suppression* – Another very useful feature allowing you to suppress zero values either before or after the main significant digits when using decimal units. For example, a 0.625 value becomes just .625 and a 1.250 becomes 1.25. Used in conjunction with the "Precision" setting to "clean up" dimension values, removing extra zeros. With the architectural units setting, the idea is the same but now "0" feet and "0" inches are suppressed.

"Angular dimensions" allow you to choose other degree measurement systems, but Decimal Degrees is sufficient for most users. Zero suppression for angles is the same concept as before, but is rarely used in this case. Your NDS dialog box should look like Figure 13.6.

➢ **Alternate Units Tab**

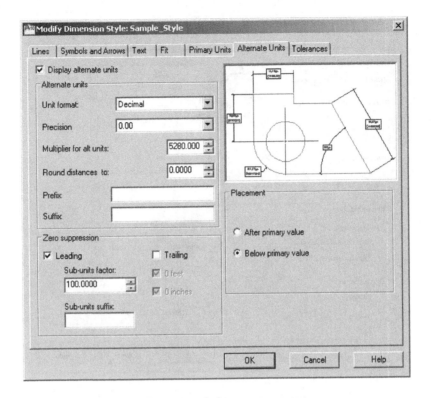

Figure 13.7 – Alternate Units Tab

This is a new tab we have not explored at all in Level 1. Alternate units are a secondary set of dimensions that reside inside parentheses attached to the main set of units. Their purpose is to present the main units in an alternate form such as Metric versus English (just one example). This is why you see a "25.4" value right away under "Multiplier for alt units:". This is how many millimeters fit into 1 inch; more on this in just a moment.

Check off the top left box called "Display alternate units". All the fields will then become editable. Take a look at all of them; most should look familiar (such as "Zero suppression" and "Prefix/Suffix") so we will not go over them again. "Unit format:", "Precision:" and "Round distances to" have also been mentioned before (reminder: don't round off without a good reason). The "Placement" category (bottom right) is also easy to understand, the choices being after or below the primary value.

Understanding alternate units then boils down to fully grasping the "Multiplier for alt units:" concept. It is a powerful and simple idea. Instead of presenting a collection of "canned" multipliers, AutoCAD allows you to set your own. So if the main unit is 1 mile, to express this in feet (as an alternate unit), type in 5280 as a multiplier. You are in no way limited as to what sort of value you can enter (inches to miles or millimeters to yards anyone?) but of course you have to know exactly *what* the multiplier is or else, as they say, "garbage in, garbage out" will result. Therefore an incorrect multiplier will get you false alternate unit readings. Explore all the features of alternate units and try entering several multipliers. For example, the number of yards equal to one mile (1,760).

In our case here, let's go with the amount of feet in a mile (5,280) and enter that as the alternate unit. Your NDS dialog box should look like Figure 13.7.

➢ **Tolerances Tab**

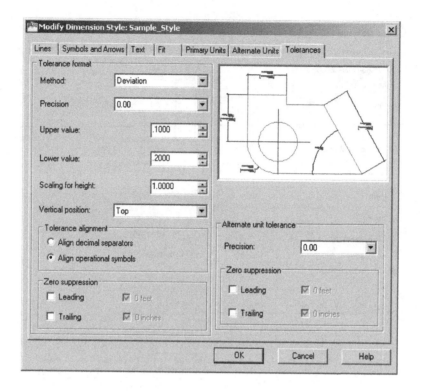

Figure 13.8 – Tolerances Tab

This final tab has also not been looked at yet. Tolerances are relevant mostly to engineering and manufacturing and less so to architecture.

Tolerances are generally defined as deviations from an ideal stated value, as used in manufacturing specifications of engineering devices (there are of course other stated definitions). The underlying need for specifying tolerances hinges on the fact that in manufacturing closer tolerances may be costlier to achieve, while larger tolerances, while cheaper, may adversely affect performance of a part. Tolerances are a way of expressing just how much a part may be off the mark and still be acceptable, based on engineering need and cost analysis. Aerospace, automotive and biomedical engineering tolerances will be closer than most consumer products for example.

Tolerances are a familiar topic to mechanical engineers and machinists. To them no further explanation is needed and AutoCAD's tolerance choices will be familiar: Symmetric, Deviation, Limits and Basic. Other options under this tab should also look familiar to all, such as "Zero suppression" and placement options.

- *Symmetric tolerances* – these indicate the overall deviation, which is symmetric for upper and lower values, therefore you will only set one (the upper value).
- *Deviation tolerance* – these indicate non symmetric deviations in either direction with a stacked +/-.
- *Limits tolerance* – these are similar in theory, and indicate the acceptable limit in one or both sets of values, but no +/- shown.
- *Basic tolerance* – Same as none, the first choice.

If the student is not currently familiar with tolerances, chances are he or she does not need them in their work situation. While this is an interesting topic to discuss in further detail, tolerances will already be familiar to those that need them, and not relevant to those who are not. We will not set tolerances in this first example.

So what did we end up with after all these settings? Well in the interest of trying out as much as possible, we didn't really create a very realistic dimension. Let's take a look at it. Click OK in the DSM box; it will disappear. Then press "Set Current" in the next box, followed by "Close". Now draw a 10" by 10" rectangle and dimension it using the basic Horizontal dimension. It should look similar to Figure 13.9.

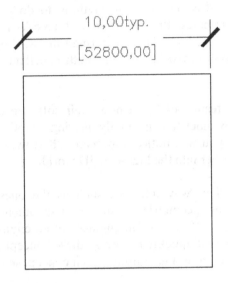

Figure 13.9 – The new dimension style

Let's try another dimension style on a hypothetical section of roadway (go ahead and draw it first). Run through all the set-up steps again using the following data, and hopefully getting a similar result to Figure 13.10.

- Architectural ticks
- Arial font, .25"
- Fit: 1.000
- Primary units: Decimal. Watch the precision!
- Suffix added on both main units (miles) and alternate units (feet)
- Alternate units – 5280 as multiplier
- No tolerances

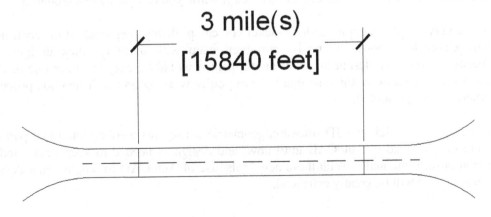

Figure 13.10 – Sample dimension

Sec 13.3 – Introduction to Constraints

AutoCAD was always a top-notch 2D drafting software application, but one with a bit of an "inferiority complex." Not in regards to any other 2D competitors, but rather toward 3D solid modeling and design software. These sophisticated products (discussed in the appendix) developed in their own world, apart from AutoCAD, via development firms that never competed against or had anything to do with Autodesk or architecture. These programs included CATIA, NX/Unigraphics, Pro-Engineer, SolidWorks and others. Their developers created quite a few revolutionary concepts and methods that transformed the way engineering design is done. Autodesk eventually jumped on-board with Inventor and later Revit; both excellent 3D products, one for engineers, the other for architects.

That left the AutoCAD development team looking at how their software could be enhanced by incorporating some of these sophisticated 3D tools without fundamentally altering AutoCAD's identity or formula for success. If you go on to study AutoCAD 3D, you will notice how many "borrowed" concepts found their way into the software as AutoCAD reached ever further into the high-end 3D world.

Some of these concepts also found their way into 2D, such as the ones we are about to discuss, namely Parametrics, which is really two topics: geometric constraints and dimensional constraints; the latter is also referred to as "dimension driven design." If you are an engineer or an engineering school student who used the previously mentioned 3D software, you will quickly recognize these concepts. AutoCAD literally took a page out of that playbook. Inferiority complex no more! Let's discuss each concept separately.

Sec 13.4 - Geometric Constraints

The fundamental idea behind geometric constraints, regardless of what software we are dealing with, is to restrict (constrain) the possible movement of drawn geometry and also force that geometry into certain positions. Ortho was a horizontal and vertical constraint on lines. So if you have two parallel or perpendicular lines, a constraint can be placed on them that will force them to always stay parallel or perpendicular, as well as force them to be that way in the first place. Expanding to more general terms, geometric constraints set "relationships" between geometry such as parallel, perpendicular, tangent, coincident, etc.

This overall concept is quite important in setting what is referred to as "design intent." If a drilled hole in a bracket absolutely has to be concentric to the bracket's corner fillet to fulfill the design intent, then that is the main driving factor in its design ("concentric" means the circles/arcs share the same center point). There may, however, be other aspects of the design, such as perhaps notches and other geometry on the bracket, which is less critical. The constraints will then "hold" the critical geometry, while you design the rest around it.

This was of course a very simple example; with a bracket you can probably keep track of the critical relationships without any help. Geometric constraints really earn their keep with more complex designs, where many constraints are needed to preserve design intent. Without them it would be easy to forget one or two important relationships, and discover later down the road that the part you're working on won't function properly because a tangency somewhere wasn't preserved.

Almost since the beginning of high-end 3D software, geometric constraints were an important part of the design approach, but were not included in AutoCAD until now, and designers needed to keep track (and check over) their geometric relationships manually. With these new tools, use of AutoCAD for engineering design (and even architecture to some extent) will be greatly enhanced.

> **Types of Geometric Constraints**

Here are the constraints available to you. We will not go through every single one, but only cover the more commonly used ones, leaving the rest for you to explore. The best way to set constraints is via the Geometric Constraints toolbar (Figure 13.11) or the Ribbon's "Parametric" tab.

Figure 13.11 – Geometric Constraints toolbar

The critical constraints we will discuss are:

- *Perpendicular* – Constrains two lines or polylines to perpendicular 90° angles to each other.
- *Parallel* – Constrains two lines to the same angle.
- *Horizontal* – Constrains two lines to lie parallel to the x-axis.
- *Vertical* - Constrains two lines to lie parallel to the y-axis.
- *Concentric* – Constrains circles, arc or ellipses to maintaining the same center point.
- *Tangent* – Constrains a tangency between curves and lines.
- *Coincident* – Constrains two points together (to be discussed along with dimensional constraints).

> **Adding Geometric Constraints**

Adding geometric constraints involves simply picking the constraint you wish to add, and selecting the first object (that will serve as the reference point) followed by the second object. The constraint will then appear next to the objects reminding you it is set. The constraint tool will also force the objects into that constraint, so if you draw two lines that are not quite parallel, once you set the constraint, they will become parallel right away.

Having read this, you may be tempted to completely do away with Ortho and even OSNAP to create accurate geometry, allowing constraints to "fix" drawn shapes. This of course isn't generally recommended, as sloppy drafting will catch up to you sooner or later (though we will try this in the next example). While geometric constraints are very useful new tools, use them when appropriate, and without forgetting or throwing away basic accuracy habits!

Let's try this out with the first constraint on the above list – perpendicular.

Keyboard: none
Cascading menus: **Parametric→Geometric Constraints→Perpendicular**
Toolbar icon: **Geometric Constraint** toolbar
Ribbon: **Parametric tab→Perpendicular**

Step 1
Draw two lines that are roughly perpendicular as seen in Figure 13.12(a).
Step 2
Select the perpendicular geometric constraint via any of the above methods

Step 3
- AutoCAD will say:

```
Enter constraint type
[Horizontal/Vertical/Perpendicular/PArallel/Tangent/SMooth/Coincident/CONcentric
/COLlinear/Symmetric/Equal/Fix] <CONcentric>:_Perpendicular
Select first object:
```

Step 4

Select the first line (the vertical one).
- AutoCAD will say: `Select second object:`

Step 5

Select the second line

Immediately after the selection of the second line, you will see it snap to perpendicular and two sets of geometrical constraints added next to each line as seen in Figure 13.12(b). Note that if you selected the bottom line first, then the vertical one would have snapped to it, not the other way around. Selection order matters.

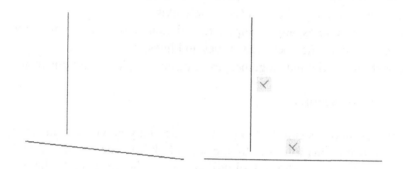

Figure 13.12 (a) and 13.12(b) – Perpendicular geometric constraints

Let's try one more, the concentric one. We'll leave the rest as an in-class exercise.

Keyboard: none	
Cascading menus: **Parametric→Geometric Constraints→Concentric**	
Toolbar icon: **Geometric Constraint** toolbar	◎
Ribbon: **Parametric tab→Concentric**	◎

Step 1

Draw an arc and a circle, one above the other as seen in Figure 13.13(a).

Step 2

Select the concentric geometric constraint via any of the above methods.

Step 3
- AutoCAD will say:

```
Enter constraint type
[Horizontal/Vertical/Perpendicular/PArallel/Tangent/SMooth/Coincident/CONcentric
/COLlinear/Symmetric/Equal/Fix] <CONcentric>:_Concentric
Select first object:
```

Step 4
Select the circle.
- AutoCAD will say: `Select second object:`

Step 5
Select the arc.

Immediately after the selection of the arc, you will see one or both snap to a concentric setting (their centers will be the same) as seen in Figure 13.13(b). Note that if you selected the arc first, the circle would have adjusted to it, not the other way around. Selection order matters.

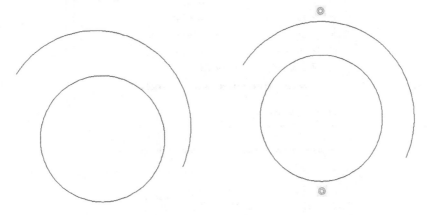

Figure 13.13 (a) and 13.13(b) – Concentric geometric constraints

For the remaining geometric constraints, try them out on your own!

> **Hiding, Showing and Deleting Geometric Constraints**

There are two ways to access these options. You can either right click on one of the constraints to reveal a small menu, as seen here in Figure 13.14, or find these same options on the Ribbon as seen in Figure 13.15 along with the rest of the constraints.

Figure 13.14 – Geometric constraints options - right click menu

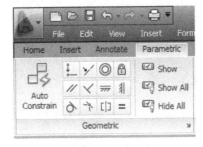

Figure 13.15 – Geometric constraints options - the Ribbon

The Constraint Settings dialog box (that can be accessed via the small drop-down arrow at the bottom right of the Geometric tab) is actually just for global settings for which geometric constraints to show. Leave all the choices selected and the transparency at 50%, as seen in Figure 13.16.

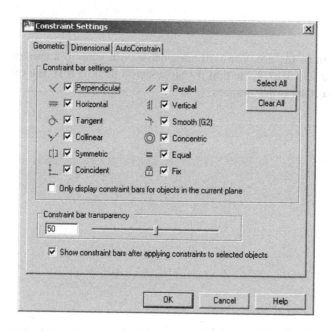

Figure 13.16 – Constraint Settings

Finally here are all six of the constraints mentioned, as you should see them on your screen after trying them out.

Figure 13.17 – Geometric constraints

Sec 13.5 - Dimensional Constraints

Since we have already discussed the basic idea of constraints, this introductory section will be shorter. The idea here is to now move from constraining pieces of geometry (and their positions relative to each other) to constraining the actual dimensions of the drawn design. So if two circles, representing drilled holes, absolutely need to be 3" apart, then constraining that dimension will ensure it won't change regardless of what other design work goes on around them.

Dimensional constraints are also a major part of 3D parametric design, and can be easily transferred to 2D work as was done here in AutoCAD. Also because the geometry has to obey the constrained dimension, one can change the geometry by simply changing the dimension value, which is quite a departure from the normal order of business with AutoCAD. This ability to change geometry by updating dimensions is called "dimension driven design" and is really the only way to change the size of an object (usually called a "feature") when using the high-end 3D design software such as CATIA or SolidWorks.

At this point you may start to see why all of the preceding is referred to as parametric design. We are designing by changing or assigning parameters to our geometry. The geometry itself is almost secondary to the relationships and data that drive its existence and creation. This type of philosophy and approach is one of several steps critical to assuring that what we have is a valid engineering design, not just a "pretty picture." The data behind the design has to make sense!

> ➤ **Working with Dimensional Constraints**

Learning the basics of dimensional constraints is straightforward. You can either add them in right away to your design or convert existing regular dimensions to constraints. Dimensional constraints have their own toolbar (Figure 13.18), or they can be accessed through the Ribbon (Figure 13.19).

Figure 13.18 – Dimensional Constraints toolbar

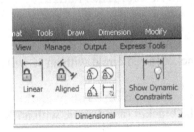

Figure 13.19 – Dimensional Constraints - Ribbon

The available dimensional constraints mirror the regular dimensions closely:

- *Horizontal*
- *Vertical*
- *Aligned*
- *Radius*
- *Diameter*
- *Angle*

Draw the following bracket (Figure 13.20), but don't dimension it, only use the values shown to create the geometry.

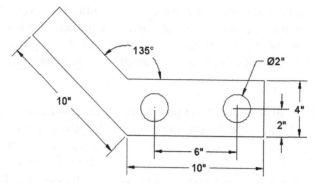

Figure 13.20 – Dimensional Constraints - Ribbon

Let's add a horizontal dimensional constraint

Keyboard: none	
Cascading menus: **Parametric→Dimensional Constraints→Horizontal**	
Toolbar icon:	
Ribbon: **Parametric tab→Linear**	

Step 1
Start up the dimensional constraint via any of the above methods
Step 2
- AutoCAD will say:

```
Current settings:  Constraint form = Dynamic
Select associative dimension to convert or
[LInear/Horizontal/Vertical/Aligned/ANgular/Radial/Diameter/Form]
<Horizontal>:_Horizontal
Specify first constraint point or [Object] <Object>:
```
Step 3
Pick one end of the bottom part of the bracket (a red circle with an X will appear).
Step 4
- AutoCAD will say: `Specify second constraint point:`

Pick the other end of the bottom part of the bracket (a red circle with an X will also appear).
Step 5
- AutoCAD will say: `Specify dimension line location:`

Locate the dimensional constraint some distance away from the part, similar to a regular dimension.
AutoCAD will display the value (d=10.0000) with a graphic of a lock as seen in Figure 13.21. Press Enter.

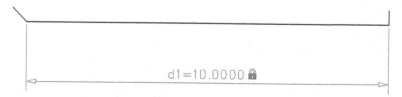

Figure 13.21 – Dimensional constraint

In a similar manner you can add additional dimensional constraints to the rest of the bracket, or convert existing dimensions (if you had some) via the convert button in the Parametric→Dimensional Ribbon tab. Notice the whole point behind this tool. The dimensional values are "locked in" as represented by the lock graphic and will not budge regardless of how other surrounding geometry may change.

Sec 13.6 - Dimension Driven Design

You can of course change the value of the dimensional constraint by just double clicking it and typing in something else (Figure 13.22).

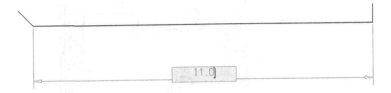

Figure 13.22 – Altering dimensional constraint

The value will force the line to move to the right, leaving a gap which is not quite what you intended. Try the same with a circle and diameter constraint (Figure 13.23).

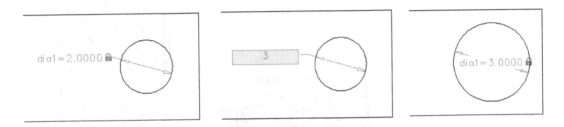

Figure 13.23 – Altering diameter dimensional constraint

Much better results; the circle changed its size! There is a very important point to understand here in order to take full advantage of these tools. *With linework and angles, dimensional constraints need to work together with geometric constraints.* Used just by themselves, dimensional constraints will only change the dimension value and drag the lines along for a ride, but if you add a coincident geometric constraint, the lines will be "glued" together so you can make entire sections of the design properly constrained. A coincident constraint will ask you to select one of the lines, then the second, and create a point where they intersect. Try this out on our bracket's bottom pieces, and try changing the bottom line's size again, leading to quite a different effect.

There is a lot more that can be written about constraints in general; it is a broad and important topic. The student is encouraged to explore further and practice setting up dimensional constraints on critical parts of a design. It will take some getting used to, and applying dimensional constraints intelligently can be awkward at first. The key is to apply these constraints only to the critical geometry, and not constrain every single line and circle on the screen. Fortunately, AutoCAD will prevent you from doing this and issue a warning that you are over-constraining, or there are conflicts, similar to the way high-end solid modeling applications do.

Level 2 Drawing Project (3 of 10) – Architectural Floor Plan

Here we will add in additional features to the floor plan, including mainly furniture and kitchen appliances. Be sure to create the appropriate layers, such as A-Chair, A-Appliances, etc. as you see fit. Dimensions for all furniture are to be reasonably approximated. The dimensions for the fireplace are given. Feel free to add more than is shown.

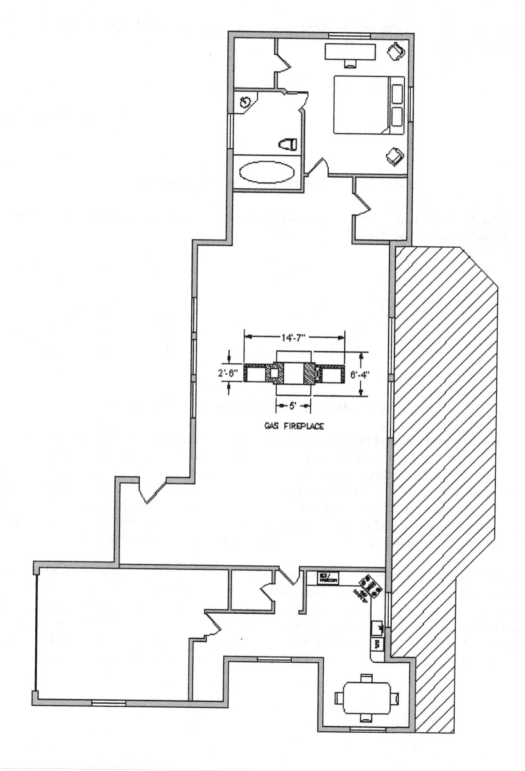

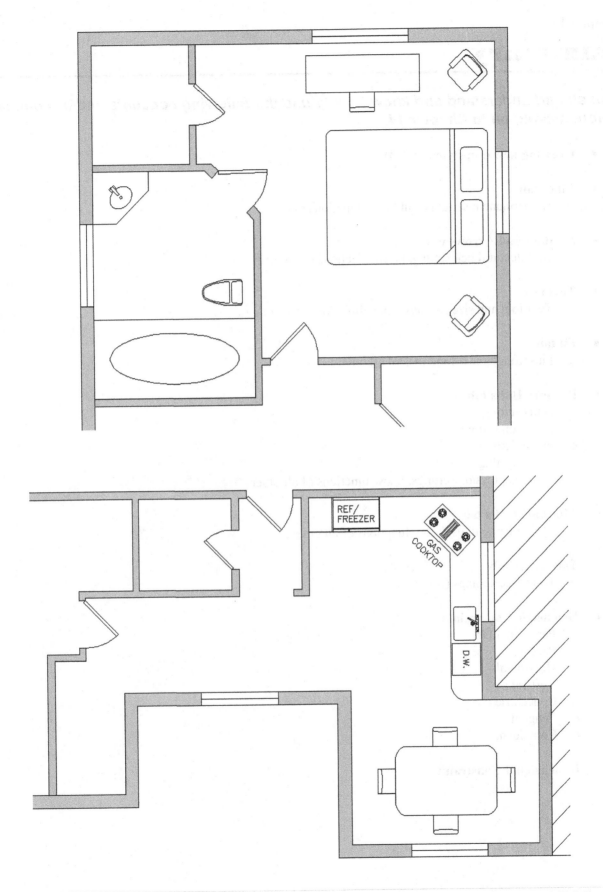

Chapter 13

Summary

You should understand and know how to use the following concepts and/or commands before moving on to Chapter 14.

- **Creating new dim style (DDIM)**

- **Lines tab**
  - No permanent settings, but know functions of all

- **Symbols and Arrows tab**
  - Architectural tick setting, know functions of all others

- **Text tab**
  - Text font and size settings, know functions of all others

- **Fit tab**
  - Fit setting, know functions of all others

- **Primary Units tab**
  - Units setting
  - Precision setting
  - Prefix setting
  - Suffix setting
  - Zero suppression settings, know functions of all others

- **Alternate Units tab**
  - Multiplier settings, know functions of all others

- **Tolerances tab**
  - Tolerance settings (if needed)

- **Geometric Constraints**
  - Perpendicular
  - Parallel
  - Horizontal
  - Vertical
  - Concentric
  - Tangent
  - Coincident

- **Dimensional Constraints**

Chapter 13
Review Questions

Answer the following based on what you learned in Chapter 13.

1) Describe the basic functions found under the Lines tab.

2) Describe the basic functions found under the Symbols and Arrows tab.

3) Describe the basic functions found under the Text tab.

4) Describe the basic functions found under the Fit tab.

5) Describe the basic functions found under the Primary Units tab.

6) Describe the basic functions found under the Alternate Units tab.

7) Describe the basic functions found under the Tolerances tab.

8) Describe the purpose of geometric constraints.

9) What are the seven geometric constraints discussed?

10) Describe the purpose of dimensional constraints.

Chapter 13

Exercises

Exercise #1 – Draw the following shape and label it exactly as shown. Be sure to note all the modifications to the dimensions including arrowheads and the presence of alternate units and tolerances. You may have to get creative with the leader. *(Difficulty level: Intermediate, Time to completion: 10 minutes)*

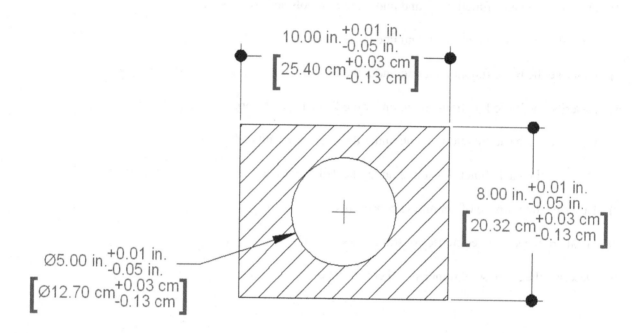

Exercise #2 – Draw a rough parallelogram shape as shown below (a). Parallel constrain all four sides, filleting if necessary, and coincident constrain all four corners. Then dimension constrain one side, using a value of 30 (b). *(Difficulty level: Easy, Time to completion: <5 minutes)*

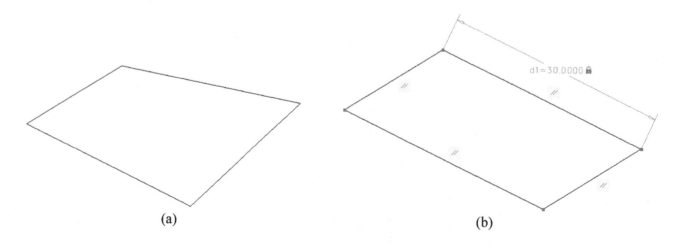

(a)   (b)

CHAPTER  14

Options, Shortcuts, CUI, Design Center and Express Tools

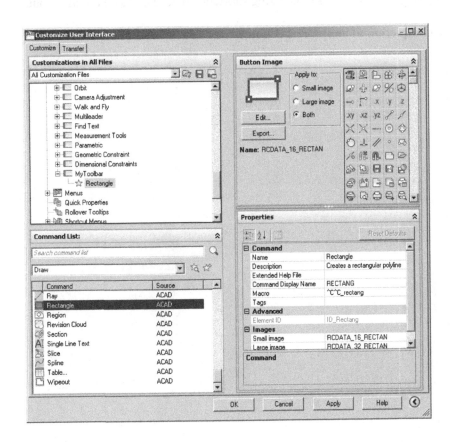

Chapter 14

Learning Objectives

In this chapter we will introduce a variety of advanced tool and discuss the following:

- Options dialog box
  o Files tab
  o Display tab
  o Open and Save tab
  o Plot and Publish tab
  o System tab
  o User Preferences tab
  o Drafting tab
  o 3D Modeling tab
  o Selection tab
  o Profiles tab
- Shortcuts and the "acad.pgp" file
- Customize Users Interface (CUI)
- Design Center
- Express Tools

By the end of the chapter you will have learned how to significantly customize your environment and speed up your work via importing pre-drawn blocks and using shortcuts to accelerate command input.

Estimated time for completion of chapter: 2-3 hours

Sec 14.1 - Options

Our topics for this chapter will be the options dialog box, followed by the concept of shortcuts (the *acad.pgp* file) and the Customize User Interface (the CUI). We'll then look at the Design Center and the Express Tools. These topics will introduce you to basic customization, greatly enhance your efficiency and hopefully lessen any remaining frustration with using AutoCAD.

The options dialog box is your access to basic AutoCAD customization, and through it you can change many essential settings to make AutoCAD behave as you want it to. Almost every user will find a few items worthy of changing, and knowing this dialog box is absolutely essential. Often while covering this topic, students remark how they have been annoyed by some feature (or lack of it) and now know how to turn it off (or on). This is exactly why we introduce this dialog box (or cover it in more detail if you have already seen it).

Note that our goal is not to go over every single item in options; it is far too big for that, and it's unnecessary anyway. Instead, we will outline the most important features and explain why they are necessary. There will be instances (in many cases) where you will be advised against changing the default values of one item or another. If you want to know more, press F1 (Help Files) while in each of the tabs for a detailed outline of every single feature. Note, however, that you should first have a clear understanding of the selected items this chapter focuses on before going into more detail.

To access the option dialog box, right click anywhere in the drawing area or command line and a menu will appear as shown below. Click the last choice: Options...

Figure 14.1 – Options...

A sizable dialog box will appear (Figure 14.2) with the following tabs: Files, Display, Open and Save, Plot and Publish, System, User Preferences, Drafting, 3D Modeling, Selection, and Profiles. With the exception of 3D Modeling (to be covered in the third book), all significant features of each tab will be introduced, though as you will see there will be limited discussion in the case of several of the tabs. Click the very first tab (Files) and let's begin.

➢ **Files Tab**

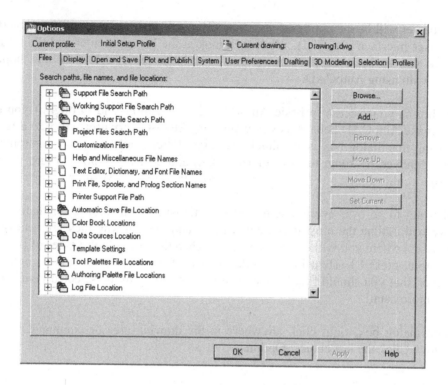

Figure 14.2 – Files Tab

This first tab will be the easiest to go over for the simple reason that we will not change anything here. What are all these listings, you ask? When AutoCAD (or any software for that matter) installs in your computer, it spreads out all over a certain area of the disk drive (where you indicated it should go upon installation) and drops off necessary files as it sees fit to do.

What you are looking at in Figure 14.2 is a collection of paths (each of which you can observe and modify by clicking the plus signs) for the locations of a variety of important AutoCAD functions – help files, automatic saves, etc. You have choices as to where you want them to go if you don't like the default locations. Some of these you may recognize, some not.

Be careful though; default settings are a good thing in this case. If you change the end location of some file, often you (or sometimes AutoCAD) may not find it, and difficulties will result, such as when you try to move the device driver files! It is strongly advised to leave all these as they are, as there really is no compelling reason to change the preset values. The author has only needed to change a few of these paths in all his years of AutoCAD use and management, and those had to do with log file locations and auto-save, so the average user will probably not have much use for this advanced tab.

If you would like a detailed description of what each path means, press F1 while you have the Files tab open.

➢ **Display Tab**

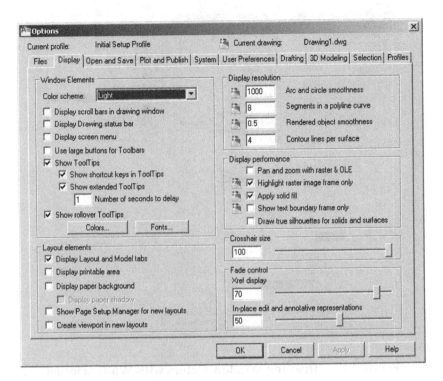

Figure 14.3 – Display Tab

This second tab will have a few items of interest, specifically **Window Elements**, **Layout elements** (on the left side of the box, top to bottom) and **Crosshair size** (on the right).

Window Elements – Under this category you do not need any of the boxes checked except the four related ToolTips boxes. The first few boxes feature color schemes, scroll bars and outdated menus that just take up screen space and do not add anything of value (though some visually impaired individuals can make use of the "large buttons for Toolbars"). ToolTips you may want to have though, to explain the meaning of new toolbars and other items.

Colors... – This refers to the color of your screen background. Generally this is colored black (as preferred by most, but not all, users). Black is a good choice for the simple reason that the dark background throws less light at your eyes from the screen, and lessens fatigue. The only catch is that printed paper output is always white, and some people prefer seeing the design as it would appear in real life (on white paper – hence the white background). Obviously it's easy to get used to the black background and pretty much disregard the "paper" aspects of the output, and this is what most users do.

Other colors are available, of course (actually all 16 million+ colors are available as a background), but as we joke in class, if you see someone using a bright neon green background, stay away...they are probably not a stable individual! A white background is of course used for screen shots throughout this textbook for clarity.

To change the color scheme for practice press the "Colors..." button and the following dialog box will appear (Figure 14.4).

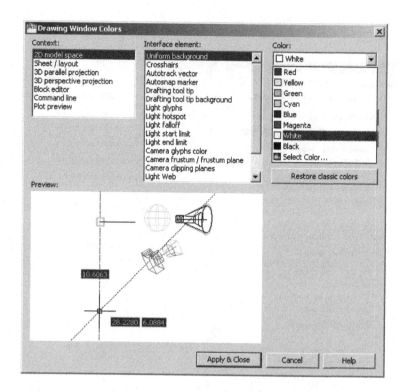

Figure 14.4 – Drawing Window Colors with White selected

Here you can select from the general "Context:" on the left (only the first choice - *2D model space* for now), and then select some element in the "Interface element" area on the right (ex: the *Uniform background*). Then pick a color you want this element to have. The figure above shows white being chosen. Note the sizable amount of choices on both the left and right columns. AutoCAD allows for the color of just about anything to be changed; browse through all the choices to get familiar. If you change so much you forget what you did, you can always press the "Restore classic colors" button seen in Figure 14.4. When done, press "Apply & Close".

Fonts… – This refers to the font present on the command line text, not the font in the actual drawing, so unless you have an objection to the Courier New font (the default), don't change anything.

Layout Elements – Here you can uncheck all of the boxes except the top one: "Display Layout and Model tabs". Generally all those options are distractions and unnecessary in Paper Space. We will discuss this in Chapter 18.

Crosshair Size – This setting is generally a personal preference. Many leave it as a "flyspeck" setting, which is 5, but it is recommended increasing the size to a full 100, as this makes life easier when you are in a Paper Space viewport, or if you're checking if a drawn line is straight (by comparing it to the straight-edge cross-hairs). In the end, it's up to you, but all screen shots in the book will feature the 100 setting.

We will not change anything under **Display Resolution**, **Display performance** or **Fade control**. The default settings are fine. If you would like a detailed description of what each selection means, press F1 while you have the Display tab open. Otherwise, check to see your settings reflect what is in Figure 14.3 and let's move on.

> **Open and Save Tab**

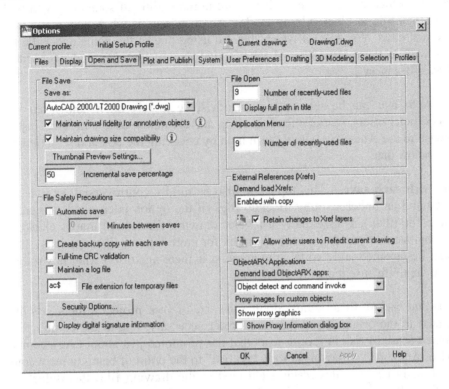

Figure 14.5 – Open and Save Tab

The Open and Save tab contains several important features worth discussing, all of them contained inside the two categories on the left-hand side, namely **File Save** and **File Safety Precautions**.

File Save - Here, let's take a look at the top option, labeled "Save as:" This drop-down menu is very important, as it allows you to permanently save your files down to an older version of AutoCAD. The significance of this cannot be overstated because you simply cannot open a file created on a more recent version using an older release. AutoCAD is an often updated software, with new releases every year (since 2004), and you can rest assured that not everyone will be running the same latest version. The author has seen AutoCAD 14 (from the mid- 90s) running on a company PC as late as 2006. While that may have been an isolated case, AutoCAD 2002 and 2004 are still in widespread use, and nothing is more frustrating to someone than getting a copy of your files for collaboration and not being able to work with them.

Autodesk of course realizes this and allows you to save files down on a case by case basis during the save procedure. AutoCAD will even automatically save one release down anyway. But to be safe (if it isn't already set this way) select the AutoCAD 2000/LT 2000 Drawing (*.dwg) choice, as seen in Figure 14.5, and that way you are assured of the file being accessible by just about everyone else. We can skip over the rest of the options under the File Save category, and move on down to File Safety Precautions next.

Automatic Save – Automatic Save is about as close to controversial as something in AutoCAD may get. The suggestion is: do not use it. Others may say, why not, it is there. The argument against it is simple; as a computer user you need to save your work often, no matter what application you are using. To rely on automatic saves is to invite trouble when you use a program that doesn't have an automatic save (or it isn't turned on). As an AutoCAD user you should be saving your work every 5-10 minutes, something students don't often do in class.

There are also some practical difficulties with automatic save. For one thing, you need to exit a command to save a drawing. AutoSave does this for you, so even if you are in the middle of something you will be kicked out for AutoSave to run (except while in block editing mode). Another issue is where automatic save files will go and what extension they will have. You can set where they go using one of the paths under the File tab, but then all automatic save files will go there regardless of what actual drawing is open and overwrite existing files. The extension of AutoSave files is .sv$.

In summary, while AutoSave can be set up, it's not advisable to rely on it, and often you will see veteran programmers and computer users ignore this feature in all computer use, and save instinctively every few minutes. If you do wish to use AutoSave, simply turn it on by checking the box and set a time threshold ("Minutes between saves") just below that.

Create backup file with each save – This topic can also be mildly controversial, as it seems like a good idea initially, yet doesn't stand up under scrutiny. Backup files (if the option is checked) are created every time you save a file. A backup file (.bak), is essentially a regular drawing file (.dwg) that is cloaked with an extension Windows doesn't recognize, but you do (see the appendix for more info on all extensions). If the .dwg file is lost, then the .bak file can be renamed and "voila", the drawing is there again. So what is wrong with this? Well, in theory nothing, but in "real life" use there are a few issues.

The main problem is that backup files literally double the size of your project folders; it's like having a copy of every drawing file next to the original, and they just aren't necessary, due to the combination of cheap storage media and a heightened sense of risk in recent years that has prompted just about every company to do daily and weekly backups. Also, AutoCAD's drawings rarely "go bad" to the point of being unrecoverable by the audit and recover commands. Finally backup files reside right next to the drawing files themselves, and if you lose the entire folder, they all go. Rarely do you lose just the drawing, and even then going to the previous night's tape backups is enough to get them back. So while some would argue that no harm is done by keeping the *.bak files, one could say: why add the extra junk to your company's server? Ultimately, the choice is up to you though sometimes company policy may dictate the use of backup files, and this will then be beyond your control.

Full-time CRC validation – CRC stands for Cyclic Redundancy Check. Most users will not need this.

Maintain a log file – Makes AutoCAD write the contents of the text window to a log file, also an unnecessary feature for most users.

Security Options... - Security Options is a relatively new feature in AutoCAD which addresses the need to secure and/or authenticate the drawings. This is not an often used feature, as most AutoCAD drawings are not top secret in nature, but the idea is intriguing and worth a look. When you click the button the following box appears.

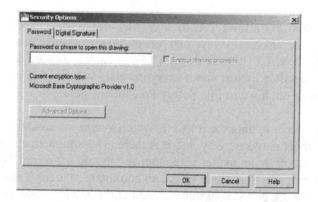

Figure 14.6 – Security Options

The first tab allows you to add a password to the drawing. Then the drawing can only be opened by someone who knows the password; though be careful – if the password is forgotten, the drawing is lost! The second tab has to do with a Digital Signature. This is in itself a lengthy topic, and is sometimes used when collaborating with others. In short, a user installs a certificate from a digital security vendor (VeriSign® in this case), which allows for a "digital signature," which in turn prevents changes to the drawing or keeps track of them. This is of course a very simplified version of the concept, but if your company already uses this, then no further explanation is necessary, and if not you will likely not care about this feature anyway.

We will not go over anything in the right-hand column of the Open and Save tab. File Open is self explanatory, and the other settings under **External References (Xrefs)** and **ObjectARX** (AutoCAD Runtime Extension) **Applications** will be left as default. If you would like a detailed description of what each selection means, pres F1 while you have the Open and Save tab open. Otherwise check to see your settings reflect what is in Figure 14.5 and let's move on.

> **Plot and Publish Tab**

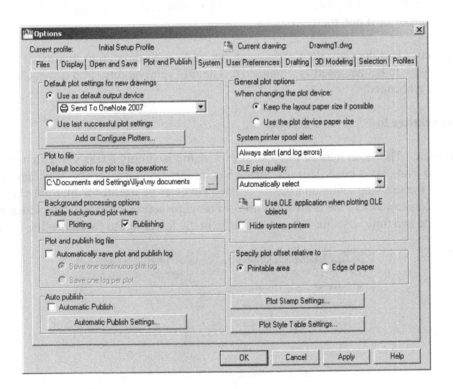

Figure 14.7 – Plot and Publish

While there is nothing critically important under the Plot and Publish tab, a few features are worth a look. We will run through everything with just general comments, while expanding on what proved to be useful over the years, and why. Starting from the top left, moving down, we have:

Default plot settings for new drawings – This is an option that selects the default plotter or adds a new one. However, this should not be done from here, rather from the Windows control panel. When new plotter/printers are added AutoCAD notices them, as well as what the chosen default is (though this is of little significance as the printer settings for each drawing are set individually). Leave everything here as default.

Plot to file - Here a drawing can be plotted to a plt (plot) file and saved in this location. While this is generally an outdated legacy command from the old days of plt files being sent (or physically carried on a floppy disk) to plotters, it has found some use today when taking drawings to a printing company (such as Kinkos®). They may not have AutoCAD (or anyone qualified to operate it), so instead they will accept the plt files, which are nothing more than ready-to-print code that can be interpreted by any plotter.

Background processing options – This has to do with being able to plot in the background while working on a drawing (with the Plotting option unchecked you must wait until the plotting is complete before continuing).

Plot and publish log file is an option that creates (or turns off) the log feature, which is nothing more than a record of the Job name, Date and time started and completed, Full file path, Selected layout name, Page setup name, Device name, Paper size name, and a few others. It is recommended to disable (unchecking) this feature unless there is a specific reason to create a record of all this information.

Auto publish – This is an option that allows you to automatically create DWF files. DWF stands for Design Web Format and is a topic all unto itself. The basic idea is to generate vector files that can be viewed by others that have the Autodesk Viewer, or if not, then through a browser.

General plot options - Here the only item of interest has to do with the output quality of OLE (Object Linked Embedded) items, a topic to be discussed in Chapter 16. The OLE plot quality drop-down choices can be adjusted depending on what is embedded.

Specify plot offset relative to – Leave as default.

Plot Stamp Settings… - This is an item of interest. A plot stamp is simply a text string that appears on all output (when activated in the Plot dialog box). This text string has information on a variety of parameters, but the ones of most interest include the drawing name and the date/time. This can be useful for internal check plots (obviously not for final output), to show where the drawing is on the company server, or to indicate which paper output is the latest based on the date and time. Use the Advanced button in the lower left to further refine the Plot Stamp.

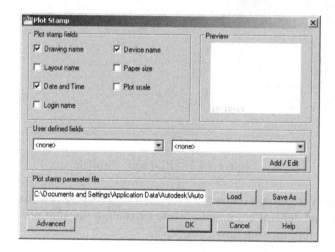

Figure 14.8 – Plot Stamp

Plot Style Table Settings… - Leave as default. These settings are addressed elsewhere.

If you would like a detailed description of what each selection means, press F1 while you have the Plot and Publish tab open. Otherwise, check to see your settings reflect what is in Figure 14.7 and let's move on.

> **System Tab**

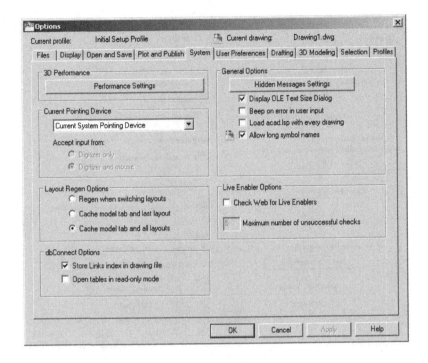

Figure 14.9 – System Tab

There will not be much to discuss here, all settings will basically remain as default.

3D Performance - We will not discuss this.

Current Pointing Device - This allows for digitizer operations, not relevant to most users.

Layout Regen Options - Leave as default.

dbConnect Options - The "db" refers to database connectivity, also not relevant to most users, but may be important in some high-end custom applications, where AutoCAD is linked to a database.

General Options - Here leave everything as default. Here resides perhaps one of the most annoying features of AutoCAD – "Beep on error in user input". For a student just starting out and making lots of errors, this may push him/her over the edge. Not recommended!

Live Enabler Options - This tells AutoCAD to check for object enablers. They can be used to display and use custom objects in drawings when the ObjectARX application that created them is unavailable. Not relevant to most users, and may sometimes cause a program stall if the internet isn't hooked up to the PC.

If you would like a detailed description of what each selection means, press F1 while you have the Systems tab open. Otherwise check to see your settings reflect what is in Figure 14.9 and let's move on.

> ➢ **User Preferences Tab**

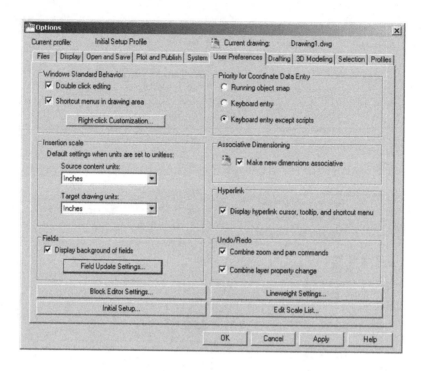

Figure 14.10 – User Preferences Tab

This tab has one very important setting under the **Windows Standard Behavior** category in the upper left. The rest of the categories are not relevant to most users, though we will briefly mention a few.

Windows Standard Behavior – This concerns speed, which is a topic that is near and dear to the heart of most AutoCAD users. As "time = $$", whatever technique accelerates their drafting is always welcome. One of the more important ones (shortcuts) is to be discussed later in this chapter. Another important "trick" is to set your mouse buttons to the most likely and needed settings for fast drafting. As such, **Right-click Customization...** is very important. Click that button and the following dialog box will appear (Figure 14.11).

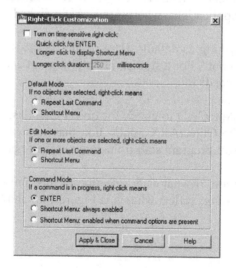

Figure 14.11 – Right-click Customization

Your mouse settings should be as shown in Figure 14.11, with the last one "Command Mode" set to ENTER. This subtle but powerful setting will speed up your basic drafting by allowing you to instantly finish commands by using the right mouse button instead of having to press Enter. Also set "Edit Mode" to Repeat Last Command and "Default Mode" to Shortcut Menu. Nothing else needs to be adjusted. Press Apply and Close.

We will briefly mention the other categories.

Insertion scale – This controls the scale for inserting blocks; leave the default values.

Fields – This sets fields preferences which are not relevant to most users; leave as default.

Priority for Coordinate Data Entry – This controls how AutoCAD responds to coordinate data input; also not necessary to change for most users.

Associative Dimensioning - All dimensions should be associative (meaning if the geometry changes so will the dimension associated with it), so leave it checked.

Hyperlink – This controls settings of hyperlinks; leave box checked. We will cover hyperlinks in the next chapter.

Undo/Redo – This controls Undo and Redo for Zoom and Pan; leave both boxes checked off.

Below these settings are four bars that you can press to set preference. They are described next.

Block Editor Settings... – This changes the environment of the block editor. You worked with the editor when you first learned dynamic blocks.

Lineweight Settings... - Displays the Lineweight Settings dialog box. These settings can be adjusted when setting up the ctb files as described in Chapter 20; not needed here.

Edit Scale List... - Displays the Edit Scale List dialog box. Additional scales are rarely needed, but can be added here if necessary.

Initial Setup – This is an interesting set of options. When you first start up AutoCAD, known as *out of the box*, it allows you to set a variety of customization settings, depending on what you are working on, architecture, engineering, etc. If you skip those steps, you have a chance to do it now as seen in the next series of screen shots.

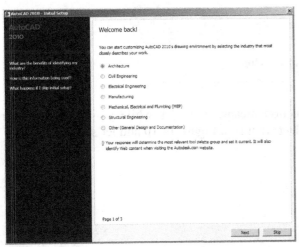

Select your industry...

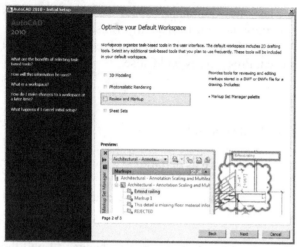

Optimize workspace…

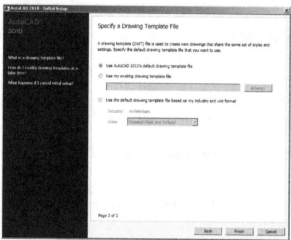

Choose template…

Each step of the way is explained by instructions/questions on the left. All of these steps are completely optional and can be customized and set in other ways. If you cancel, you will see this:

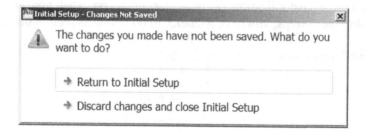

If you would like a detailed description of what each selection means, press F1 while you have the User Preferences tab open. Otherwise check to see your settings reflect what is in Figure 14.10 and let's move on.

> **Drafting Tab**

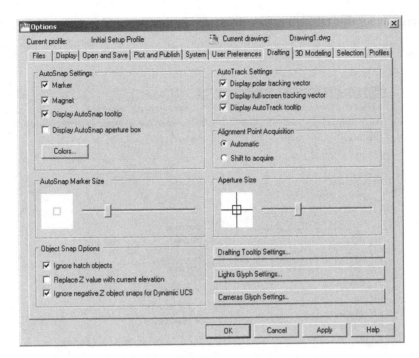

Figure 14.12 – Drafting Tab

The entire left side of this tab is devoted to snap point markers, those yellow colored geometric objects that represent endpoints, midpoints, etc.

AutoSnap Settings – Leave as default, including colors, unless you prefer something different.

AutoSnap Marker Size – Leave as default, unless you have trouble seeing them or find them too big.

Object Snap Options – Leave all as default.

AutoTrack Settings – These have to do with AutoCAD's tracking tool OTRACK which will be briefly looked at in Chapter 15 along with other miscellaneous topics. Leave settings as default.

Alignment Point Acquisition – Leave as default.

Aperture size – This is perhaps the one item that may occasionally be adjusted to suit the user's taste, but generally keep it the same size as the AutoSnap marker.

Drafting Tooltip Settings… - These feature minor nuanced adjustments to drafting tooltips (such as distance or angle displays) and their visibility to the users (the transparency). Leave all as default.

Lights Glyph Settings… and **Cameras Glyph Settings…** – Advanced setting, not relevant at this point; leave as default.

If you would like a detailed description of what each selection means, press F1 while you have the Drafting tab open. Otherwise check to see your settings reflect what is in Figure 14.12 and let's move on.

> ### 3D Modeling Tab

Settings and adjustments under this tab will not be discussed in this textbook.

> ### Selection Tab

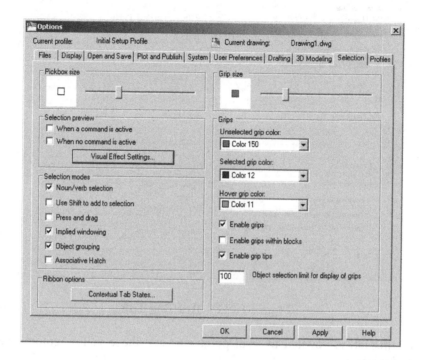

Figure 14.13 – Selection Tab

This entire tab is devoted to minor adjustments to your drafting environment, which may or may not be important depending on your personal tastes and style of drafting.

Pickbox Size – This is simply what the cursor becomes when you are in the process of picking something. Set it too small and you'll have trouble selecting the geometry. Set it too big and you will select more than you wanted. Generally leave it at the default size shown above.

Selection Preview – This is a relatively new feature that causes objects to "light up" when you mouse over them. What exactly constitutes "light up" and a few other items is the business of the "Visual Effect Settings…" button. When you press it, you see the following (Figure 14.14).

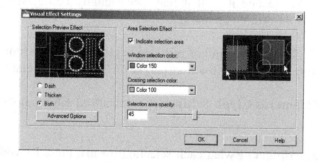

Figure 14.14 – Visual Effects Settings

Here you can set whether you want the lighting up effect to be dashed linework and/or thicker lines (as if a highlighter magic marker went over them), or both. On the right is an interesting setting that has to do with the inside colors of the familiar window/crossing. You can change those colors as well as the intensities. You can eliminate them as the colors are really unnecessary; what's important is whether the selection rectangle is dashed or solid.

Selection Modes – This is just more adjustments to object selection. Leave all as default as seen in Figure 14.13.

Ribbon options – Some options regarding contextual tab states. Leave as default.

Grip Size – This should be self explanatory. Leave the size as default unless you prefer bigger or smaller grips.

Grips are yet more grip settings for colors and other adjustments. There is really no reason to change grip color. Blue means cold (unselected), red means hot (selected) and hover can be any color (default green is just fine). Leave the other settings as default (you want grips enabled usually, but not inside blocks). Leave the maximum amount of grips that can be shown at 100 so as to not overwhelm the drawing if you select more.

Really most of these settings can be left alone. They are included for the benefit of the handful of people that may want to change something and feel that with a $4000 software program, they ought to be able to alter *anything*. If you would like a detailed description of what each selection means, press F1 while you have the Selection tab open. Otherwise check to see your settings reflect what is in Figure 14.13 and let's move on.

> **Profiles Tab**

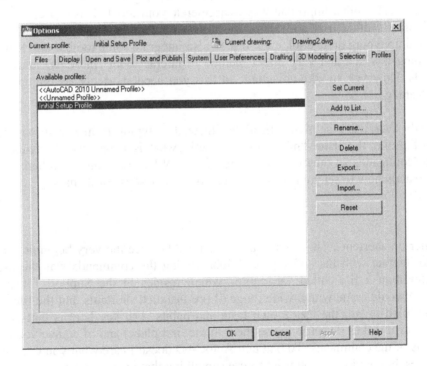

Figure 14.15 – Profile Tab

This tab is simply there to save your personal settings, so if someone else using your computer and AutoCAD prefers other settings (maybe that neon green background color), that person can set his or her own and save them as well. Simply press the Add to List… button and enter a name for the profile as seen below and press Apply & Close as seen in Figure 14.16.

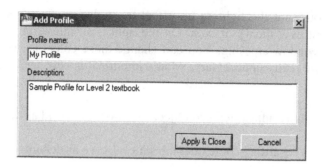

Figure 14.16 – Add Profile

The profile will be added to the list (as many as you want can be added), and can be restored by selecting it and pressing Set Current. Profiles can be exported (as ARG files), imported, renamed and deleted as needed. Profiles were more relevant in the days when PCs were more expensive and each one in a company may have had a few users, but some people still use them to avoid office arguments about exactly how AutoCAD "should look" if one was to "borrow" a PC for the day, change AutoCAD to suit their taste, and forget to reset it to how the original owner/user likes it.

Sec 14.2 - Shortcuts

Recall that it was mentioned in the previous pages how important speed is to an AutoCAD designer. One of the tools to help you work faster was already mentioned – the mouse right-click settings under the User Preference tab. Shortcuts, however, are another important way to accelerate your input. This is useful only if you type. If you are a die-hard Ribbon or toolbar user and abhor the keyboard, then you won't have a need for them. Many students, however, end up typing quite a lot, even if they also integrate some toolbar and Ribbon use. The reasons are quite simple; typing is a fast and very direct method to interact with AutoCAD and remains quite popular despite other methods. However, to take advantage of this you need shortcuts, as typing out the full command defeats the speed advantage.

You may have already been using shortcuts all along. Instead of typing in **arc**, you may have just pressed the letter "a", instead of **move**, the letter "m". This is exactly what is meant by shortcuts; it is the process of shortening the typed commands so they can be entered faster. What we want to do here is put all this on solid footing and outline exactly how to set your own shortcuts to make best use of them.

> **PGP File**

Here is a basic primer on shortcuts. They were part of AutoCAD since the very beginning, as typing commands was the best way to interact with the software and abbreviating the commands was the logical next step. The abbreviations are stored in a file called "acad.pgp" which resides in the Support folder of your AutoCAD installation. AutoCAD would come with a wide range of pre-installed shortcuts, but the file was easily found and was meant to be modified to suit the user's own taste and habits. The key here is that you need to have your shortcuts memorized (or it makes no sense to use them in the first place) and of course you need to pick the most efficient abbreviations of the commands. To that end AutoCAD doesn't care what you use - it can be numbers, or names of your relatives, but obviously you want to use something that makes sense to you and shortens the typing effort. Ultimately the first letter, first and second or first and last is used for all but a few of the commands.

Here are some suggestions for the Modify commands:

Erase – **e**
Move – **m**

Copy – **c**
Rotate – **ro**
Scale – **s**
Trim – **t**
Extend – **ed**
Fillet - **f**
Offset - **o**
Mirror - **mi**

The idea here is to abbreviate the most often used commands in the shortest way possible (such as when there is a conflict – circle is "**ci**", copy is just "**c**" because copy is generally used more often). Sometimes doubling up of letters is OK such as with Array ("**aa**") because "**a**" is taken by arc, and so on. Listed in Figure 14.17 is the author's acad.pgp file. It is presented here only for illustration purposes and may or may not match what you may want in yours, though it can be a good starting point to creating your own.

A,   *ARC
AA,   *ARRAY
B,   *BHATCH
BL,   *BLOCK
C,   *COPY
CI,   *CIRCLE
CH,   *PROPERTIES
D,   *DTEXT
DI,   *DIST
DDO,   *OSNAP
E,   *ERASE
EL,   *ELLIPSE
ED,   *EXTEND
EX,   *EXPLODE
F,   *FILLET
I,   *INSERT
ISO,   *LAYISO
L,   *LINE
LI,   *LIST
LA,   *LAYER
LF,   *LAYFRZ
M,   *MOVE
MI,   *MIRROR
MM,   *MATCHPROP
ML,   *MLINE
O,   *OFFSET
PP,   *PURGE
PL,   *PLINE
PE,   *PEDIT
Q,   *QSAVE
R,   *REGEN
RO,   *ROTATE
RE,   *RECTANG
S,   *SCALE
SP,   *SPLINE
T,   *TRIM
U,   *UNDO
UU,   *UCSICON
W,   *WBLOCK
X,   *EXTRUDE
XL,   *XLINE
Z,   *ZOOM

Figure 14.17 – Sample acad.pgp file

> ➤ **Altering the PGP File**

So how do you find and alter the pgp file to suit your tastes? Prior to AutoCAD 2004 you had to go look for it in the "C:\Program Files\AutoCAD 2002\Support" folder (for ACAD 2002 in this example). You no longer have to do this, and can access the file automatically through the drop-down menus as follows: **Tools→Customize→Edit Program Parameter (acad.pgp)**. The pgp file will then appear on your screen (opened in a Notepad file – which is just a simple ASCII text editor if you never used it before). Scroll down the file until you see the rather lengthy built-in listing of shortcuts. At this point you may do one of three things:

A) Leave the entire file as it is, and use it without modification.
B) Erase and/or change the shortcuts for commands that you either don't use or would like to have modified.
C) Erase the entire list and start fresh, entering the shortcuts as you would like them to be, one by one. If this is the case be sure to follow the format given where you enter the shortcut, then a comma, then a few spaces, and then the command preceded by a "*".

When done save the acad.pgp file and close out of Notepad. At this point for the changes to take effect you can either restart AutoCAD, or (if you are in the middle of a drawing and don't want to do that) type in **reinit**, and press Enter. The following dialog box will pop up (Figure 14.18).

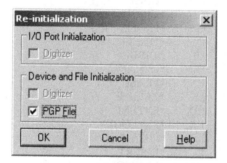

Figure 14.18 – Re-initialization of the pgp file

Check off the PGP File box and press OK. The pgp file is ready to go and you can use it right away. To add or subtract commands just repeat the steps at the beginning of this section, save the pgp file and re-initialize again.

It is strongly recommended that you at least try to modify and use the pgp file. It is almost always faster to type the shortcuts than to click icons. You can eventually learn to type quickly and without even glancing at the keyboard (similar to a good secretary or court reporter), assuming you don't already type fast, as many of my students do. Another advantage is that by typing commands you free up space on your screen where toolbars would normally go.

Sec 14.3 - Customize User Interface (CUI)

The customize user interface dialog box (referred to from here on as just the CUI) is the next item of interest in this chapter. The CUI has to do with changing the look and functionality of menus and toolbars (among other items), and as such can be of importance to a small set of users that may need to do this for aesthetic or functional reasons.

The CUI is an advanced and somewhat mysterious concept, and is not one taken advantage of by most users as either they don't know much about it or don't wish to change their toolbars or menus. A few though will find this customization ability of great interest and spend time customizing the look of AutoCAD to the point of almost having AutoCAD look like another program entirely.

We will walk a middle ground here and present some CUI basics, demonstrating the creation of a new custom toolbar (something the author had an occasional use for) and leaving the student with some CUI ideas to pursue further if he or she is interested in doing so.

Let's introduce the basic CUI as it opens in default mode. Type in **cui**, press Enter (or alternatively use the Ribbon's Manage tab→User Interface) and after a few seconds the following dialog box (Figure 14.19) will open up. Press the expansion button (small arrow, bottom right) to make it full size if it does not do this automatically.

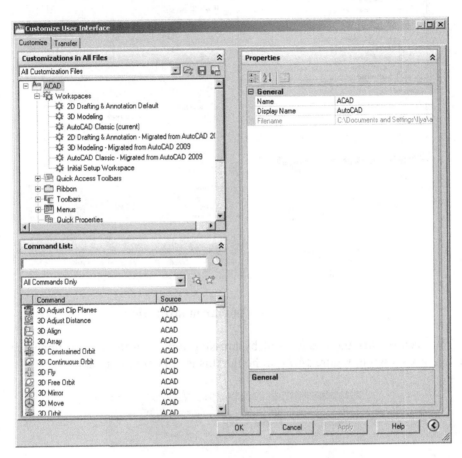

Figure 14.19 – The CUI

In general, the CUI is your window into customizing virtually anything in AutoCAD that you can push, select or type in. The full list of candidates is in the upper left of the CUI box – namely Workspaces, Toolbars, Menus, Dashboard Panels and all the other choices.

Just to get our feet wet, let's create a custom toolbar that only has the buttons you want and none of the ones you never use (great for 3D where toolbar use is more of a necessity than in 2D).

1) In the upper left-hand corner find the Toolbars category and right-click, selecting New Toolbar. Once clicked, a Properties Palette will appear on the right. Give your new toolbar a name in the name field (under General).
2) On the bottom left begin selecting the commands you want your new toolbar to have (you can filter the commands to find them easier), and drag them into the new toolbar name on the upper left. As soon as you begin doing this, a Toolbar Preview and Button Image category will appear on the right-hand side as seen in Figure 14.20 when "rectangle" was added as the first command.

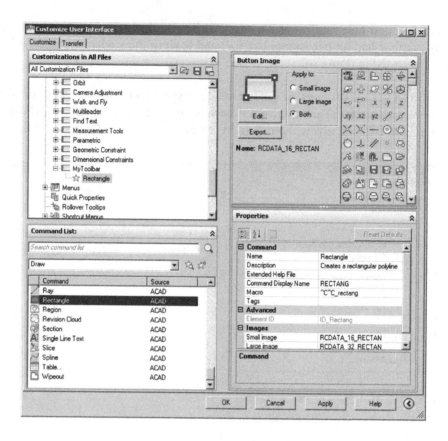

Figure 14.20 – Toolbar modification

Interestingly, you can even modify the image itself by pressing Edit… in the Button Image category which will take you to the dialog box shown in Figure 14.21, although this is not commonly done.

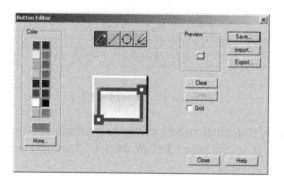

Figure 14.21 – Button Editor

Finally after adding a few more commands (Revcloud, Circle and Spline), pressing Apply and OK, the new toolbar was born, as shown in Figure 14.22; this is certainly a good trick if you like toolbars and want to only see what you use on yours.

Figure 14.22 – The newborn toolbar

This is of course just the tip of the iceberg. You can modify all the rest of the choices. Shortcut Keys under the "Keyboard Shortcuts" are quite useful. This refers to shortcuts such as CTRL+C for Copy Clip (also known as Accelerator Keys). You can set up your keyboard to accelerate many commands. Here are just some of the preset ones (Figure 14.23).

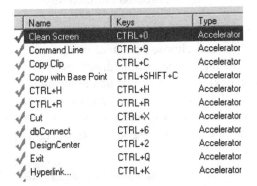

Name	Keys	Type
✓ Clean Screen	CTRL+0	Accelerator
✓ Command Line	CTRL+9	Accelerator
✓ Copy Clip	CTRL+C	Accelerator
✓ Copy with Base Point	CTRL+SHIFT+C	Accelerator
✓ CTRL+H	CTRL+H	Accelerator
✓ CTRL+R	CTRL+R	Accelerator
✓ Cut	CTRL+X	Accelerator
✓ dbConnect	CTRL+6	Accelerator
✓ DesignCenter	CTRL+2	Accelerator
✓ Exit	CTRL+Q	Accelerator
✓ Hyperlink...	CTRL+K	Accelerator

Figure 14.23 – Accelerator Keys

It may be worth your time to look through the help files and investigate the CUI further; it really is a bottomless pit of customization, though be careful not to change something just for the sake of change, but rather only if you feel it may help your productivity.

Sec 14.4 - Design Center

You will probably guess what the Design Center (DC) is as soon as you open it by using the cascading menus **Tools→Palettes→DesignCenter** (or typing in **dc and pressing Enter**). Be sure to dock yours to the left side of the screen and expand the two vertical categories to equal size for good visibility as shown in Figure 14.24.

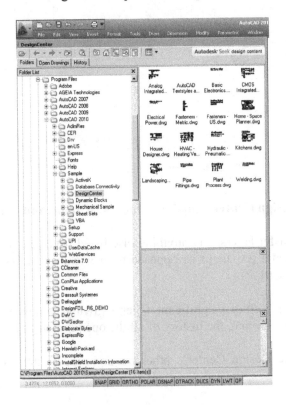

Figure 14.24 – Design Center

What you are seeing here (far left window) is basically the Windows Explorer that is used regularly to view files (such as when you click on My Computer). This is a big clue to the first of two functions of the DC, which is to view the contents of one drawing on the left, and, as we will soon see, transfer parts of that drawing into the active drawing on the right. Let's give this a try using the process outlined next.

By default the DC will open the //Sample/DesignCenter file folder on the left, revealing a collection of sample drawings. You can of course browse for and choose an actual work drawing you created (that's the whole point after all), but for our purposes right now one of the sample drawings is just fine. Go ahead and pick one, let's say the HVAC one. Clicking the plus sign next to the file name and expanding the folder reveals a collection of useful data such as Blocks, Dimstyles, Layers, Layouts, and others. Now click a category such as Blocks. A collection of blocks present in this drawing will be revealed in the middle window, next to the file tree. Click and drag any block from the choices given on the left into the drawing space on the right as seen in Figure 14.25.

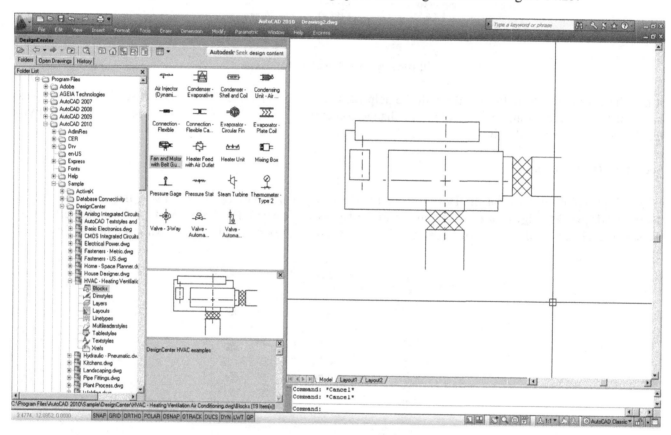

Figure 14.25 – Design Center transfer of blocks (Ribbon and toolbars turned off)

In the same manner you can transfer layers, and anything else listed from one drawing into another. This is very useful to use when you only need a few items as opposed to everything, and is a great visual alternative to the familiar Copy/Paste (Ctrl_C and Ctrl_P).

The next two tabs of the DC are not all that significant. Open Drawings simply looks at the content of the current active drawing, and History is exactly that, the history of all the open drawings this far in the drawing session.

Next we will take a look at the Design Center's other major purpose, importing predrawn blocks into your drawing.

Take a look at DC Online. Note, however, that this tab does not appear automatically in AutoCAD 2009 and 2010. You need to install the CAD Manager Control Utility at the time of AutoCAD install or later on, and check off "Enable DesignCenter (DC) Online". Assuming this has been done (and that you have an internet connection at your computer), click the tab and browse the large collection of blocks found on Autodesk's website for you to use for free. Here is what it looks like with the 2D Architectural folder opened to Bathrooms→Bathtubs, and one of the bathtubs brought into the active drawing on the right by clicking and dragging. Note that, depending on your internet connection, you need to give the system a moment to download the symbol. A preview and some information will appear in a smaller bottom window.

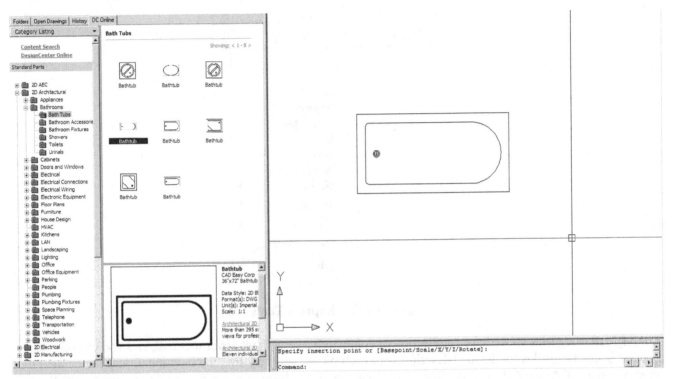

Figure 14.26 – Design Center Online

The collection of blocks is quite extensive. Take some time to look through most of them. Besides Architectural blocks there are also Mechanical, Electrical and Manufacturing, each with extensive subfolders of items. The blocks are generic in nature and may only be useful in some situations (others may call for more vendor-specific blocks). In any case it's good to know there is an available library if the need arises, especially one that is completely free.

Sec 14.5 - Express Tools

Express tools are a set of commands grouped together under the Express drop-down menus. These commands add extra functionality to AutoCAD, and while some of them are redundant or not that useful, others became so indispensable that they were incorporated into regular AutoCAD menus and command sets. It really is a mixed bag, and our goal here will be to highlight only the most useful ones, leaving the rest for you to explore if needed.

Express Tools were originally called Bonus Tools and got their start around Release 14 as a collection of useful routines and small programs. Where they originally came from is a matter of debate, but Autodesk likely acquired them from other third party developers or possibly regular programming-savvy users who felt that AutoCAD just didn't have that perfect command and developed their own as "plug-ins."

However it happened, these tools found their way to Autodesk and the company began to offer them as an "extra" to AutoCAD (not technically part of the core software). Sometimes these tools were offered for sale, sometimes for free to VIP subscribers and more recently for free to everyone, though they often still needed to be installed in a separate and deliberate step while installing AutoCAD itself.

The first step in learning Express Tools is to check that you have them (you should with AutoCAD 2010) and we will proceed from here on with the assumption that all is well and you are ready to go over them. Here is what the top-most listing looks like when dropped down from the top cascading menu (Figure 14.27).

Figure 14.27 – Express Tools

Although all categories above have a collection of commands, we will not cover them all, and you can refer to the Express Tools Help files for detailed description of each. Often you may find the functionality of the express tool already a part of AutoCAD in some other way, shape or form. Here is a run-down of the most useful Express Tools.

> **Blocks**

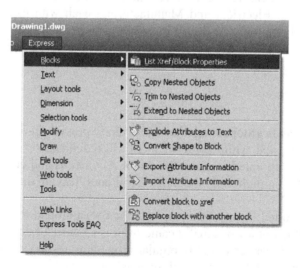

Figure 14.28 – Express Tools (Blocks)

Copy Nested Objects – This command is useful for plucking out nested items, such as those embedded in Xrefs (a topic to be covered later) or blocks. These objects are otherwise stuck inside and not easily accessible for copying. The command line equivalent for this is **ncopy**, and does the same thing.

Explode Attributes to Text – When you explode regular attributes (a Chapter 19 topic), they do not turn to regular text, instead taking the shape of attribute definitions. This useful express tool will force them to revert to ordinary text so you don't have to retype anything.

Nothing else from Blocks will be covered.

> **Text**

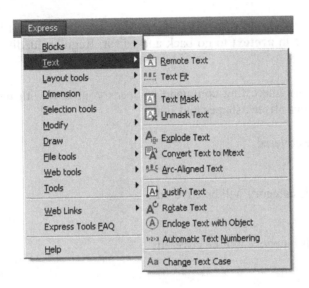

Figure 14.29 – Express Tools (Text)

Convert Text to Mtext – What you see is what you get with this command that converts lines of text to a paragraph of Mtext. It's faster and more efficient than copying the text in one line at a time or redoing Mtext completely.

Arc-Aligned Text - This is a unique command used to tackle a problem that has never been addressed by AutoCAD prior to Express tools: how to enter text that stretches around an arc. The procedure is to draw an arc and select this tool (or just type in **arctext**), then select the arc. The following dialog box will appear (Figure 14.30).

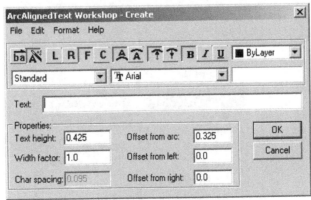

Figure 14.30 – Arc Aligned Text

Enter the actual text and adjust parameters such as fonts, size and anything else you want (they all have explanatory tooltips if you float your mouse over them) and finally press OK when done. You will have the following (or something similar).

Figure 14.31 – Arc-Aligned text

If the text is too squeezed or spread out, adjust the sizing. Note that the text can be over or under the arc. Also remember that you need to type in **arctext** to go back and edit it. Regular ddedit or double-clicking won't work here.

Enclose Text with Object – This interesting command encloses your text with rectangles, slots or circles. Run through the menu options, setting offsets, shapes and sizes.

Nothing else from Text will be covered.

> **Layout tools**

Nothing from this Express tools category will be covered.

> **Dimension**

Nothing from this Express tools category will be covered.

> **Selection tools**

Nothing from this Express tools category will be covered.

> **Modify**

Flatten objects – This is a great tool for 3D work, and will flatten (convert) a 3D image into a 2D one. Nothing else from this Express tool category will be covered.

> **Draw**

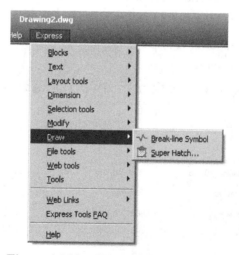

Figure 14.32 – Express Tools (Draw)

Break-line Symbol – Students have found the break line symbol one of the most useful Express tools, because everyone seems to hate drawing one from scratch. Here the process of making one is as easy as clicking two points and indicating a midpoint for the actual break line. Try it by selecting the command and specifying a first and second point along a line; then, using the MIDpoint OSNAP, place the break symbol in the middle.

Super Hatch... - Be careful with this one. A super hatch creates hatch patterns out of ordinary images, by basically copying them many times over to fill an area. This looks interesting, but really slows down the computer as graphics are copied over and over.

> **File tools**

Nothing from this Express tools category will be covered

> **Web tools**

Nothing from this Express tools category will be covered

> **Tools**

Nothing from this Express tools category will be covered

> **Layer Express tools**

The Express tools just covered are the most useful ones to the average designer, and also the ones the author used the most over many years of varied experience. You should still however review the others at your own pace just to know what is available. You will find a lot of overlap and redundancy between what is still in the Express tools and what has been incorporated into regular AutoCAD. Autodesk doesn't like to throw much away.

The Express tools are also missing something very important, perhaps the most important ones of all – the Layer Express tools. They were once part of the gang, but their importance has been acknowledged by an upgrade to permanent status within regular AutoCAD menus. Although there are a bunch of them (most are redundant), we will focus on the primary three that are of significant importance:

- Layer Freeze (command line - **layfrz**)
- Layer Isolate (command line – **layiso**)
- Layer Walk (command line – **laywalk**)

Besides typing, these tools are also accessible via the drop-down menus under **Format→Layer tools** as seen in Figure 14.33.

Figure 14.33 – Layer Express Tools

Layer Freeze

In a complex drawing, layers may number in the dozens or even hundreds. Any tool that can simplify dealing with them is a welcome addition. Layer freeze allows you to freeze a layer merely by selecting it. This bypasses the tedious process of finding out what layer an object is on (if not known), then freezing it manually via the LA dialog box or the toolbar. Simply type in **layfrz** or select **Format→Layer tools→Layer Freeze**.

- AutoCAD will say: `Select an object on the layer to be frozen or [Settings/Undo]:`

Pick an object that represents the layer you want frozen and it will disappear. This can be repeated as often as needed, and is a good way to clear a lot of junk from a busy and dense drawing.

Layer Isolate

Layer isolate allows you to isolate a layer merely by selecting it. The command will turn off (not freeze in this case) all the other layers, leaving the selected one by itself. This bypasses the tedious process of doing it manually. Simply type in **layiso** or select **Format→Layer tools→Layer Isolate**.

- AutoCAD will say:
  ```
  Current setting: Hide layers, Viewports=Vpfreeze
  Select objects on the layer(s) to be isolated or [Settings]:
  ```

Pick an object that represents the layer you want isolated, press Enter and all others will disappear. It is another good way to clear a dense drawing.

Layer Walk

This is perhaps the most interesting one of all the layer controls and is a good way to shuffle through an entire drawing layer by layer. Layer Walk presents a list of layers and isolates one of them. Then you use the arrow keys on your keyboard to select your way down the list. AutoCAD will turn layers on or off in succession, allowing you to inspect each layer, one at a time (walk through them). To try this, open up any file that has a number of layers. Then type in **laywalk** and press Enter or select **Format→Layer tools→Layer Walk...**

A dialog box will appear listing all the layers that are visible from layer "0" until the last one, as shown on a sample file in Figure 14.34. Click on layer "0"; it will highlight and all the rest will disappear. Use the arrow keys to cycle down through each layer, and notice how each one appears and disappears as its turn comes up in the listing.

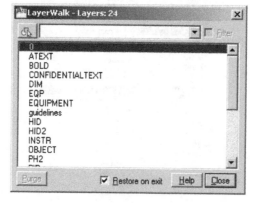

Figure 14.34 – Layer Walk

This is a great tool to review all the layering on an unfamiliar drawing as well as one of your own to insure accuracy. You will notice that along the way many AutoCAD designers will let their layering accuracy slip as the project gains complexity, and by the end layers are sloppy, with objects residing where they aren't supposed to be.

After inspecting all the layers, you can either stop on any one of them for editing, or if the Restore on exit selection box is checked, then all the layers will come back on when you press Close.

Most of the other layers commands are redundant as you can already perform these functions in other familiar ways (via the LA dialog box or toolbar), so we will not discuss them further, but review them at your own pace.

Level 2 Drawing Project (4 of 10) – Architectural Floor Plan

Here we will create a rudimentary electrical plan using outlets, lights, switches and wiring. Create the appropriate layers, which should at a minimum include:

E-Switch
E-Outlet
E-Light
E-Wiring
E-Panel

Some of the symbols are shown below connected by a wire. They are reasonably close to the standard you may find in architectural design, but that of course can vary, especially in regards to how architects choose to represent lighting. Even with switches and outlets there are numerous types (dimmer, 3-way, duplex outlet, GFI, etc.), so this only scratches the surface. Be sure to follow the standard at your company. The plan is of course greatly simplified from the original and does not follow strict electrical guidelines; it is primarily an exercise for you to become familiar with electrical drafting.

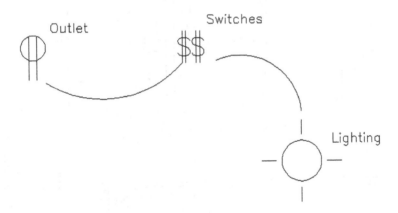

Lay out the electrical devices as seen in the floor plan on the next page. Some of the wiring runs use spline.

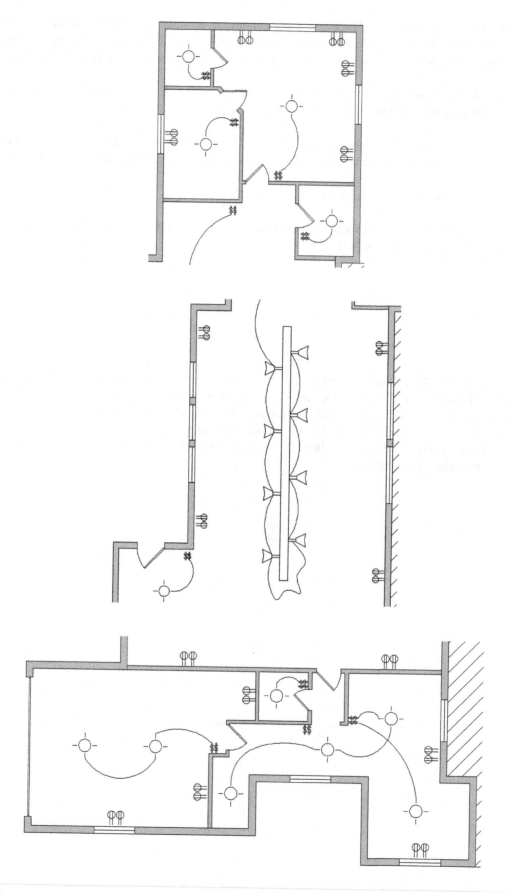

Chapter 14
Summary

You should understand and know how to use the following concepts and/or commands before moving on to Chapter 15.

- **Files tab**
- **Display tab**
- **Open and Save tab**
- **Plot and Publish tab**
- **System tab**
- **User Preference tab**
- **Drafting tab**
- **3D Modeling tab**
- **Selection tab**
- **Profiles tab**
- **Shortcuts**
- **CUI**
- **Design Center**
- **Express Tools**

Chapter 14
Review Questions

Answer the following based on what you learned in Chapter 14.

1. What is important under the **Files** tab?

2. What is important under the **Display** tab?

3. What is important under the **Open and Save** tab? _

4. What is important under the **Plot and Publish** tab?

5. What is important under the **System** tab?

6. What is important under the **User Preference** tab?

7. What is important under the **Drafting** tab?

8. What is important under the **Selection** tab?

9. What is important under the **Profiles** tab?

10. What is the "**pgp**" file? How do you access it?

11. Describe the overall purpose of the **CUI**.

12. Describe the purpose of **Design Center**.

13. Describe the important **Express Tools** that were covered.

Chapter 14

Exercises

Exercise #1 – Based on what you learned about the Options dialog box, set up your own profile. Be sure to include or at least consider all the variables and settings discussed. Save the profile as "Profile_1_Your_Name"
(Difficulty level: Easy, Time to completion: 10 minutes)

Exercise #2 – Based on what you learned about the "pgp" file, create your own based on whatever abbreviations you prefer. Be sure to utilize the commands shown in Figure 14.18 as well as adding any of your own that you may want included.
(Difficulty level: Easy, Time to completion: 20 minutes)

Exercise #3 – Based on what you learned about CUI, create your own toolbar that features the following commands icons:

Line
Circle
Rectangle
Move
Copy
Rotate
Offset

(Difficulty level: Easy, Time to completion: 15 minutes)

Exercise #4 – Open up Design Center and open the DC Online tab. Go to the Architectural 2D folder and download a full set of furniture and appliances for a small apartment including bed, tub, chairs, sink, stove, refrigerator, etc.
(Difficulty level: Easy, Time to completion: 15 minutes)

Exercise #5 – Review the express tools including creating another arc aligned text and break line. Open up any multi-layer file and review the layer tools: layfrz, layiso and laywalk.
(Difficulty level: Easy, Time to completion: 20 minutes)

LEVEL 2

CHAPTER 15

Miscellaneous Topics

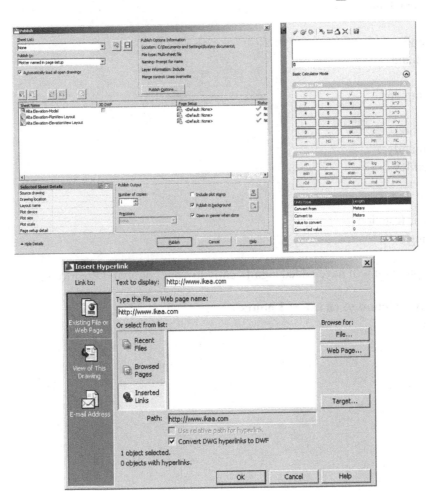

Chapter 15

Learning Objectives

In this chapter we will continue to introduce a variety of advanced tool and discuss the following:

- Align command
- Audit/Recover commands
- Break/Join commands
- CAD Standards
- Calculator command
- Defpoints concept
- Divide/Point Style commands
- Donut command
- Draw Order
- eTransmit command
- Filters
- Hyperlinks
- Lengthen command
- Object Snap Tracking
- Overkill
- Point command
- Publish command
- Rasters
- Revcloud command
- Sheet Sets
- Selection methods
- Stretch command
- System Variables
- Tables
- Tool Palette
- UCS/Crosshair Rotation
- Window Tiling

By the end of the chapter you will have added significantly to your command repertoire.

Estimated time for completion of chapter: 3 hours

Sec 15.1 - Introduction to Miscellaneous Topics

This will easily be the most diverse and unique chapter you will read in the book. This text was written from the start to simplify and isolate core topics and essential ideas, without cluttering the themes with additional AutoCAD commands and concepts. There are several more "big ideas" left, such as Xrefs, Paper Space and Attributes. Those will be addressed in upcoming chapters. However that leaves out a great deal of useful concepts and commands that didn't quite fit in anywhere else. These bite-sized chunks of information form the basis of this chapter. Many of these topics are just random commands, standing alone, not part of any other major concept, some are part of a larger family, but all are useful to one extent or another and are necessary for a well rounded knowledge of AutoCAD.

It is in this chapter that we will really look into the "nooks and crannies" of AutoCAD and try to cover just about anything you may find browsing through the menus. Obviously this is not an exhaustive list, some items have been left out due to marginal usefulness or just space limitations, but the important ones are for the most part here. The list is in alphabetical order. Go through it at your own pace, noting how each concept or command can help you in your design work. There is much to discover.

> **Align Command**

This command does exactly what it advertises, namely aligns one object to another. In Figure 15.1 we have a rectangle drawn and rotated to some random angle, and below it a line. While one can always measure the angle of the rectangle and then use the rotate command, there is a much easier and accurate way to align the rectangle to the line (or vice versa).

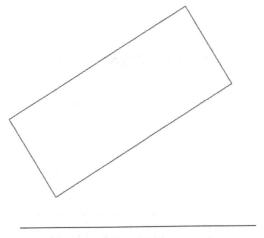

Figure 15.1 – Alignment objects

Type in **align** and press Enter. You can also use the cascading menus **Modify→3D Operations→Align**. There are no toolbar or Ribbon equivalents.

- AutoCAD will say: `Select objects:`

Select the object to be moved into alignment (the rectangle). Press Enter.

- AutoCAD will say: `Specify first source point:`

Pick the lower left corner of the rectangle using endpoint osnap (point 1 on Figure 15.2)

- AutoCAD will say: `Specify first destination point:`

Pick the left point of the line using endpoint osnap (point 2 on Figure 15.2)

- AutoCAD will say: `Specify second source point:`

Pick the lower right corner of the rectangle using endpoint osnap (point 3 on Figure 15.2)

- AutoCAD will say: `Specify second destination point:`
Pick the right point of the line using endpoint osnap (point 4 on Figure 15.2)
- AutoCAD will say: `Specify third source point or <continue>:`
You have no need of a third set of points, so press Enter.
- AutoCAD will say: `Scale objects based on alignment points? [Yes/No] <N>:`

There is something interesting you can do here. Not only can you align the rectangle to the line but also scale it up to the line if needed by selecting `Yes`. We won't try this now, but go back after learning the basic align command and try this variation of it. For now just press Enter to accept the default `No` option. The rectangle will be aligned to the line as in Figure 15.3.

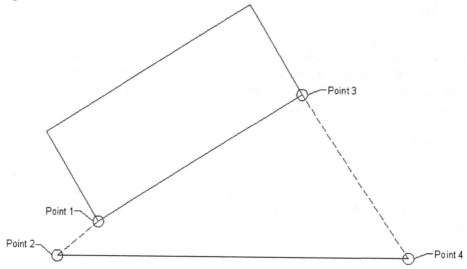

Figure 15.2 – Alignment – Points

Figure 15.3 – Alignment – Final Results

> **Audit/Recover Commands**

Though rare, AutoCAD does lock up and crash (usually a hardware/memory issue) but can also be software related. In the event of a crash the drawing is auto saved but may be corrupted when you try to open it up again. Sometimes also drawings just cause AutoCAD to behave oddly for no specific reason (i.e: commands don't work as expected or just fail). In either case the drawing will need to be audited and/or recovered.

Auditing is the lower-level check. AutoCAD evaluates the integrity of the drawing and corrects *some* errors. You just simply type in **audit** and press Enter or alternatively use the cascading menus **Application menu→Drawing Utility→Audit**. There is no toolbar or Ribbon equivalent. AutoCAD will ask about fixing any detected errors (say "yes") and will cycle through drawing entities and in a few seconds generate a log similar to what you see next (assuming nothing wrong was found or fixed as seen here).

```
Auditing Header
Auditing Tables
Auditing Entities Pass 1
Pass 1 1100 objects audited
Auditing Entities Pass 2
Pass 2 1100 objects audited
Auditing Blocks
10 Blocks audited
Total errors found 0 fixed 0
Erased 0 objects
```

Of course if something was found, fixes would be reported. Sometimes however audit is not enough and you must recover the drawing. This is not as drastic as it sounds. The recover command is powerful and almost always brings the drawing back as a matter of routine. The author can count on one hand the amount of drawings that were lost forever due to corruption in 12 years of CAD design. And even in those cases, going to the previous nights backups would have restored the drawing, minus a few hours of work.

To use this feature, open a blank drawing file, type in **recover** and press Enter or alternatively use the cascading menus **File→Drawing Utilities→Recover**. There is no toolbar or Ribbon equivalent. You will be prompted to browse and find the unfortunate drawing. Once you select it and press Open, recover will run and generate its own log as seen here (assuming nothing was found as the author ran both audit and recover on clean files to generate these logs).

```
Drawing recovery.
Drawing recovery log.
Scanning completed.
Validating objects in the handle table.
Valid objects 755  Invalid objects 0
Validating objects completed.
Salvaged database from drawing.
Auditing Header
Auditing Tables
Auditing Entities Pass 1
Pass 1 900 objects audited
Auditing Entities Pass 2
Pass 2 900 objects audited
Auditing Blocks
19 Blocks audited
Total errors found 0 fixed 0
Erased 0 objects
```

Following the log, you will see this message box:

If something however *was* wrong, recover should fix it and the drawing will open as normal. Immediately save it!

> ➢ **Break/Join Commands**

The break command can take a geometric object and put a break in it (remove a section another words), or if applied at an intersection with another object, break that object at the intersection without actually removing any material. Join command is the exact opposite. It will add material in to close the break, but will only work in certain circumstances. Let's examine each command closer.

Break - Method 1
Draw a line of any size or direction and start up break via any of the following methods.

Keyboard: Type in **break** and press Enter.
Cascading menus: **Modify→Break**
Toolbar icon: **Modify** toolbar ☐
Ribbon: **Home tab→Modify** ☐

- AutoCAD will say: `Select object:`
Go ahead click to select the line at some random spot. It will turn dashed.
- AutoCAD will say: `Specify second break point or [First point]:`
Pick another nearby point on the line, and a section between these two points will be removed (Figure 15.4).

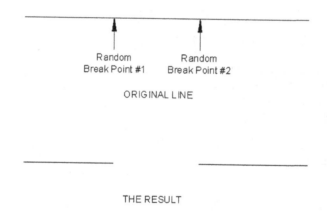

Figure 15.4 – Break command

In the previous method you had no control over the first point of the break. When you selected the line the first point was chosen then. However, as shown in the next method, you can select the line to be broken, then using the First point option specify exactly where to start the break. We will use this to our advantage to create four lines out of two as seen next.

Break - Method 2
Draw a horizontal line of any size, followed by a perpendicular vertical one that is of equal size. They should cross perfectly in the middle. You will have a plus sign. One good trick to do this is to first draw a circle and just connect the four quadrant points with two lines as seen here:

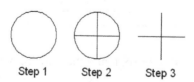

Start up the break command via the previous methods.
- AutoCAD will say: `Select object:`

Go ahead click to select the horizontal line at some random spot. It will turn dashed.
- AutoCAD will say: `Specify second break point or [First point]:`

Type in **f** for `First point`.
- AutoCAD will say: `Specify first break point:`

Using OSNAPs select the midpoint of the horizontal line.
- AutoCAD will say: `Specify second break point:`

Once again, using OSNAPs select the same midpoint of the line.

Now repeat the above steps with the vertical line. The end result will be four lines where there was only two just moments earlier as seen when the grips are revealed. All four are broken at their intersection points.

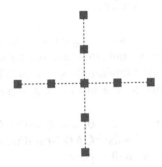

Join

The join command is included here because it is the exact opposite of break. It will close the gap between two coplanar lines. Note the important disclaimer however. The join command will not unite two lines that are not on the same plane. This means that if they are not pointing in the exact same direction and are not literally "one after the other" then join will not work.

Draw two coplanar lines (or use break to put a gap in) as seen below:

Keyboard: Type in **join** and press Enter.
Cascading menus: **Modify→Join**
Toolbar icon: **Modify** toolbar ━╍━
Ribbon: **Home tab→Modify** ━╍━

Start up the join command via any of the above methods.
- AutoCAD will say: `Select source object:`

Select the first line.
- AutoCAD will say: `Select lines to join to source:`

Select the second line.
- AutoCAD will say: `Select lines to join to source:`

You're done with the two segments so press Enter.
- AutoCAD will say: `1 line joined to source`

A solid line will take the place of the two segments.

> ➢ **CAD Standards**

This is a rather obscure procedure that may be valuable to a few organizations, and the author has on occasion recommended it to prime contractors who often find their design files bouncing around from subcontractor to subcontractor. After all the respective additions to the drawings are finished they are sent back to the main contractor. There it is found that the standards have deteriorated as everyone did their own thing in regards to layers and text/dim styles. A way to identify and clean up the mess would be useful, hence the existence of this tool.

The CAD Standards procedures can be a bit confusing, and the process does not guarantee effortless and perfect results. Depending on the damage, you may still need to do manual clean up of layer properties and errant fonts, etc. The best defense is to hold all others who may work on your drawings to a high standard from the start, and outline what is acceptable and what isn't. Obviously this can be challenging in just one office, never mind implementing it for designers at other remote locations.

Here's is the basic idea. Come up with one file that is in great shape. This can be a regular job that is singled out for its perfect or near perfect use of company standards and is of the highest quality and integrity. Then do a "save-as" for this job under a "dws" (**DraWing Standards**) extension. This is your standards file. All others will be compared to it as they come in from subcontractors or just other designers.

To implement it, open up a job that just came in and is of questionable CAD quality. Type in **standards** and press Enter or alternatively use the cascading menu **Tools→CAD Standards→Checks**. There is no toolbar or Ribbon equivalent. The Configure Standards dialog box will appear.

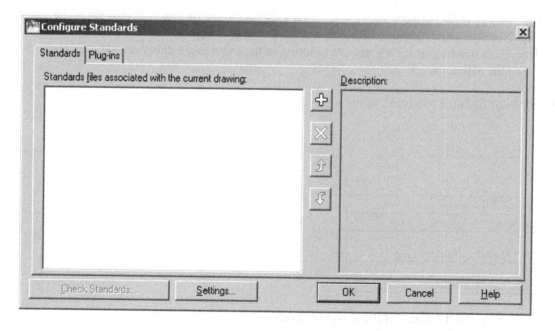

Figure 15.5 – Configure Standards

Next use the plus (+) sign to open up the saved dws file. The file will appear in the window on the left with some descriptions on the right. The Check Standards button (lower left) will also light up, at which point press it to execute the check. The check standards dialog box will appear as seen in Figure 15.6.

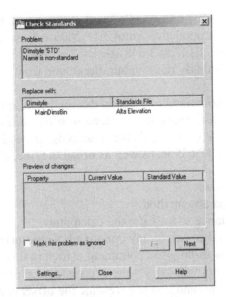

Figure 15.6 – Check Standards

Next you will need to run through the fixes, item by item. Continue pressing the Fix button as AutoCAD runs through the fixes and matches up text, layers and other items. When done the following Checking Complete notice will display.

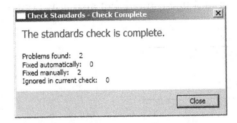

Figure 15.7 – Checking Complete

> ➤ **Calculator Command**

The calculator is a surprisingly powerful and versatile tool within AutoCAD, and one which is often ignored by users. The function's power and versatility should come as no surprise, as AutoCAD itself is highly mathematical software, a calculator in itself, performing complex operations "under the hood" as you draw, copy and rotate objects.

The calculator function can perform all regular and scientific calculations you would expect from your handheld device, but in addition can also assist in construction of geometry, and even able to make use of AutoLISP, a programming language. We will not go in depth here by any means, but will provide one example of construction assistance using the transparent line command calculator, and one example of the actual graphic calculator tool (QuickCalc) to do unit conversions. The student is encouraged to read the help files for a wealth of additional information.

First a brief overview of the **cal** command. This was the original version of the calculator. It's a command line tool enabled by typing in **cal** and pressing Enter. AutoCAD will then say: >> Expression: at which pint you can now enter some math function using the familiar characters such as +, -, * and /. Pressing Enter again will yield an answer.

Not too long ago the limit of the calculator was +32,767 and -32,768 as the largest displayable final integer, but that restriction was lifted somewhere around the AutoCAD 2005 release and you can now work with integer numbers between +2,147,483,647 and -2,147,483,648 (that's 2 to the 15th and 31st power, respectively, if you *really* wanted to know where those seemingly random numbers came from).

Let's use the previously described cal command to assist in design, because crunching numbers alone can be done by your handheld or Windows calculator. Suppose you need to offset a line one-half of some number, but you don't know what that value is (say ½ of 17.625" for example). Instead of reaching for that calculator, let AutoCAD do both; calculate what ½ of 17.625 is, as well as offset the line, all in one command. To do this:

1) Draw a line of any reasonable size.
2) Start up the offset command via any method.
3) You don't know the offset distance so let's use the calculator.
4) Type in **'cal** (the apostrophe makes cal, or any command, temporarily transparent, meaning you can enter it without exiting out of another in-progress command, which in this case is "offset"). Press Enter.
5) The >>>> Expression: prompt will ask for the expression. Enter 17.625/2 and press Enter.
6) Your answer (8.8125) will appear, and you may resume the offset command.
7) Continuing as usual, select the line and pick a direction to complete the offset and Esc to exit out.

This was a very simple case of how the calculator assists in design tasks. It can do a lot more in that regard and you are encouraged to examine the help files for additional examples. We will now however move on to briefly introduce the QuickCal.

The calculator function is not just command line limited, but also now comes in a graphic version.

Keyboard: Type in **quickcalc** and press Enter.
Cascading menus: **Tools→Palettes→QuickCalc**
Toolbar icon: **Standard** toolbar
Ribbon: **View tab→Palettes**

Here is what you will see (Figure 15.8).

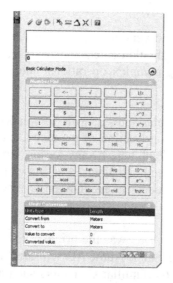

Figure 15.8 – QuickCalc

The Number Pad and Scientific keypads are self explanatory. If you ever used a handheld calculator, you should be right at home with this one. Variables, at the bottom, will not be covered, but Units Conversion deserves a special mention. It is a VERY useful tool for converting between a wide variety of Length, Area, Volume and Angular values. Simply select from the fields on the right (drop down arrows will appear) and enter the values to convert just below that. A converted value will appear according to what you requested.

➢ **Defpoints Concept**

Defpoints is not a command, rather a concept that pops up often and bears a short mention. Defpoints stands for Definition Points and they appear in the drawing (and mysteriously in the Layers dialog box – where most users first notice them) as soon as you add any dimension to your drawing. What are they exactly? They are the tiny points where a dimension attaches itself to the object being dimensioned. The Defpoints layer doesn't go away easily, and usually prompts questions from students as to its nature and intent. The layer fortunately does not plot, and you can pretty much disregard its existence if you choose. One benefit of it not plotting is that you can put any objects that you don't want to see on paper on it (such as viewports as will be discussed in Chapter 18). In fact that is exactly what many users do, though a dedicated VP layer is recommended instead. In conclusion…don't worry about Defpoints and continue on.

➢ **Divide/Point Style Commands**

Divide does exactly what it advertises; it divides an object into equally spaced intervals. Note however that the object isn't actually broken into those segments (that's for a break command to do) but rather just divided by equally spaced points. Then you can use these points as guides for further work. Let's give this a try. Draw a horizontal line of any size as seen below.

Type in **divide** and press Enter or alternatively use the cascading menu **Draw→Point→Divide**. There is no toolbar or Ribbon equivalent.
- AutoCAD will say: `Select object to divide:`

Select the line
- AutoCAD will say: `Enter the number of segments or [Block]:`

Enter some value; 8 for example, and press Enter.

The line will look the same with no apparent divisions. It seemed like the command hasn't worked. It actually worked just fine, and the line is divided into the eight segments using seven points as expected. The problem is that you can't see the points; they are obscured by the line, and we need to change the point style to something more visible. Using the cascading menu, select **Format→Point Style…** You will see the following (Figure 15.9).

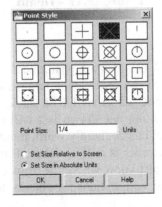

Figure 15.9 – Point Style

Select a point style that isn't too complex; the "X" format in the top row is a popular one. Press OK and the divided line will appear as you were expecting it, divided into the eight visible segments (Figure 15.10).

Figure 15.10 – Divided line

➢ **Donut Command**

While this command may not be used all that often, it's still worth a mention. It creates exactly what is advertises; a donut of varying size. Broken down in to its components, a donut is two circles of different size, one inside the other, and a fill. You have control over the sizes of both circles, and a result the size of the donut.

Keyboard: Type in **donut** and press Enter.
Cascading menus: Draw➔Donut
Toolbar icon: none
Ribbon: none

Start up the donut command via any of the above methods.
- AutoCAD will say: `Specify inside diameter of donut <0.5000>:`
You can accept this value by pressing Enter or type in another (we'll accept the .5 here).
- AutoCAD will say: `Specify outside diameter of donut <1.0000>:`
Once again you can accept this value or type in your own (we'll accept the 1.0 here).
- AutoCAD will say: `Specify center of donut or <exit>:`
Click anywhere on the screen and the donut will appear as seen in Figure 15.11.

Figure 15.11 – Donut

➢ **Draw Order**

Draw order refers to the stacking order of objects that are positioned one on top of another. Another words, which one is on top as well as what is the rest of the order. This feature is commonly used in most graphic design programs, and has an occasional use in AutoCAD as well. The stacking order is usually not of major concern in day to day basic drafting. If items intersect, the one drawn last will be on top, and in cases of two intersecting lines, the intersection point will be one pixel in size, hardly worth mentioning. Sometimes however, such as when two or more hatch patterns are stacked, the order is critical. We will use this feature later in this chapter to great effect when we discuss Wipeout. For now just know where the draw order commands are. They are found in the cascading menu under **Tools➔Draw Order**, with options for objects, text and dimensions.

"Bring to Front" and "Send to Back" send objects strictly to and from the top and bottom of the order while "Bring Above Objects" and "Bring Below Objects" are more selective, moving items one step at a time. With these four tools you can arrange any amount of objects in any order.

> ➤ **eTransmit Command**

The purpose of this command is to assist with e-mailing of drawings. The problem is that when drawings are sent out, items can be inadvertently omitted. These omissions may include something major like associated Xrefs, to something secondary such as fonts and styles. As a project grows more complex it's easy to forget to send Xrefs, and even easier to forget about the unique fonts, styles and ctb files that have been generated over time. The recipient of the email may have only some of these, and what they will download and reconstruct on their end may not be quite what you intended for them to see.

eTransmit is a "go-getter" of sorts. It sniffs out every associated file and everything else that may be relevant to recreate the file(s) in a new location as originally intended. Then it packages everything together in a folder, and sends it out. While veteran users have remarked that an experienced and aware designer needs to always have a clear idea of what to send (eTransmit is a recent invention after all), it is still a valuable tool for novice and expert alike.

So what is (and isn't) included in the eTransmit transmittal? Here's a partial list of what *is* included. You should know of most of these files. If not refer to the appendix.

*.dwg - Root drawing file and any attached external references.
*.dwf - Design Web Format files that are attached externally to the root drawing or Xrefs.
*.xls - Microsoft Excel spreadsheet files that are linked to data extraction tables.
*.ctb - Color-dependent plot style files used to control the appearance of the objects in the drawings of the transmittal set when plotting
*.dwt - Drawing template file that is associated with a sheet set
*.dst - Sheet set file
*.dws - Drawing standards file that is associated to a drawing for standards checking

Here's a partial list of what is *not* included. Don't worry if you do not recognize all of these formats. You will know most of them by the end of Level 2.

*.arx, *.dbx, *.lsp, *.vlx, *.dvb, *.dll - Custom application files
(ObjectARX, ObjectDBX, AutoLISP, Visual LISP, VBA and .NET).
*.pat - Hatch pattern files.
*.lin - Linetype definition files.
*.mln - Multiline definition files.

To use etransmit, save your work but keep the file open. Type in **etransmit** and press Enter, or alternatively use the cascading menu **File→eTransmit...** There is no toolbar or Ribbon alternative. The following dialog box will appear (Figure 15.12).

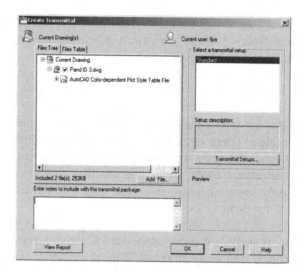

Figure 15.12 – Create Transmittal

The procedure is highly automated, but you do have to select a few parameters. The current drawing that is open is the one that will be packaged together. Some of the included items are shown in the file tree. Expanding the + signs will show specific files (which you can deselect if desired). You also add more drawings to the transmittal using the Add File… button. If you're fine with just the current drawing then press the Transmittal Setups… button. You will see the following dialog box (Figure 15.13).

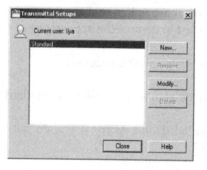

Figure 15.13 –Transmittal Setups

Let's create a new setup as an exercise. Press the New…button and the following dialog box will appear (Figure 15.14). Give the transmittal a name and press Continue.

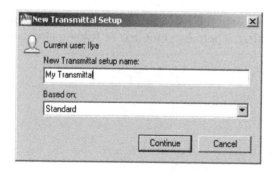

Figure 15.14 – New Transmittal Setup

The following dialog box will then appear (Figure 15.15).

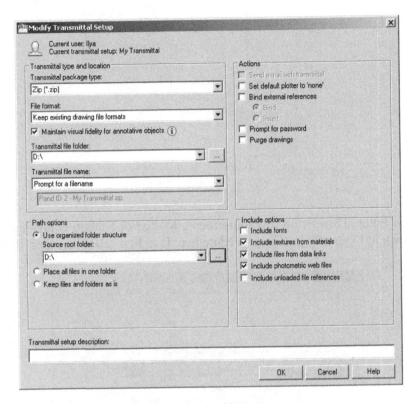

Figure 15.15 – Modify Transmittal Setup

All you really have to do here is visually check all the settings. Looking at the fields from top to bottom we see the Zip option for actually sending the files (recommended), followed by a variety of information related to formats, names and locations. Finally below that we have a number of checked off options. You should keep the files in one folder, and include everything listed in the first four options. The last four options, in contrast, may be unnecessary. Use your best judgment.

When done press OK, then Close in the next dialog box (Figure 15.13) and finally OK in the main setup box (Figure 15.12). You will be asked where to put the zip file and for any last minute name changes. Pressing Save will drop the file into its place, and it is ready for emailing.

> **Filters**

Filters were briefly discussed in the context of layers (Chapter 12). While the idea is roughly similar, the application is slightly different, so let's review from the beginning.

Filters are nothing more than a method to isolate and select objects by using their features or properties. So for example you can isolate and select all the circles by virtue of them being circles, and find green objects because they all have green in common. You can also add together properties and create requests like: "find me circles that are red only" and so on. This is a good tool for quickly finding stuff in a complex drawing.

To experiment with filters, let's first draw a simple drawing which is just a collection of shapes. Draw the following set of circles, rectangles and arcs. Then change the shapes' colors to blue or red as shown next.

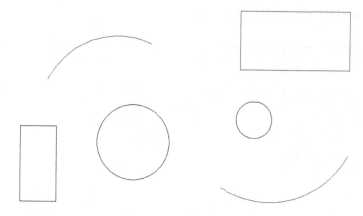

Type in **filter** and press Enter. There is no cascading menu, toolbar or Ribbon equivalent. The object selection filter will appear as shown in Figure 15.16. The procedure here is to first select the filter, then add it to the list and finally apply it.

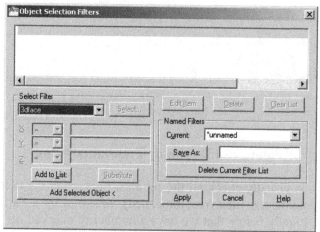

Figure 15.16 – Object Selection Filter

Select arc from the Select Filter drop down menu on the left. Then press the Add to List: button just below that. For good measure let's also add the color red. Select color from the same Select Filter menu. You will notice at that point the Select button will light up. Go ahead and select the color red. Then press Add to List:. Your Filter should look like this now:

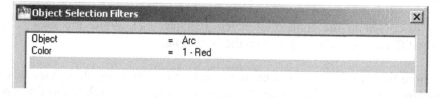

Now press the Apply button. AutoCAD will ask you to select objects. Select everything by typing in "all". Every red arc in the drawing will become dashed after the first Enter. Press Enter again and the grips will come on. Now type in Erase and press Enter. All the red arcs will disappear as seen in Figure 15.17(a) and (b).

Filters can be saved for future use via the Save As: button and the listing can be modified or deleted via the Edit Item, Delete and Clear List buttons.

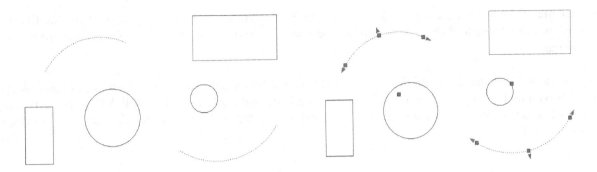

Figure 15.17(a) and (b) – Red Arcs selected by Filter

> **Hyperlinks**

The hyperlink command attaches a hyperlink to either an object or text. During this process you are asked to specify what this hyperlink will open when activated. It can be a website (usually) or a document (sometimes). In either case, it's a great command to quickly access additional useful information. Someone for example, can see a drawing of a product and follow the link to a website from the manufacturer, or call up an Excel file that lists all the specs of a design.

To try out the hyperlink features, draw a symbol for a chair and a website link text as seen below. Make a regular block out of the chair.

WWW.GOOGLE.COM

We would like to link the chair (an *object*) to a furniture website (IKEA® for example) and link the Google text (a *text string*) to the Google website. Type in **hyperlink** and press Enter or alternatively use the cascading menus **Insert→Hyperlink**. There is no toolbar or Ribbon equivalent. AutoCAD will ask you to select objects. Go ahead and pick the chair block. The following dialog box will then appear (Figure 15.18).

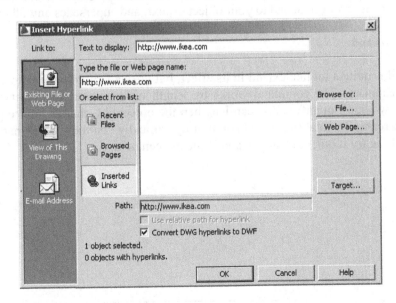

Figure 15.18 – Insert Hyperlink

Enter the name of the desired website in the appropriate field. That would be where it says "Type the file or Web page name" not the "Text to display". Not much else needs to be done, so press OK. Repeat the process for the Google text as well.

Hover your mouse over the chair and then the Google text and notice something new. There is a hyperlink symbol that appears with text prompting you to hold down the Ctrl key and click to follow the link (Figure 15.19). If your computer is currently connected to the internet, go ahead and do that in both cases (chair and Google) to bring up the respective websites.

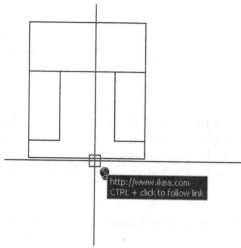

Figure 15.19 – Hyperlink

There is an interesting variation to this as mentioned earlier. You do not have to attach a website to the link. You can also attach a regular document, so it is called up when the hyperlink is clicked as before. Pretty much anything can be attached if it can be found and the computer can run the application; Word, Excel, PowerPoint documents, pictures, videos, you name it.

To do this, follow the same steps as before, but this time select the "File…" button under the "Browse for:" header on the right side of the dialog box. Find the file of interest, double click it and OK the Hyperlink dialog box. That's it; the document is now attached to your object or text, and supersedes any other links you may have.

➢ **Lengthen Command**

The lengthen command is quite a lot like extend in its effect, but goes about doing it differently. It's worth adding to your AutoCAD vocabulary. The command works by lengthening non-closed objects, such as lines or arcs (circles and rectangles wouldn't qualify). You can lengthen the objects by change in size [DElta], by percentage added or subtracted [Percent], by total final size [Total] or dynamically "on the fly" [Dynamic]. All this applies to angles (of arcs) as well. To try it out, draw an arc and a line as seen next.

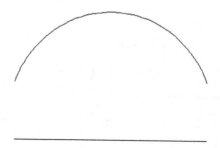

Type in **lengthen** and press Enter. Alternatively use the cascading menu **Modify→Lengthen**. There is no toolbar or Ribbon equivalent.

- AutoCAD will say: `Select an object or [DElta/Percent/Total/DYnamic]:`

Pick the `Percent` choice by typing in **p** and pressing Enter.

- AutoCAD will say: `Enter percentage length <100.0000>:`

Enter a value such as 150 and press Enter.

- AutoCAD will say: `Select an object to change or [Undo]:`

Pick the line and it will lengthen 150%. Press Esc to exit the command.

The process is essentially the same with the arc and the other lengthening options. Repeat the command several times and run through each option.

> **Object Snap Tracking**

Object snap tracking is a drawing aide that allows you to find various OSNAP points of an object and connect to them without actually drawing anything that touches that object; a sort of remote OSNAP. This function can be activated when the OTRACK button is pressed in at the bottom of the AutoCAD screen and will be deactivated when the button is pressed back out. The best way to see the purpose and function of this command is to try it. Draw a series of rectangles or squares spaced some distance apart as shown below. You don't have to number them; that is only for identifying purposes.

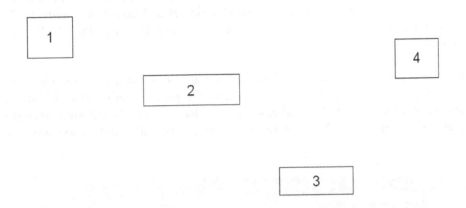

What we want to do is connect a series of lines from (and to) the midpoints of each of these shapes via right angles (not directly). An example of this may be a cable diagram or an electrical schematic. One way would be to start a line from the midpoint of the right side of shape #1, stop somewhere randomly and repeat the process from the top of shape #2. Then you can use trim, extend or fillet to connect the lines together. This however is cumbersome and slow.

A better way is to begin the line from the right side midpoint of shape #1 as before (turn on ORTHO also), but then activate OTRACK, and *without clicking*, position the mouse over the midpoint of the top of shape #2. A straight dashed *vertical* line will appear. Follow that dashed line back up until the mouse meets the straight *horizontal* dashed line. Then and only then, click once and continue to draw the line to shape #2. This is shown in Figure 15.20. Notice what OTRACK does. It sensed where that midpoint of shape #2 was and gave you a position to confirm and click before proceeding. Complete similar lines from the bottom of shape #2 to the top of shape #3 and then #4.

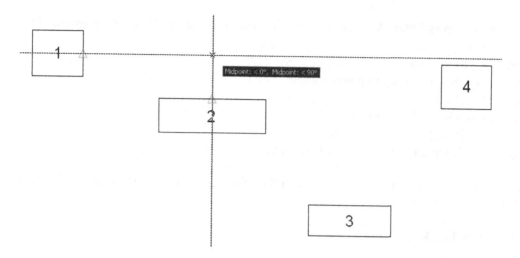

Figure 15.20 – OTRACK

> ➤ **Overkill Command**

This interesting command is technically part of the Express Tools under <u>M</u>odify→Delete <u>d</u>uplicate objects, but it's more entertaining to type in its rather dramatic other name. What overkill does though is quite useful. It deletes duplicate objects that overlap each other. In the 2D flat world of AutoCAD it can be quite challenging and time consuming to find these overlaps, especially if similar shapes and colors are involved. With overkill you simply select the offending objects and AutoCAD figures out which to keep and which to delete, effectively merging them all into one.

To try it out, draw 5 lines one on top of another. Although it may not be obvious to a novice, an experienced user will notice these lines are a bit darker and "fuzzy", indicating the presence of overlap; it's subtle but there. Now type in **overkill** and press Enter. AutoCAD will ask you to select objects. Select all 5 lines using a window or crossing (not one at a time) and press Enter again when done. The following dialog box will appear (Figure 15.21).

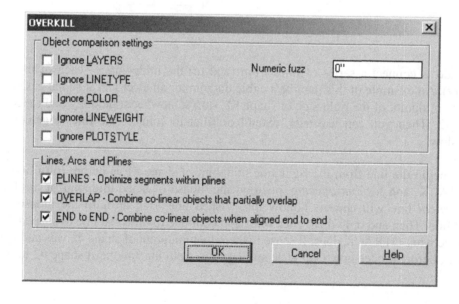

Figure 15.21 – Overkill

You can leave all the defaults as they are, pressing OK to have all the lines merge into one line. This is probably the simplest application of the overkill command. However what if the lines are on different layers?

In this case you have to check off Ignore LAYERS and the command will merge the layers into one. Numeric fuzz full works by comparing the distances between nearly overlapping objects and acts on them if the distance falls inside the fuzz value. Overkill has other useful features as well, so go through the command referring to the Express Tools help file for more info.

➢ **Point Command**

You have already seen and used points when we discussed the divide command. In that case points were created as a means to divide an object. To see them clearly you needed to modify the form of the point, from a dot to something more visible using **Format→Point Style...** The point command can be used to create points from scratch as described next.

Keyboard: Type in **point** and press Enter.
Cascading menus: Draw→Point→Single Point
Toolbar icon: **Draw** toolbar ___
Ribbon: **Home tab→Draw** ___

Start up the point command via any of the above methods.
- AutoCAD will say: Current point modes: PDMODE=3   PDSIZE=0.0000
                                           Specify a point:

Click anywhere and a point will appear.

➢ **Publish Command**

The publish command is a convenient way to print multiple sheets (Layouts) all at once. This command depends on knowing the basics of Paper Space, so you may want to come back after covering that topic in Chapter 18. Assuming however you have done so, the idea here is quite simple. When a file with multiple Layouts is opened, the publish command sees them and allows you to print them all at once, as opposed to clicking on each tab one at a time and using the plot command. In Figure 15.22 the author has opened a file called Alta Elevation (a recent job that features a design spread out over two Layouts – a Plan View and an Elevation View), then started up the Publish command via any of the following methods.

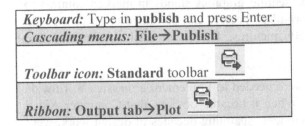

Keyboard: Type in **publish** and press Enter.
Cascading menus: File→Publish
Toolbar icon: **Standard** toolbar ___
Ribbon: **Output tab→Plot** ___

The following dialog box will then appear (Figure 15.22). Note that you can expand it if it isn't by pressing the Show details button on the lower left.

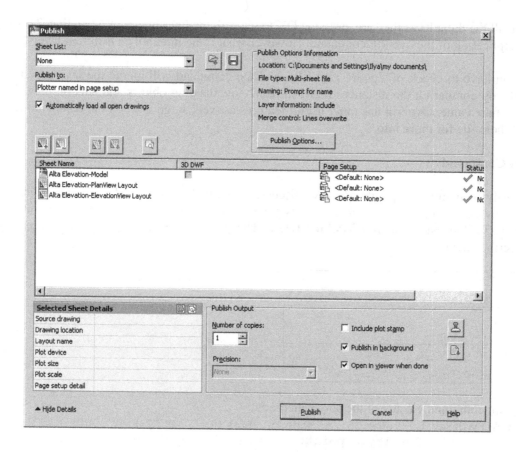

Figure 15.22 – Publish Command

The dialog box features some options to fine tune publishing, such as adding additional sheets, plot stamp, how many copies and a few others. All you have to do at this point is just press the Publish button on the bottom and AutoCAD will print out 1 copy of every single Layout the file contains. It's quick and easy and saves time. Go ahead and explore this very useful command on your own for additional features.

> **Raster**

Raster is a term you may hear of in AutoCAD circles, and it is not so much a command as a concept in graphic design. A Raster Graphic is nothing more than a bitmap which in turn is a grid of pixels viewable on any display medium, such as paper or computer monitor. This method of presenting graphics stands in marked contrast to Vector Graphics which are based on underlying mathematical formulas. Vectors can be scaled up or down with no loss of clarity, bitmaps cannot. So raster images are basically pictures. These can easily be inserted into AutoCAD as covered in a previous chapter.

Raster graphics hold a unique place in AutoCAD history. They were needed to implement a massive worldwide conversion from paper-based hand drawn designs to CAD files. When it became obvious that Computer Aided Design was the future, millions upon millions of hand drawn documents spanning decades of design work, were scanned into the computer, became backgrounds and were patiently re-drawn. Government and private industries around the industrial world collectively spent large sums of money and many man-hours to transform these archives into viable CAD files for future use. Most of this work was done during a ten year span 1988-1998, though some of it was started prior and some work is still on-going. The author recalls putting in some significant screen time doing these conversions early in his career and for a new AutoCAD user this was excellent, if boring, practice: scan, draw, print, and check. Repeat.

Raster scanning software can be quite sophisticated. At the time we used regular large sheet scanners which simply pulled up the graphic on the screen. We then tugged at it until it was the right size (by comparing a drawn line to a scanned known dimension). Later we acquired more sophisticated software that actually recognized shapes and attempted recreating these pieces of geometry to somewhat mixed results. Any smudges, imperfections or odd shapes could potentially confuse the system. Noise level (sensitivity) settings were used to control how much was picked up, and it was usually better to set it lower so you could add what was missing rather than constantly erase accidentally picked up junk. As an AutoCAD user today you may never need these tools but it is good to know what they are just in case.

> **Revcloud Command**

This is a command that thousands of architects were screaming for. Revision Clouds, or Revclouds for short, are used to indicate changes on a drawing. You can draw an ellipse, rectangle or circle just as well, but the unmistakable cloud shape is noticed most easily. To draw one of these by using the arc command would be certainly quite tedious. Let's give it a try the right way.

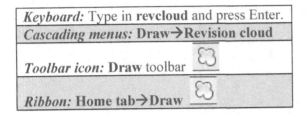

Keyboard: Type in **revcloud** and press Enter.
Cascading menus: **Draw→Revision cloud**
Toolbar icon: **Draw** toolbar
Ribbon: **Home tab→Draw**

Start up the revcloud command via any of the above methods.
- AutoCAD will say:
  ```
  Minimum arc length: 0.5000 Maximum arc length: 0.5000
  Style: Normal
  Specify start point or [Arc length/Object/Style]
  <Object>:
  ```
Click anywhere once.
- AutoCAD will say: `Guide crosshairs along cloud path...`
Draw a circular shaped revcloud connecting the last arc to the first one (no extra clicking needed).
- AutoCAD will say: `Revision cloud finished.`
Your result is should look somewhat similar to Figure 15.23 (you may have to zoom in or out to see the revcloud).

Figure 15.23 – Revcloud

The command had several options. For example if you do not connect the last arc to the first and instead just right-click, AutoCAD will ask you: `Reverse direction [Yes/No] <No>:` Here you can actually turn the revcloud "inside out", though this is rarely done. You can also change the size of the cloud's arcs by selecting the Arc length option, and even turn objects (like circle and rectangles) into revclouds using the Object option. Take the time to experiment with all of these.

> **Sheet Sets**

Like many new features in AutoCAD, they are not revolutions, rather evolutions of previous ideas. So it goes with Sheet Sets. This tool, introduced in AutoCAD 2005, was meant to combine some aspects of Paper Space Layouts and the Publish command and a few new twists, to create an "all under one roof" method of setting up printing of a project. The idea is to set up sheet sets (1 Layout = 1 Sheet) that define exactly what is needed to plot out the entire job correctly. These sheets can come from different drawings and even different jobs, and are grouped together by their respective categories. The idea is that they are now easier to handle and organize.

While the author has seen some reluctance in the industry to using a tool that somewhat duplicates existing functionality (even if new features are added), a few organizations, not surprisingly ones that produce jobs that span dozens of pages, took to this idea. We will give a very brief overview, and the student can then decide if they want to pursue this further.

To create a sheet set, you need to set it up. Go to **File→New Sheet Set...** and the following box will appear:

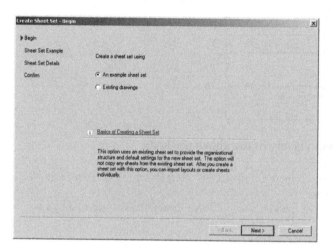

Figure 15.24 – Create Sheet Set-Begin

Select the first choice, "An example sheet set" and press Next>. The following will appear:

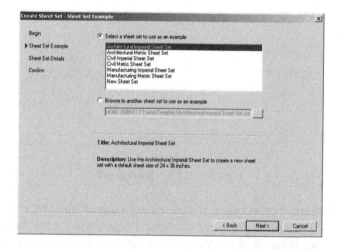

Figure 15.25 – Create Sheet Set-Sheet Set Example

Select the appropriate sheet set – Architectural Imperial for example, and press Next>. The following will appear:

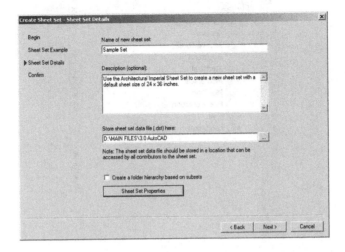

Figure 15.26 – Create Sheet Set-Sheet Set Details

Here you can give the set a name and location. Press Next>, and the following will appear:

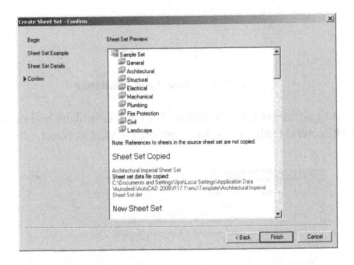

Figure 15.27 – Create Sheet Set-Confirm

The various sheets sets sorted by category are visible. Press Finish and you will see the Sheet Set manager appear as seen in Figure 15.28.

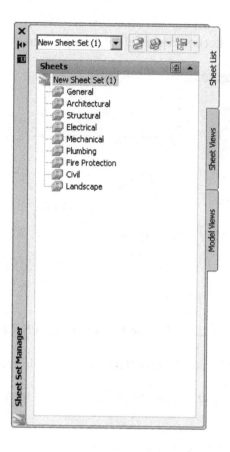

Figure 15.28 – Sheet Set Manager

This was all just to set up the basic sheets (placeholders). They are still blank and need drawings associated with them. Right-click on the Architectural folder and select "Import Layouts as Sheet…". The following will appear:

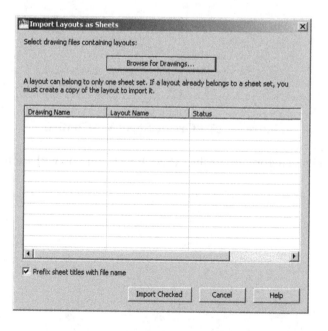

Figure 15.29 – Import Layouts as Sheets

You now have to browse for the right drawing that is associated with the Architectural layout. Continue down the categories until all are filled. This is just the basic idea of sheets sets. Its an extensive tool that can span many more pages of descriptions, and you should explore some of it on your own.

> **Selection Methods**

What we will outline here is a few other methods of selecting items that you may have not yet seen; namely the Fence (F), Window Polygon (WP) and the Crossing Polygon (CP). These are all applicable when selecting objects for a number of operations such as erase, move, copy, trim and others.

To try out window and crossing polygons, draw a set of circles as seen below and execute the erase command. Then when it's time to select objects, type in **wp** for window polygon and press Enter.

- AutoCAD will say: `First polygon point:`

Click anywhere.
- AutoCAD will say: `Specify endpoint of line or [Undo]:`

Continue to "stitch" your way around the circles you wish to select, and right-click when done. Notice the difference here between the standard window selection tool: (strictly rectangular in nature), and the flexible "any shape you want it" window polygon. Its action of course is similar; everything it encloses will be selected.

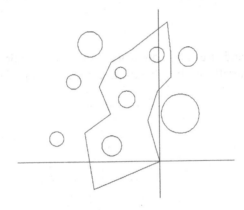

Figure 15.30 – Window Polygon

The crossing polygon works the same exact way except anything it touches gets selected, much like the regular crossing operates.

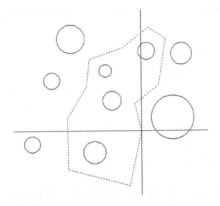

Figure 15.31 – Crossing Polygon

The fence option works by having a drawn line be the cutting tool during a trim command, and select multiple lines to be trimmed. Draw one horizontal line and a bunch of vertical ones as seen below. Then execute the trim command, but when it's time to select the lines to be trimmed, type in **f** for fence, then click twice to draw a line *across* the lines to be trimmed, and press Enter to make them disappear.

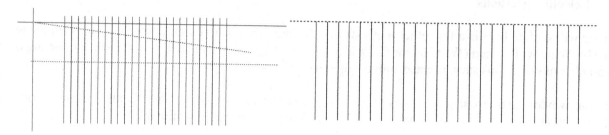

Figure 15.32 – Fence command and the results

In case you've discovered it on your own, yes, you can accomplish the exact same thing by using a regular crossing to eliminate the lines once a cutting edge is selected. However the fence command comes in real handy when going around corners as the flexible line-based approach can bend around any shape, while the rigid rectangular shaped crossing cannot.

> **Stretch Command**

This is an easy command to learn and use, and can work wonders for quick shape modifications. The stretch command uses the crossing tool to literally stretch shapes. Create two rectangles as seen below:

We want to stretch the right sides of both of both rectangles some distance. Start up the stretch command via any if the following methods.

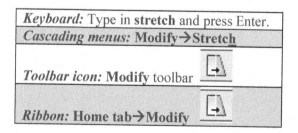

Keyboard: Type in **stretch** and press Enter.
Cascading menus: **Modify→Stretch**
Toolbar icon: **Modify** toolbar
Ribbon: **Home tab→Modify**

Using a *crossing* (this won't work with the window selector) select the first quarter or so of the rectangles as seen in Figure 15.33.

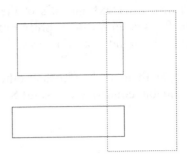

Figure 15.33 – Stretch in progress 1

They will become dashed. Press Enter and click anywhere to begin stretching them as seen below. It will simplify matters if you have Ortho on, and all OSNAPs off.

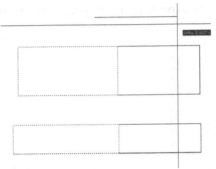

Figure 15.34 – Stretch in progress 2

Finally click elsewhere and the rectangles will have a new length. You did not have to explode and redraw them or use grips. You can use the Displacement option to key in specific X,Y, or Z distances, or stretch to another OSNAP point. Note that stretch doesn't work on partially selected circles (it will move, not stretch fully selected ones).

> **System Variables**

There isn't much to say about system variables except just to make you aware of what they are, so if they come up in reading of other texts or the AutoCAD help files, you know what you are dealing with. System variables are Yes/No, On/Off or 0,1,2,3 types of commands that are used to indicate a *preference or a setting*. AutoCAD has over 500 of them for every conceivable variable (something that changes) in the software, and listing them all is futile, though if you wish to see them all type in **setvar**, press Enter, then type in **?**, and press Enter twice. Many system variables are set graphically via a dialog box, and the user may not even be aware or know it's a system variable they are changing. Perusing the help files and AutoCAD literature, especially on an advanced level, you will find many references to these. Knowing system variables is important for those wishing to learn advanced customization (such as AutoLISP).

Here's a useful example that is typed in on the command line. Suppose you were mirroring text and the new set is backwards as seen below.

hello   ollǝɥ

That isn't quite right, as you want text to be readable even after mirroring, so some sort of system variable (SV) needs to be set. The responsible SV is called **mirrtext**, and it's of the On/Off variety represented by 0 and 1. Somehow it got set to 1, and needs to be 0. Type in mirrtext, press Enter, type in the new value of 0 and press Enter again. Try the mirror command again - it should work fine now.

As mentioned before, this is just one example, there are hundreds of others. Keep an eye out for them. After years of using AutoCAD you will amass a formidable collection of useful SVs that can resolve some thorny problems for less experienced designers.

> **Tables**

If you haven't yet used tables, here's a quick primer on them. They are pretty much what you'd expect, a bunch of rows and columns, with a header on top, ready for you to enter information into. You can specify the number of rows and columns and add/take away at a later time as well. Tables are quite reminiscent of low-tech Excel spreadsheets and can be useful in certain situations (like when you need data displayed in a hurry).

We'll cover some basics here and let you explore more on your own. Start up the table command via any of the following methods.

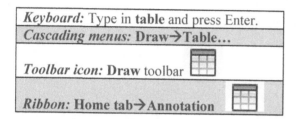

Keyboard: Type in **table** and press Enter.

Cascading menus: **Draw→Table...**

Toolbar icon: **Draw** toolbar

Ribbon: **Home tab→Annotation**

You will then see the following (Figure 15.35).

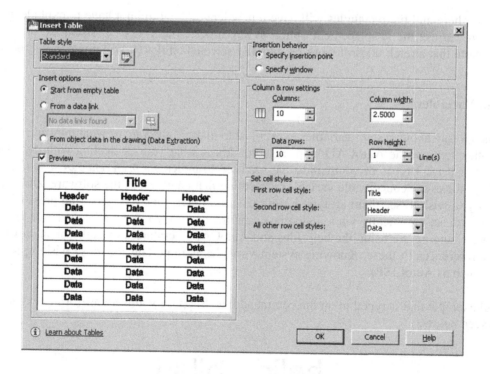

Figure 15.35 – Insert Table

Under the "Column & row settings:" set your columns and rows as well as their respective width and heights (we'll do a 10 × 10 here). When done press OK and you will be asked to click where you want the table to go. When you position it you will see the following (Figure 15.36).

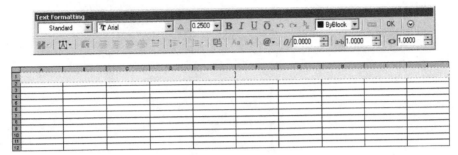

Figure 15.36 – Insert Header

Here you are being asked to set up the header. Type in "Sample Data Table" and press OK. You will see the completed table as seen next.

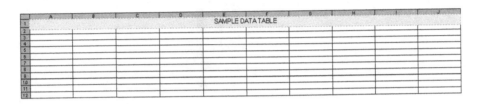

You can now fill in the table with data by clicking in each individual cell and typing in values as seen below:

There is more to this command, such as sizing and modification of the table, as well as some ability to link up with Excel spreadsheets, so go ahead and explore it further. You can also set up styles of tables by first using the **tablestyle** command, which will give you the following dialog box for "designing" a table (Figure 15.37).

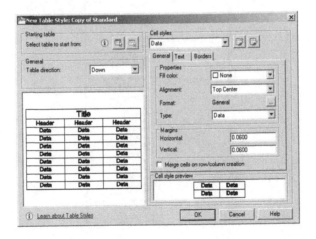

Figure 15.37 – Table Style

> ➤ **Tool Palette**

This useful palette, shown in Figure 15.38 can be accessed by typing in **toolpalettes** or alternatively via cascading menu **Tools→Palettes→Tool Palettes** or by pressing **Ctrl+3**.

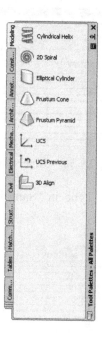

Figure 15.38 – Tool Palette

What the tool palette represents is a *collection of commonly used commands and predefined blocks*. What you actually see when the palette comes up may be different from the previous figure, as it depends completely on what menu is pulled up or what tab is clicked. In all cases, to use this palette, just click on a command (or click and drag) and it will be executed as though you typed or selected a toolbar icon. To access more menus, simply right click on the grey "spine" The following menu will appear (Figure 15.39):

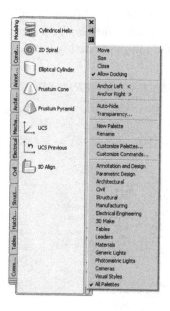

Figure 15.39 – Tool Palette, Menu

You can select the various menus at the bottom, and they will appear with their associated tabs. The rest of the choices (top half) are self explanatory. There is more to this palette however. First, select All Palettes at the bottom of the list you were just looking at. Then go to the bottom of the tabs on the *left* and click. Another extensive menu will appear as shown in Figure 15.40.

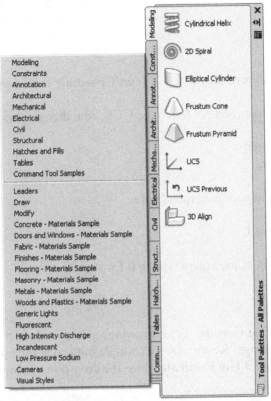

Figure 15.40 – Tool Palettes, Tabs Menu

Take the time to go through each of the menu choices. The amount of available blocks is quite extensive. Although we will not go into this further, you may also customize what tabs you have available by creating your own, and adding custom commands and blocks. This is similar to creating new toolbars using the CUI as covered in Chapter 14.

> **UCS/Crosshair Rotation**

Rotating the crosshairs or the UCS icon is a very useful trick for working with tilted shapes. Not everything you work on will be perfectly horizontal and vertical; some designs will be rotated at some angle and it would be nice to align the crosshairs to them. This will allow for easier drawing and better orientation for the designer. Commands such as Ortho can now be implemented at the new tilted angle.

Rotating the UCS icon is *not* the same as rotating just the crosshairs, but we will not split hairs (so to speak) over the conceptual differences and go over both methods as if they are the same. To rotate the UCS icon, which will by default rotate the crosshairs, simply type in **ucs**, press Enter, and type in **ob** for the OBject option. Then, assuming you have a rotated design you are working with, click a line that best represents the angle of the design's rotation and the UCS icon/crosshairs will align itself to that line (it will actually stick to it). To revert back to normal, retype **ucs**, press Enter and type in **w** for world and press Enter. The UCS icon/crosshairs will return to normal.

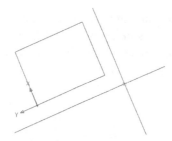

Figure 15.46 – UCS Icon/Crosshair Rotation

If you just want to rotate the crosshairs and not touch the UCS icon, then do this:

1) Type in **snap** and press Enter.
2) Type in **ro** and press Enter.
3) For base point, type in **0,0**.
4) For angle of rotation type in **per** for perpendicular.
5) Select an angled line
6) Turn off the SNAP command.

To get back to normal, repeat all six steps, except type in **0** for step #4.

➢ **Window Tiling**

This is just a simple tool for allowing viewing of two drawings side by side in a convenient manner. Open up two drawings and press one of the drawings' Restore Down (middle button) at the upper right, so they are tiled one on top of another. Then select **Window→Tile Vertically** from the drop-down menu as seen next.

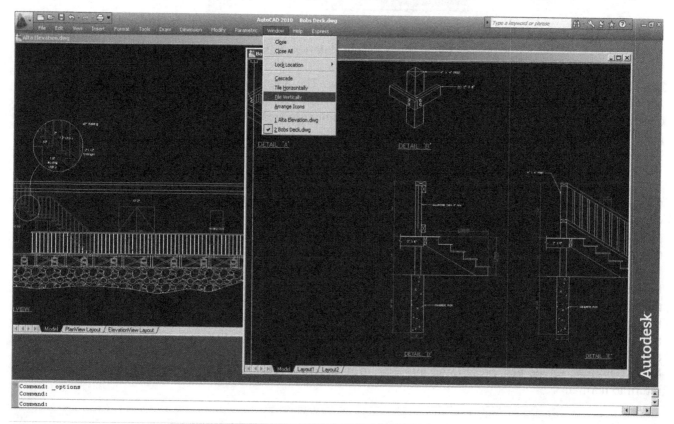

After the button is clicked, you will see this (Figure 15.41).

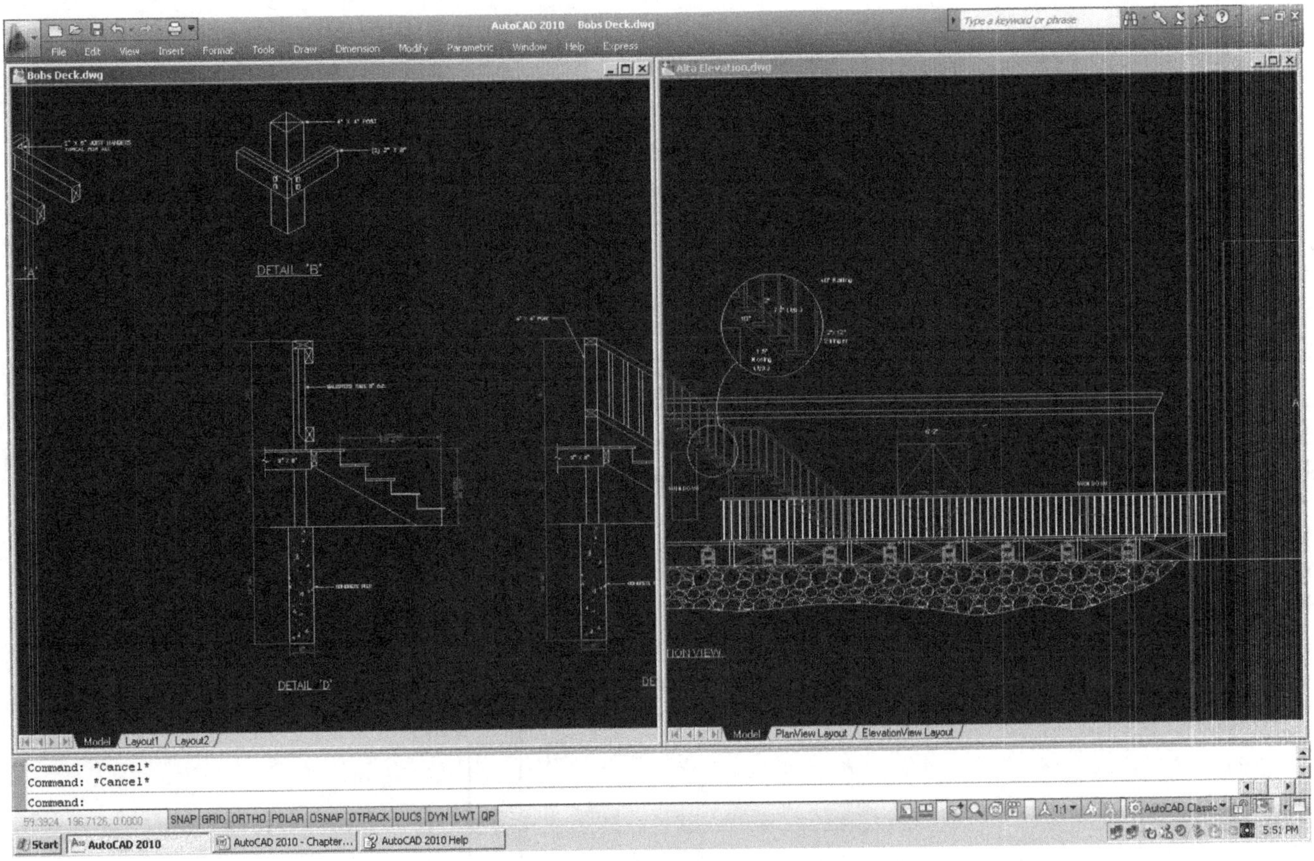

Figure 15.41 – Vertical Tiling

Both drawings will be lined up neatly on the screen. You can now click in and out of each and use copy/paste commands to transfer items if necessary. The same thing can of course be done for a horizontal tiling.

> **Wipeout Command**

For reasons not immediately apparent, but to be explained in detail, this command is one of the most important in 2D in producing drawings that not only possess extraordinary clarity but are easy to work with and modify. When used properly, wipeout is truly an "insider's" secret that is not known by beginner students, and is one of the reasons why drawings done by AutoCAD experts have that intangible "something" that sets them apart from those done by less experienced designers.

First let's define what a wipeout is. It is simply a tool that creates a mask (or screen) that blocks out the viewer's ability to see other drawn geometry that the wipeout covers. To use it, draw a shape of some sort. Then type in **wipeout**, press Enter and click your way around all or part of the shape. When you press Enter or right-click, the shape will disappear behind the wipeout as seen in Figure 15.42.

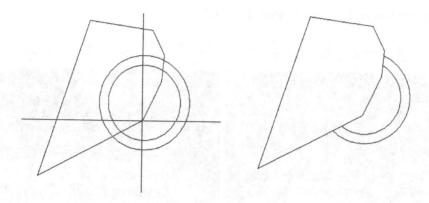

Figure 15.42 – Creating wipeout and the result

Notice the wipeout itself is still visible (something that will be addressed later), but the double circle shape *has* been partially blocked as expected. Practice using this command a few times just to become familiar.

Remember that a wipeout itself is an object. It can be moved, erased, copied, rotated, scaled and just about anything else you can do to a regular geometric object. You can also click the wipeout (click on the lines, not the blank center) and activate its grips. Then the grips can be used to reshape it into any shape whatsoever.

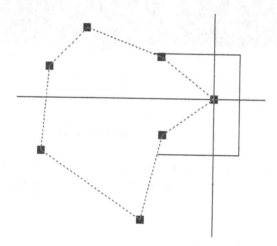

Figure 15.43 – Wipeout grips

So what is so important about wipeout that qualified it for such commentary as seen a few paragraphs ago? Well, this command allows us to do something very special in 2D work. We can give our designs *depth*.

When you look around you in the real world, objects are generally not transparent. So if you are looking at a bus in front of a building, part of the building is obscured by the bus. If another car were to park between you and the bus, it would obscure part of the bus, and so forth. In 3D, this isn't a problem at all. All objects are solid, and while they can be viewed in transparent wireframe form, a click of the button gives you rendered models that *create an instant illusion of depth*. This is important, as it matches our everyday experience and we quickly identify what we are looking at. In the image of Figure 15.44, the chair and the computer monitor all block a part of the computer desk. What is in front of what is easy to tell in the rendered image on the left, but not as easy with the wireframe one on the right.

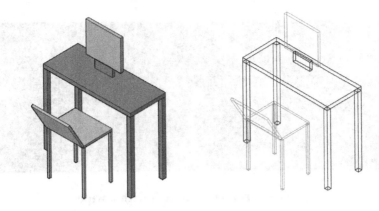

Figure 15.44 – Solid and wireframe

Let's try the same thing with the bus and car.

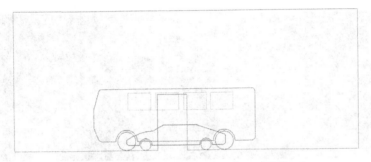

Figure 15.45 – Bus, building and car, Wireframe

Which is in front of which in Figure 15.45? You really can't tell without shading the image as in Figure 15.46.

Figure 15.46 – Bus, building and car, Solid

In 2D though we have a problem: we cannot shade in the conventional sense! Everything is drawn using lines and nothing is solid, even though it may be in real life. In a plan view of an architectural floor plan this is not an issue as we are looking straight down, but what about in elevation view such as the images above? A section through a room in elevation view may feature furniture obscuring walls and walls obscuring other items behind them. Or in electrical engineering, we have equipment on a rack obscuring mounting holes. To provide realism and a sense of depth, we need to somehow obscure our view of items not seen. Now, yes you can trim linework behind the object, but what if the object moves? You now have a gaping hole. We need a way to temporarily obscure and hide geometry that is tied directly to the object doing the hiding, so if it moves the linework reappears. Wipeout to the rescue!

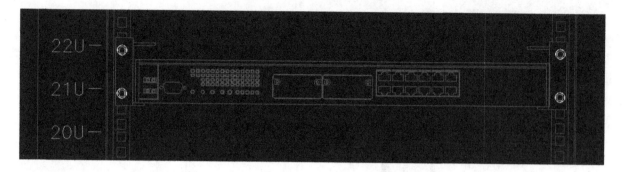

Figure 15.47 – A router on a rack with a wipeout

In Figure 15.47 notice how the equipment bolted to the rack obscures part of the rack. Now if the equipment is moved such as in Figure 15.48 (notice the numbers to the left change) the wipeout moves with it and it acts like a solid object.

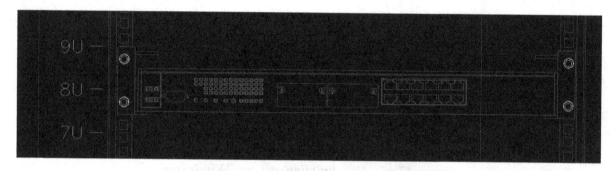

Figure 15.48 – Equipment relocated

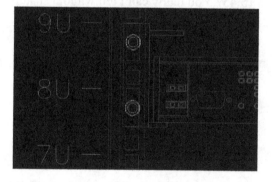

Figure 15.49 – Same equipment with no wipeout

So in summary, when a wipeout is attached to an object, it will prevent anything behind it from being seen, exactly as in real life. So a bookshelf with a wipeout attached will block the wall behind it and electrical equipment will block the mounting holes it is attached to. Move the object and the wipeout moves, also blocking out whatever you happen to be blocking. This is the technique that the pros use to give drawings a uniquely realistic look in 2D design and you should learn this as well. Let's go through some details.

The essential steps of making this technique work include the following:

Step 1
Create the drawing of what you are designing – such as this rack equipment.

Step 2
Using wipeout, click to trace the outline of the drawing.

Step 3
Using **Tools→Draw Order→Send to Back** put the wipeout behind the design.

Step 4
Make a block out of the design + wipeout.

The object has now been "solidified" and can be placed in front of other objects. If you create more objects in this manner, you can simply use the Draw Order tool to shuffle them around top to bottom as needed (as long as they are all blocks). In this manner you can create a rather impressive sense of depth and clarity to an otherwise busy drawing.

Take a look at Figure 15.50. It is a larger section of the same cabinet shown in previous figures. It could not have been done this way without extensive use of the wipeout command. Notice how it is easy to tell which equipment is in front and which is behind the rack. There are in fact several layers of objects – the equipment bolted to the front of the rack, then the rack itself, then the equipment in the back of the rack.

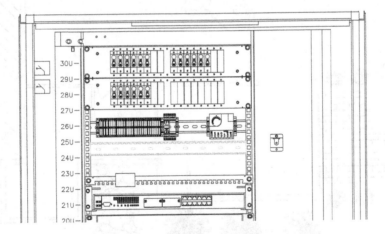

Figure 15.50 – Equipment with wipeout

The final item to mention with wipeouts is the fact that they themselves can be hidden (while their ability to hide objects still remains in effect). To do this, simply type in wipeout, press Enter and select **f** for Frames in the submenu. AutoCAD will then say: Enter mode [ON/OFF] <ON>: If you type in **off**, AutoCAD will turn off all wipeouts on the screen, while preserving their effect on other objects.

Level 2 Drawing Project (5 of 10) – Architectural Floor Plan

Here we will first add some landscaping using advanced lines that were learned in Chapter 11. You will use spline, xline and a variety of hatches to create shrubbery, trees, a driveway and two walkways as seen in the plan below. Create the appropriate layers before starting the design (L-Driveway, L-Tree, etc.).

Then add layer states that you learned about in Chapter 12. Create three layer states called Electrical (with all the E - layers), Architectural (with all the A- layers) and Landscape (with all the L- layers).

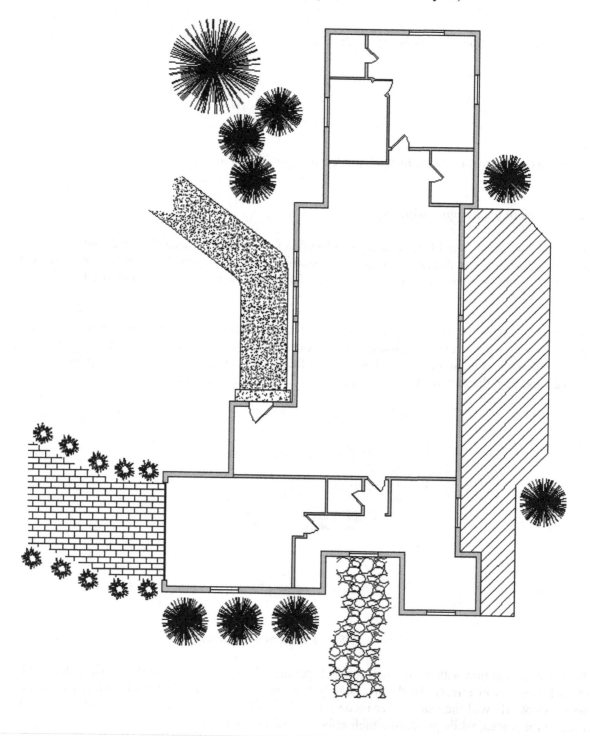

Chapter 15

Summary

You should understand and know how to use the following concepts and/or commands before moving on to Chapter 16.

- Align command
- Audit/Recover commands
- Break/Join commands
- CAD Standards
- Calculator command
- Defpoints concept
- Divide/Point Style commands
- Donut command
- Draw Order
- eTransmit command
- Filters
- Hyperlinks
- Lengthen command
- Object Snap Tracking
- Overkill
- Point command
- Publish command
- Rasters
- Revcloud command
- Sheet Sets
- Selection methods
- Stretch command
- System Variables
- Tables
- Tool Palette
- UCS/Crosshair Rotation
- Window Tiling

Chapter 15
Review Questions

Answer the following based on what you learned in Chapter 15.

No review questions are assigned to this chapter. However, be sure to go through each of the commands in details again to thoroughly understand them. Specifically you should review the following new commands and concepts as these are easy to practice and are more hands-on than theoretical. Several of the following commands will be the basis for the four exercises to follow.

1) Align
2) Break/Join
3) Chamfer
4) Divide/Point Style
5) Donut
6) Hyperlinks
7) Lengthen
8) Revcloud
9) Stretch
10) Table
11) Wipeout

Chapter 15
Exercises

Exercise #1 – Draw the following two rectangles according to dimensions given. Pedit the width to .25" and shade each separately with a solid hatch, color #9. Then use **align** to place the 10" box on top of the 1' box.
(Difficulty level: Easy, Time to completion: 5 minutes)

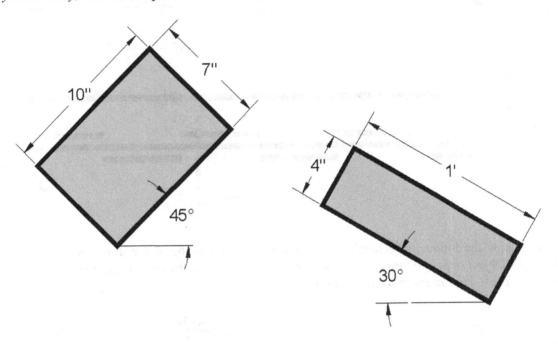

Exercise #2 – Draw the following set of lines and **chamfer** them to the dimensions shown. Then pedit the result to a thickness of .25".
(Difficulty level: Easy, Time to completion: 5 minutes)

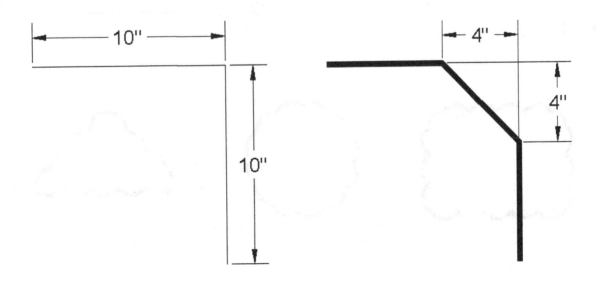

Exercise #3 – Draw the following line and **divide** it into seven sections using six points of the style shown. Then pedit the line to a thickness of .25" and break it using the **break** command (First point option) at each of the points. Finally offset each alternate section .5" up and .5" down as shown.
(Difficulty level: Easy, Time to completion: 5 minutes)

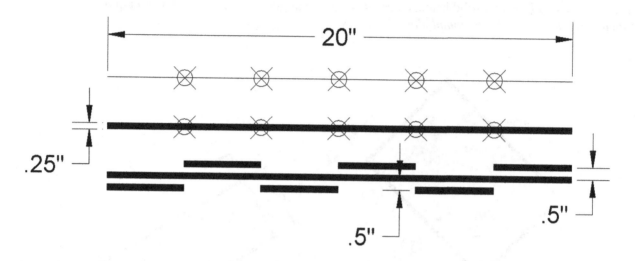

Exercise #4 – Draw the following rectangle, circle and triangle according to the dimensions shown. Then use **revcloud** and the Object option to turn each into a revcloud. Finally pedit the new shapes to a thickness of .25".
(Difficulty level: Easy, Time to completion: 5 minutes)

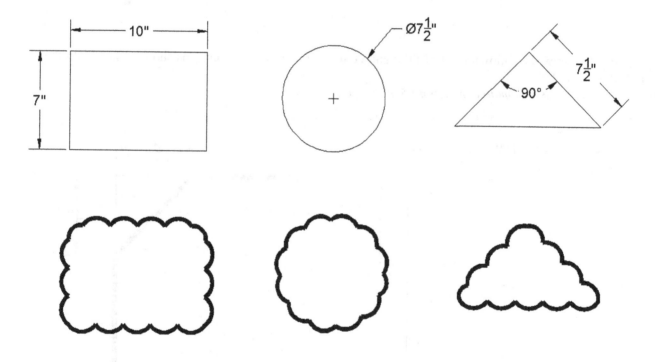

CHAPTER  16

Importing and Exporting Data

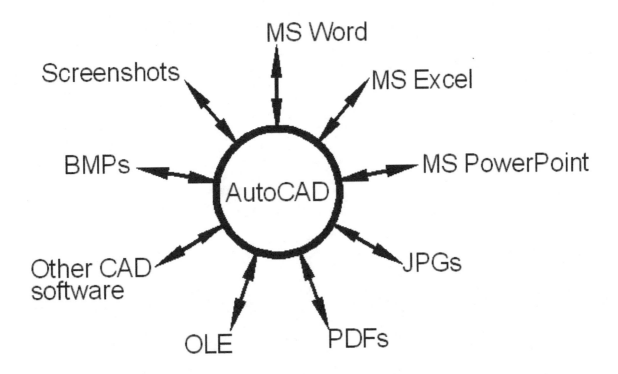

Chapter 16

Learning Objectives

In this chapter we will learn how AutoCAD interacts with other software that you are likely to use in the course of design work. We will cover the following topics:

- AutoCAD and MS Word
- AutoCAD and MS Excel
- AutoCAD and MS PowerPoint
- AutoCAD and JPGs
- AutoCAD and PDFs and Screenshots
- AutoCAD and other CAD software
- Exporting and SaveAs
- Importing and OLE

By the end of the chapter you will be able to smoothly import and export data between a variety of common office and design applications.

Estimated time for completion of chapter: 2 hours

Sec 16.1 - Introduction to Importing and Exporting Data

AutoCAD and the designers that use it usually do not work in a "software vacuum". They typically interact with not only AutoCAD but also a number of other applications, some of them generic and used by many other professions, and others unique to their own field. As such, there is often a need to either import data *into* or export data *out of* AutoCAD. Examples are numerous, such as an architect who wishes to transfer notes typed in MS Word into his project details page in AutoCAD or an engineer that needs to drop in an Excel spreadsheet. Just as often, AutoCAD drawings may need to be inserted into MS Word reports, PowerPoint presentations and other applications. You may then need to create PDFs of your drawing or insert a PDF or an image into one.

Swapping files between the various software applications has in the past been occasionally tricky business. Companies are under no obligation to make such tasks flow smoothly, as it is not their business what else you may have on your PC (which may include software that is their direct competitor). Of course, such thinking doesn't score any points with the customer, therefore as a rule developers try hard to make sure there are as few compatibility issues as possible. Today such tasks flow much smoother but still require knowledge of procedure that can be unique for each application.

The sheer quantity of software on the market is staggering, so we will limit the discussion to:

- Software or files that are used or encountered by virtually everyone who uses a computer (Word, Excel, PowerPoint, JPG and PDF), or

- Software that has special relevance to AutoCAD users (such as other CAD programs)

We will focus on three distinct tasks: how to *import* files *from* these applications, how to *export* AutoCAD files *into* them (analyzing each software pairing bidirectionally if you will), and then conclude with a brief overview of the Export/Insert feature and OLE.

Sec 16.2 - Importing/Exporting to and from the MS Office Applications

We will begin with MS Word, Excel and PowerPoint as those are probably the most important and most relevant to the majority of users. Throughout, unless noted otherwise, we will make extensive use of Copy/Paste. These "copy/paste to clipboard" tools are some of the most useful in computing, allowing you to shift around just about anything from one spot to another. The keyboard shortcut is **Control-C** for copy and **Control-V** for paste, or you can just right-click and select copy and paste.

Be aware that the "copy" isn't AutoCAD's standard copy command, so don't use that. Yes, some students mixed these up, so let's be clear from the start. Open up Word, Excel and PowerPoint and keep them handy as we go through each one.

> **Word into AutoCAD**

Example of use: You would like to bring in extensive typed construction notes onto one of the pages of the AutoCAD layout.

Procedure: Go ahead and click-drag to highlight the text in Word first and right-click or Control-C to copy to clipboard. Go to the AutoCAD screen, but do **not** right-click and paste the text in. If you do this, it will appear as an embedded object (with a white background) and will not be editable. Instead, create an **mtext** field of an appropriate size, and paste the copied text into it. The result will be easily editable text that can be further formatted. In Figure 16.1, the top line is an embedded field, and the bottom line is the correct mtext.

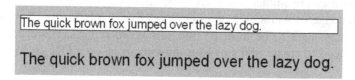

Figure 16.1 – Proper insertion of Word text

> ➤ **AutoCAD into Word**

Example of use: You are preparing a report and would like to include some of the graphics spread out among the text to enhance clarity or as supporting documentation.

Procedure: This will be a direct application of the copy/paste function *or* the "PrtScn" (print screen) key. With copy/paste you no longer have to change the background color of AutoCAD before insertion (developers finally conceded to the fact that yes, most paper is white and the black AutoCAD background just doesn't look right – an automatic color change is performed), but be careful with older AutoCAD releases!

Once inserted, some cropping will be necessary to get rid of excess white space. Often the print screen button followed by pasting in is easier, and the newer 2007 Word brings the cropping tool up right away when you click the pasted-in screenshot. Indeed, this was the initial method for importing screenshots throughout this textbook.

A typical view of this procedure is shown in Figure 16.2.

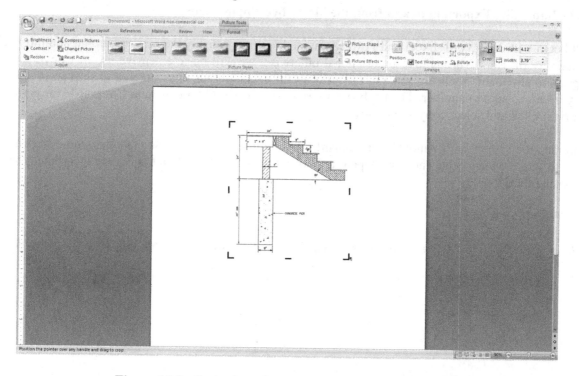

Figure 16.2 – Insertion of an AutoCAD design into Word

> **Excel into AutoCAD**

Example of use: You would like to bring in a spreadsheet into your electrical engineering drawing that details the type of connections and wiring used in a design.

Procedure: Once again, you will make use of Copy/Paste. In Excel, create a small group of cells, number them (adding borders if desired), then highlight all, right-click copy; go into AutoCAD and right-click paste. As soon as you do this, the following dialog box will appear (Figure 16.3).

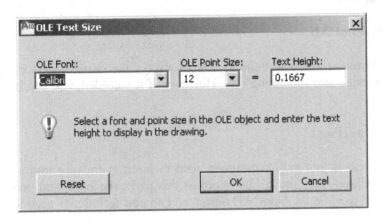

Figure 16.3 – OLE Text Size

This is the OLE (Object Linked Embedded) Text Size dialog box, which allows you to alter the text height of your inserted spreadsheet. The exact size is not too important for now (you can easily scale objects up or down anytime after insertion), but the concept of OLE is important and will be discussed again soon. For now, press OK and you will see the following (zoom in if needed).

1	2
3	4
5	6
7	8
9	10

Figure 16.4 – Embedded Excel cells

This is your basic Excel insertion. The cells are regular objects (very much like blocks), and you can erase, copy and move them as needed. You can also scale them by clicking once and moving around the resulting grip points. You cannot mirror or rotate the cells however.

A major question usually arises at this point from students. Are the inserted Excel cells linked to Excel, allowing for changes and updates? They are not, in the true sense of the word. What you can do is double click these inserted cells and Excel will be called up temporarily for editing purposes, but any changes made to the original file after accessing it *from* Excel (not AutoCAD) will not be reflected in AutoCAD.

Having truly linked files is a very useful idea. The author has worked with some fancy Excel/AutoCAD interfaces where extensive Excel data was used to generate corresponding AutoCAD drawings, but this trick requires additional programming and extensive customization, and is beyond the scope of this book.

> **AutoCAD into Excel**

This action basically pastes an image into Excel with the same Copy/Paste procedure as for Word. This however is rarely done as Excel is primarily for spreadsheet data, and unless you need a supporting drawing there is just no overwhelming reason to do this.

> **PowerPoint into AutoCAD**

This action is not very common as PowerPoint is almost always the *destination* for text and graphics, not an intermediate step to be inserted into something else. However, if it's needed, you can insert slides into AutoCAD via regular Copy/Paste.

> **AutoCAD into PowerPoint**

This makes a lot more sense as you may sometimes want to insert a drawing into a PowerPoint presentation. Fortunately PowerPoint behaves very much like Word in this particular case, and one Copy/Paste followed by some cropping is all it takes to insert an AutoCAD design and no new techniques need to be learned.

Sec 16.3 - Screen Shots

A Screen Shot or Screen Capture refers to capturing a snapshot of everything on your screen as seen by you, the user, and sent to memory, to be inserted somewhere. This is done by simply pressing the Print Screen button which on most computers is called "PrtScn" or "SysRq" or sometimes just F12. You can capture what is on your screen this way and insert it into Word or PowerPoint, and then crop out what you don't need to see. While many computer users may know of the "print screen" function to actually print something, they may overlook this useful trick for inserting data into other applications.

Sec 16.4 - JPGs

This file compression format (used most often with photographs) can be inserted directly into AutoCAD and is a great idea if you would like a picture of what it is you have drafted next to the drawing. Examples include an aerial photo of a site plan for architectural and civil applications and a photo of an engineering design after manufacture for as-built or record drawings.

Inserting JPGs is very easy. Click on a file you want to insert and right-click copy it. Then simply right-click paste into AutoCAD. A few things happen at this point.

- AutoCAD will ask for an insertion point:
  ```
  Specify insertion point <0,0>:
  ```
  Click anywhere you like.
- AutoCAD will then ask for image size:
  ```
  Base image size: Width: 0.00, Height: 0.00 Inches
  Specify scale factor or [Unit] <1>:
  ```
  Scale the jpg by moving the mouse to any size you wish (or type in a value).
- AutoCAD will then ask for the rotation angle.
  ```
  Specify rotation angle <0>:
  ```
  If you want the image rotated, enter a degree value; otherwise press Enter.

The result of a jpg insertion is shown in Figure 16.5.

Figure 16.5 – Embedded JPG

The image can be erased, copied, moved, rotated, mirrored and just about anything else you can do to an element. You can also do some very basic image adjusting by *double-clicking* on the picture. A dialog box shown in Figure 16.6 will appear, allowing you to adjust Brightness, Contrast and Fade, as well as reset everything. Any changes will apply only to that jpg, not copies of it (if other copies exist).

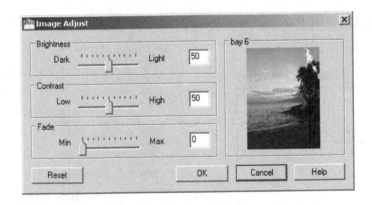

Figure 16.6 – Image Adjust

Sec 16.5 - PDFs

The Portable Document Format (PDF) is of course Adobe's popular format for document exchange. AutoCAD drawings can be easily converted to PDF by simply printing to them (in other words, selecting PDF as the printer – assuming Acrobat is installed). The print function will execute, and AutoCAD will ask you where you want the file saved. Select the destination and your PDF file is created.

Another way to generate a PDF, especially if you need to customize the output, is via the Ribbon's Output tab→Export option. If you select the PDF option a "Save as PDF" dialog box appears and allows some tinkering with options and output to finesse the results, as seen here in Figure 16.7.

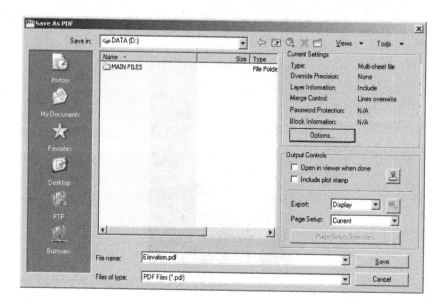

Figure 16.7 – Save as PDF

As of AutoCAD 2010 you can now bring in a PDF file as an underlay. This means it will come into AutoCAD as a background which you can see but not edit. This is not unlike raster images (mentioned in the previous chapter) and is quite useful if all you have of a design is a PDF. After insertion you can draw over it, or use it as a supporting image, such as a key plan. To bring in the PDF use the cascading menu Insert→PDF Underlay... You'll then be prompted to look for the file you want to insert, and upon selection you will see the following (Figure 16.8).

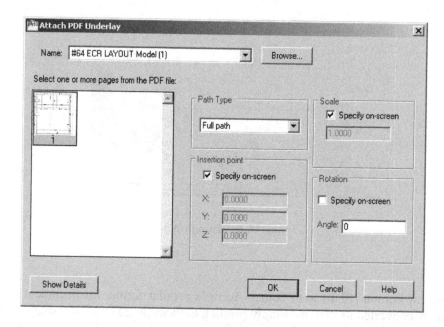

Figure 16.8 – Attach PDF Underlay

Once you press OK, the PDF will insert into the AutoCAD file and prompt you for an insertion point and scale factor. Once the image is embedded, you can click on it and the Ribbon will change to show you some of the image editing options as seen here in Figure 16.9, with "Change to Monochrome" and some Fade and Contrast adjustments among the more useful tools.

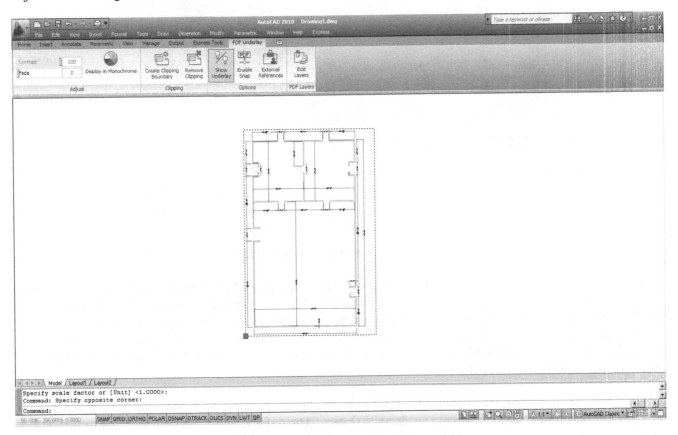

Figure 16.9 – PDF Underlay Ribbon menu

Sec 16.6 - Other CAD Software

Initially, as AutoCAD gained market dominance, it did not really attempt, nor had any incentive, to "play nice" with other CAD software on the market. Being the 2D industry leader allowed AutoCAD to get away with not accepting any other files in their native format. It was their way or the highway, but that slowly changed. One major concession in regards to importing and exporting was with MicroStation, AutoCAD's nemesis and only serious competitor. MicroStation always opened and generated AutoCAD files easily, but not the other way around. Finally in recent releases AutoCAD listed the "dgn" file format as something you can import. You were then able to bring in MicroStation files, more or less intact. In AutoCAD 2009 the settings box was expanded and that was carried over to 2010 as seen here in Figure 16.10. You are now able to compare layers and other features side by side and make adjustments so the target file is similar to the original.

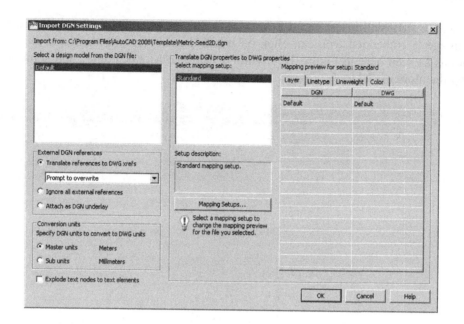

Figure 16.10 – Import DGN Settings

To access this, use the cascading menu **File→Import…** and search for the MicroStation dgn file you're interested in importing. Another way around all of this is just to have the MicroStation user generate the AutoCAD file. You can also export dwg files directly to dgn as well by using the cascading menu **File→Export…** and selecting dgn as the target file. You will then see the same settings dialog box of Figure 16.10. Of course MicroStation will accept the AutoCAD file no problem, so this may not be necessary. If you would like to learn a bit more about MicroStation in general see the appendix.

Let's expand the discussion to other CAD software. What should someone do if they are using another CAD program and need to exchange files with an AutoCAD user? The dxf file is the solution as it is a format agreed upon by most CAD vendors as the "go-between" that can be understood by all applications for easy file sharing. AutoCAD will easily open a dxf or generate its own dxf files for any drawing. This will be discussed momentarily.

A few observations on sharing files: among architects, the issue is not much of a problem as AutoCAD is the main application in this profession and competitors such as ArchiCAD, will easily open AutoCAD's files and send a dxf right back if needed. In engineering however, AutoCAD has a lesser presence. While some electrical and civil engineers use AutoCAD, many others, especially mechanical, industrial, aerospace, automotive and naval engineers, must instead use 3D modeling and analysis software such as CATIA, NX, Pro/Engineer, SolidWorks and many others.

These applications do not interact well with AutoCAD. Their "kernel" (the software's core architecture) is significantly different in design and intent from AutoCAD's ACIS 3D kernel. They generate native files for internal use and unfamiliar (to AutoCAD users) files such as IGES and STEP for sharing and for analysis or manufacturing purposes. Some of the previously mentioned software *can* generate dxf files, but those files are "flattened" and what you get is a 2D snapshot of what used to be a 3D model. Better than nothing, but not terribly useful for further work in AutoCAD. It is two different worlds indeed. AutoCAD can also now import images from CATIA and SolidWorks using the OLE command (to be discussed soon), but once again they are also just flattened images.

This can all admittedly be a somewhat confusing scenario. Because there are so many different CAD packages on the market, an AutoCAD designer needs to be aware and up to date on what others in their industry are using, and think on their feet when it comes time to exchange files, as there usually *is* a way. Let's now take a look at the specifics of exporting, importing and the OLE command to explore the available options.

Sec 16.7 - Exporting and the Save As Feature

As just mentioned with a MicroStation example, exporting data can be done by selecting **File→Export...** from the cascading menu. What you will see is the following dialog box (Figure 16.11).

Figure 16.11 – Export Data

Your drawing can be exported to the following formats:

- **3D DWF** – This is to convert the drawing to web format for use on viewers.
- **Metafile** – This is to convert the drawing to a Windows Meta File, a Microsoft graphics format for use with both vectors and bitmaps.
- **ACIS** – ACIS is the 3D solid modeling kernel of AutoCAD as well as a format a drawing can be exported to. Few other software currently use the ACIS kernel (SolidEdge comes to mind), as Parasolid is the industry standard, so this is a rarely used export format.
- **Lithography** – STereoLithography files (*.STL) are the industry standard for rapid prototyping and can be exported from most 3D CAD applications including AutoCAD. Basically, it's a file that uses a mesh of triangles to form the shell of your solid object, where each triangle shares common sides and vertices.
- **Encapsulated Post Script** – EPS is a standard format for importing and exporting PostScript language files in all environments, allowing a drawing to be embedded as an illustration.
- **DXX Extract** – DXX stands for Drawing Interchange Attribute and is a DXF File with only block and attribute information. Not relevant to most users.

- **Bitmap** – This will convert the drawing from vector form to bitmap form, with the resulting pixilation of the linework. Not recommended.
- **Block** – This is another way to create the familiar block covered in Level 1.
- **V8 DGN** – MicroStation Version 8, as described earlier.

The Save As dialog box can be accessed any time you select **File→Save** or **File→Save As...**from the drop-down cascading menus. You can also type in **saveas**. In either case you will get the following dialog box (Figure 16.12).

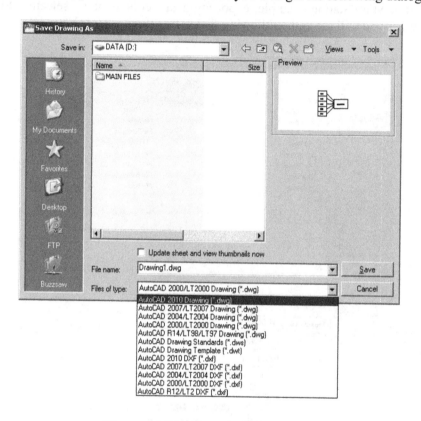

Figure 16.12 – Save As Dialog Box

The extensions present here are described in detail in the appendix, but in summary you can save your drawing as an older version of AutoCAD (an important step that is automated via a setting in Options, as covered in Chapter 14) or as a dxf as described earlier. Notice the dxf files can also be created going back in time, all the way to Release 12 (that was a mostly DOS-based AutoCAD). The other extensions, dws and dwt, are used less often and are also mentioned in the appendix.

Sec 16.8 - Inserting and OLE

The final discussion of this chapter will feature the Insert menu option in general and the OLE concept/command in specifics. The Insert cascading menu is shown in Figure 16.13.

Figure 16.13 – Insert Menu

Most of what is featured in this drop-down menu you may already be familiar with, or it will be covered soon as a separate topic:

- Block insertion is the same as typing in **insert** (from Level 1)
- Hyperlinks and Raster Images are covered in Chapter 15
- External References is all of Chapter 17
- Layouts is a Paper Space topic from Chapter 18

Other items we will not cover (and you will rarely need):

- Fields
- 3D Studio, ACIS, Drawing Exchange Binary, etc.

Let's then focus on OLE or Object Linked Embedded. OLE is a way to embed (insert) non-native files into AutoCAD on a provisional basis, meaning these files are not part of AutoCAD (they can't be anyway as they are of non-AutoCAD formats), they are merely dropped in and are visually present. In principle you can drop in just about anything, with just about any results. Some files will sit nicely, others won't appear and others may even cause a crash. Let's take a look at the dialog box.

Select OLE Object… from the Insert menu and the following dialog box will appear (Figure 16.14).

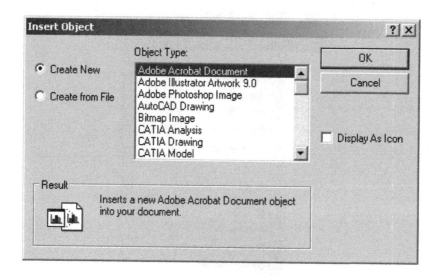

Figure 16.14 – Insert Object

These, in theory, are the various software file items that can be inserted into AutoCAD. Looking through it top to bottom, you can see that a wide variety of Adobe, Microsoft, and even Dassault (they own CATIA and SolidWorks) files can be inserted. You can also, in principle, insert video and audio clips (one can envision music files playing upon opening of an AutoCAD drawing!), and even Flash animations. Simply pick what it is you are looking to insert and select "Create from File". Some of the more useful OLE insertions are related to the Microsoft products listed, but much of that topic has already been covered earlier in this chapter. To insert a file, select "Create from File" and browse for the file you are interested in.

Level 2 Drawing Project (6 of 10) – Architectural Floor Plan

Here we will add furniture to the living room area, and a variety of hatch patterns to the rest of the rooms. Create a carpeting layer A-Carpet for those hatches. The rest of the furniture can go under the existing furniture layer. Some furniture dimensions are given, and others can be assumed. Note how some of the furniture and all the doors are frozen to allow for the carpeting hatches. Some of the hatch patterns used include:

- AR-HBONE
- NET
- HOUND

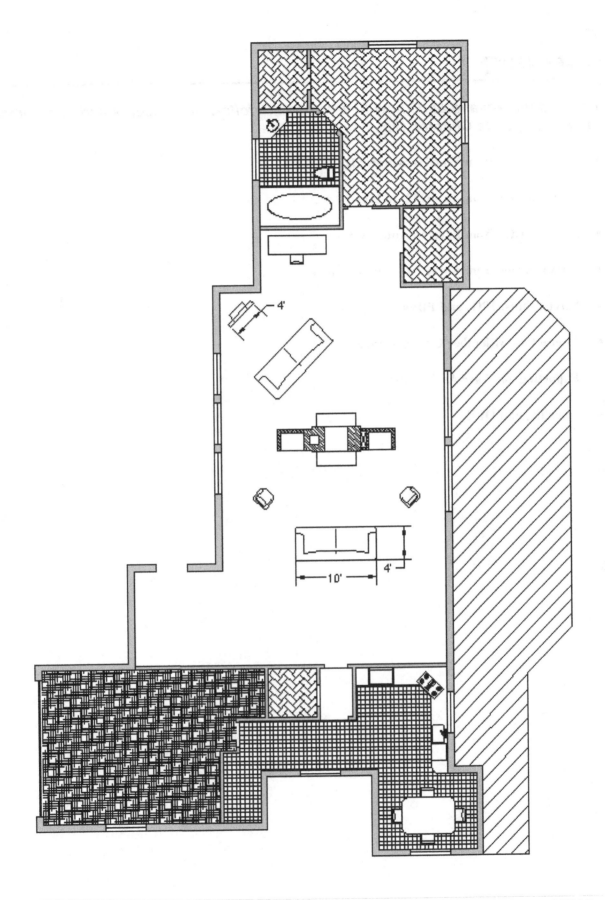

Chapter 16

Summary

You should understand and know how to use the following concepts and/or commands before moving on to Chapter 17.

- Data transfer from Word to AutoCAD

- Data transfer from AutoCAD to Word

- Data transfer from Excel to AutoCAD

- Data transfer from AutoCAD to PowerPoint

- Screen shots, JPGs and PDFs

- Interaction with other CAD software

- Exporting and Save As features

- Inserting and OLE

Chapter 16
Review Questions

Answer the following based on what you learned in Chapter 16.

1. What is the correct way to bring Word text into AutoCAD?

2. What is the procedure for bringing AutoCAD drawings into Word?

3. What is the correct way to bring Excel data into AutoCAD?

4. What is the procedure for bringing AutoCAD drawings into PowerPoint?

5. What is the procedure for importing JPGs and generating PDFs?

6. Describe the basic ideas of interacting with other CAD software?

7. What is Exporting and Save As?

8. What is OLE?

Chapter 16
Exercises

Exercise #1 – At your own pace review everything presented in this chapter.
(Difficulty level: Easy, Time to completion: 15 minutes)

1. Correctly import an MS Word paragraph in to AutoCAD.
2. Correctly import MS Excel data into AutoCAD, and scale it up and down.
3. Correctly import an image from AutoCAD into MS Word using:
   a) Print Screen function
   b) Copy and Paste
4. Export a drawing to "dxf" format.

CHAPTER  17

External References (XREFs)

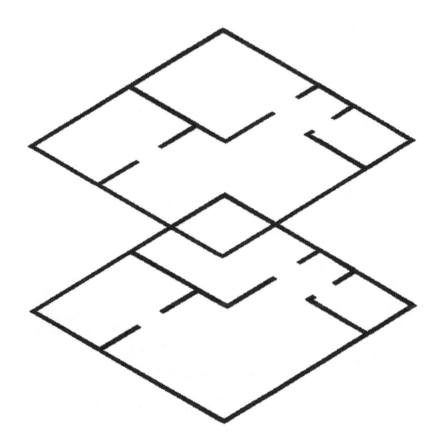

Chapter 17

Learning Objectives

In this chapter we will introduce and thoroughly cover the concept of External References (Xrefs). We will specifically introduce:

- The primary reasons for using Xrefs
- Loading Xrefs
- Unloading Xrefs
- Binding Xrefs
- Updating and Editing Xrefs
- Layers in Xrefs
- Multiple Xrefs

By the end of the chapter you will be well versed in applying this critical concept to your design work.

Estimated time for completion of chapter: 1-2 hours

Sec 17.1 - Introduction to Xref

External reference, or Xref for short, is a critically important topic for architectural AutoCAD users, and very useful knowledge for all others. Along with Paper Space (next chapter), Xref forms the core of Level 2, and is considered an advanced application of the software. Like many advanced topics, you will need to understand fundamentally what it is and when to use it. The "button pushing" or the mechanics of how to make it work is the easy part. In this introduction, let's first define what an Xref is, and why there is a need for this rather clever concept that has existed since AutoCAD's early days.

What is an Xref? As defined, an Xref is a file that is electronically attached (referenced) to another file. It then appears in that file as a fully visible background that you can use to position new design work against. This Xref is not really part of the file, it is merely "paper clipped" to it and cannot be modified in a traditional way; it merely serves as a background. The word Xref is also an action verb, as in "I need to Xref that file in".

Why do we need an Xref? What is the benefit? The preceding paragraph was just a strict definition and may not have yet illuminated the real reason for Xref. To see why we have this concept in AutoCAD let us propose the following scenario:

You are designing a tall office building. It is boxy in shape (similar to the former World Trade Center towers), and as such the exterior doesn't change from floor to floor. The interior, however, does. The first floor is a lobby, the second is a restaurant, the third is storage, and floors four through thirty are all office space. In short, for a designer, once the exterior is done, the rest is all interior-space work. Thus going from floor to floor, the exterior design file can be Xref'd in and just "hang out," providing a background for placement of interior walls and other items. This dramatically reduces the size of each file as a copy of the exterior is merely referenced, and it is never part of the interior files. Got 100 floors to do? No problem, one Xref file of the exterior is all that's needed.

- *Xref benefit #1: Reduce file size by attaching a core drawing to multiple files.*

If this were all there was, Xref would be a neat trick, and would have surely fallen out of favor as computer speed, power and disk storage space increased over the years. However, the preceding was just a warm-up. The main reason for the Xref concept is *design changes*. If you need to change the shape of the exterior walls, you only need to do it *once*, and the change will propagate through all files that are using that Xref. Think about the implications for a moment. In the old hand drafting days, each floor had to be drawn separately. Imagine now a last minute change to the exterior design. Hundreds of sheets would have been updated by hand. Score one for AutoCAD!

- *Xref benefit #2: Automated design change updates to core drawings.*

Need another benefit? How about security. The Xref file can be placed in a folder that is only accessible to the structural engineer and lead architect, not the interior designer or electrical engineer. Granted, this "need to know" basis for access may not be necessary, but it's good to have a way of keeping someone from shifting structural walls to accommodate the carpeting.

- *Xref benefit #3: Critical file security.*

Hopefully you are now sold on the benefits of an Xref. So how is this actually done in practice? The Xref is typically not just the exterior walls but also columns and the interior core (elevators and stairs) as those also tend to not change from floor to floor. The designer creates these files, naming them something descriptive such as "ExteriorWallsRef", "ColumnsCore", etc. Then the Xrefs are attached together and then to the first floor file (or to the first floor file individually – more on that later).

The first floor file is the active file of course, and is named "Floor_One_Arch". At first it will be blank, showing just the Xref. Then the interior design is completed. Then the designer opens another blank file, attaches the same Xref or set of Xrefs and names this file "Floor_Two_Arch" and so on. Xrefs can be attached, detached, refreshed and bound to the main drawing file. We will cover all these shortly. We will also cover some Xref related topics such as layering and methods to edit Xrefs in place.

Sec 17.2 - Using Xrefs

To demonstrate the features of the Xref command we will first need to have a building exterior file to work with. You can select one of your own, or one assigned by your instructor if taking a class. Alternatively (and for good practice), draw the floor plan in Exercise #1 at end the end of the chapter. We will be using that floor plan for the rest of the Xref discussion. In Figure 17.1 the plan is rotated horizontally and the left corner moved to point 0,0. All the dimensions are frozen and the file is saved as ExteriorWalls.dwg.

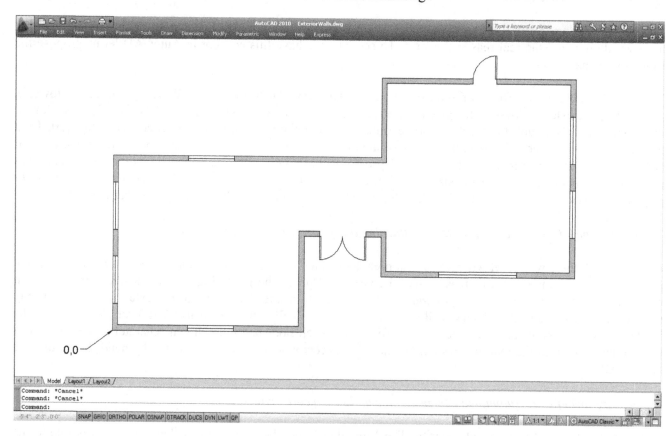

Figure 17.1 – The Xref file

There is, as of now, nothing special about this file. We still need to Xref it to the working file.

Open up a blank file and save it as Floor_One_Layout. Then type in **xref** and press Enter. Alternatively you can use the cascading menu **Insert→External References…** The following palette will then appear (Figure 17.2).

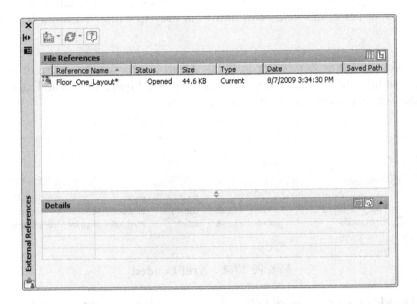

Figure 17.2 – The Xref palette

At the upper left is a white rectangle with a paper clip (hence the earlier paper clip reference). That is the "attach drawing" button. You can click the down arrow to reveal further attachment options (Image, DWF, DGN), but we don't need those. Simply click on the paper icon and the Attach Reference File browsing window opens. Browse to find the Xref file and click Open. The External Reference insertion box appears as shown in Figure 17.3.

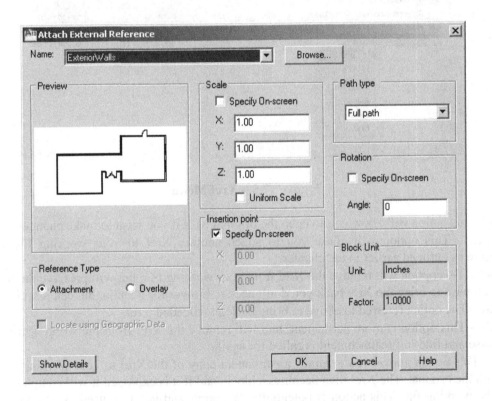

Figure 17.3 – External Reference

Examine what is featured here, most of which you should be familiar with (such as Insertion Point, Scale and Rotation). Uncheck Insertion Point; the fields will turn from grey to white. We want the Xref to insert at 0,0,0. Finally press OK.

The Xref will appear with its lower left corner at 0,0,0. Zoom out to see the entire drawing if not apparently visible. Notice the new addition to the Xref palette as seen below in Figure 17.4. The ExteriorWalls file is visible and indicated as Loaded with the appropriate date, time and path.

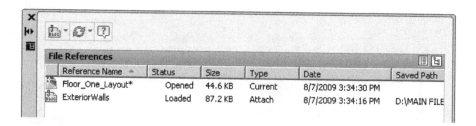

Figure 17.4 – Xref Loaded

You can now proceed to add new design work to the file, as the Xref procedure is complete. Let's skip ahead then and explore some of the other available options now that the Xref is loaded. By right-clicking the XrefExteriorWalls paper icon, you will get access to the following important menu (Figure 17.5).

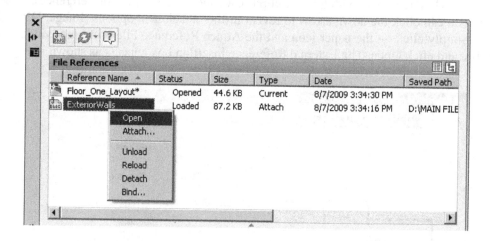

Figure 17.5 – Xref Menu

- **Open** – This option will allow you to open the original Xref if you wish to make changes.
- **Attach...** - This option will allow you to attach another Xref to your working file with the same procedure as outlined earlier in the chapter.
- **Unload** – This option will remove the Xref from your working file, but will preserve the last known path, so if you wish to bring the Xref back you don't need to browse for it again, but can just click Reload.
- **Reload** – The option referred to above to bring back an unloaded Xref.
- **Detach** – This option is more permanent, and will remove the Xref from the working file, requiring you to browse and find it if reattachment is called for again.
- **Bind** – This option allows you to attach a permanent copy of the Xref to your working file and sever all ties with the original. The Xref then becomes a block, and if it is exploded it will completely fuse together with the working file. This action is potentially dangerous and is rarely used as you are now stuck with the Xref permanently. Two possible uses may include permanent drawing archiving and emailing the drawing set (so as to not forget the Xrefs), though e-transmit can take care of that.

Sec 17.3 - Layers in Xrefs

There is a very logical problem that creeps up when using Xrefs. The Xref file itself has layers, often many of them. There may also be several Xrefs in one active drawing. And of course the drawing itself may have numerous layers of its own. It is almost guaranteed that on a job of even medium complexity, layer names can and will be duplicated. This of course is not allowed in AutoCAD, and some method is needed to separate layers in the Xref from mixing and being confused with layers in other Xrefs or the main design drawing. The solution is quite simple. In the same manner that people have first and last names to tell them apart (especially helpful if the first names are the same), so do layers when Xrefs are involved.

All layers that are part of a certain Xref will have that Xref's name preceding the layer name with a small vertical bar (and no spaces) between them. All layers that are native to the working file will of course have no such prefix.

This simple method ensures no direct duplication by having the Xref file name act as the "first name" and the layer name as the "last name." An example is shown below in Figure 17.6. New layers have been added in the host file. Notice the small vertical bar in the Xref layer name and the inclusion of the Xref file name.

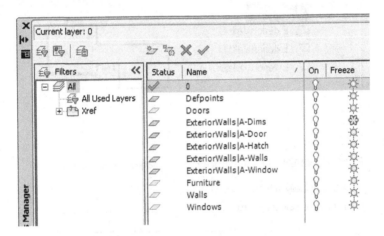

Figure 17.6 – **Xref Layering**

When an Xref is bound to the drawing the layer's vertical bar changes to a 0 set of symbols, indicating they are now bound and part of the working drawing as seen in Figure 17.7. You should be familiar with this if you ever examine a bound set.

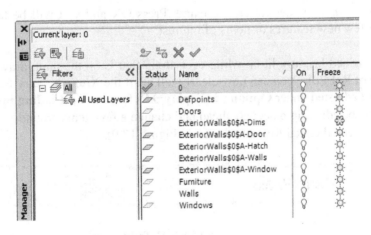

Figure 17.7 – **Xref Bound Layering**

Sec 17.4 - Editing and Reloading Xrefs

There was once a time that Xrefs could not be changed in place. This means you had to open the original Xref file to edit it, and couldn't do this from the main working file. This made sense as you generally want to limit access to the original Xref so it cannot be casually edited.

All things (and software) change and a number of releases ago AutoCAD started allowing Xrefs to be edited in place using a technique similar to editing of blocks (somewhat diminishing the security benefit of Xrefs as mentioned earlier). Let's give it a try. Double click any of the lines making up the Xref while in the main file. The Reference Edit will pop up as seen in Figure 17.8.

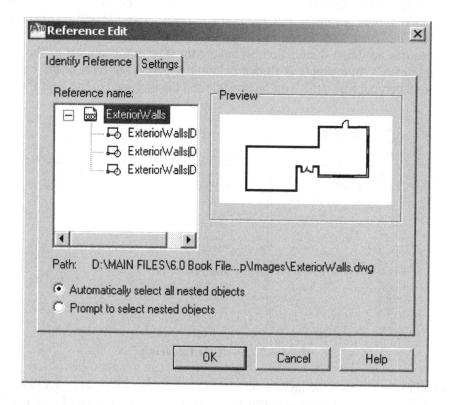

Figure 17.8 – Xref Bound Layering

The Xref you clicked on is listed under Reference name. Press OK and you will be taken into Reference Edit mode. You may notice a few new features or items of interest:

- The Xref has remained brightly lit up while everything that is on the main drawing file has faded down. This is a feature of Ref Edit. You can now clearly tell apart the Xref geometry from the rest. The intensity of the fade can be adjusted under **Options→Display→Reference Edit fading intensity**.
- The Xref can now be edited in place; go ahead and change a few items around.
- A new tool bar appeared called Refedit (shown in Figure 17.9).

Figure 17.9 – Refedit

When you have finished editing the Xref, press the last button all the way to the right called Save References Edits. A warning icon will pop up; press OK and you are done. The Xref has been edited in place.

Just as a reminder, this procedure is only for editing the Xref while you have the main working file open, and what you just did can be also done in a far simpler manner by just opening the Xref as its own file (thought with Ref Edit you have the advantage of seeing the internal design while editing). In fact, when changes are extensive the designer may just open the Xref itself and work on it all day, without resorting to Ref Edit. Every time they save their work the changes become permanent and accessible to any other designers that may be working on a job that uses that Xref.

A good question may come up at this point. How do the other designers know that the Xref changed so they can adjust their interior work accordingly? Many releases ago the only way to know was for the person working on the Xref file to simply tell others via a visit to their cubicle/office, phone or email. The other users would then press Reload All References as seen in Figure 17.10.

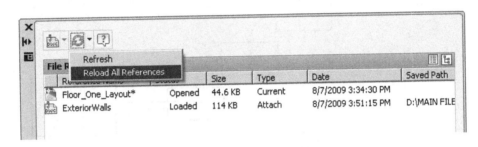

Figure 17.10 – Reload Xref

Of course you can still do this, and designers often refresh every few minutes on their own if they know someone is actively working on the Xref. However, a new tool appeared recently. It is simply a "blurb" announcement that appears in the lower right corner of AutoCAD's screen as shown in Figure 17.11. Press the blue link and the Xref will reload and refresh automatically.

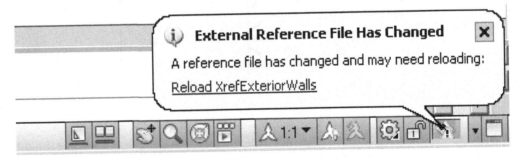

Figure 17.11 – Reload Xref announcement

One final brief topic in this chapter concerns how to chain together Xrefs if there is more than one of them. Some methods are suggested next.

Sec 17.5 - Multiple Xrefs

As mentioned earlier in the chapter, many complex drawings use more than one Xref. Typically one may find the exterior walls, columns and core as three separate Xrefs. The ceiling grid (RCP) can also be Xref'd in for extensive ceiling work, as can the demolition plan and many other combinations.

There are essentially two options. Let's say you have five Xrefs total to work with. You can attach them all one to another and then to a master working file (daisy chain), or one at a time as shown in Figures 17.12 and 17.13.

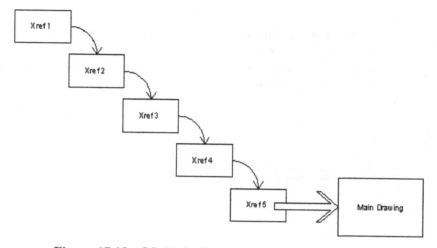

Figure 17.12 – Multiple Xref Attachment – Method 1

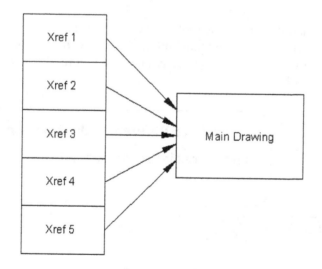

Figure 17.13 – Multiple Xref Attachment – Method 2

Which method you use is up to you (there may be company policy or a convention in use at your job). Each method has minor pros and cons and you should be expecting either one to be used in the industry.

One final word on when NOT to Xref, as this feature gets abused occasionally. Xref'ing of title blocks, furniture and text is not what Xref is for. It is intended for major pieces of infrastructure, not secondary objects. The logic behind Xref in the first place is to attach objects that aren't likely to change and can serve as the backdrop. Use the tool wisely.

Sec 17.6 – Ribbon and Xrefs

Some Xref functions can be accessed via the Ribbon, and we will discuss them separately in this section. You can find them under the Insert, Reference tab as seen in Figure 17.14.

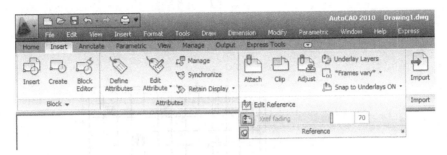

Figure 17.14 – External References via Ribbon - 1

One important control is the "Xref fading" slider you will see toward the bottom of the tab. This will control the darkness of the Xref. Some users prefer the Xref to be dark, similar to the surrounding design, while some prefer it very light, so they clearly know what is an Xref and what is the design.

Finally the Ribbon will switch to a dedicated Xref tab if you click the Xref as seen here in Figure 17.15, though there is much duplication of the previous tab's commands.

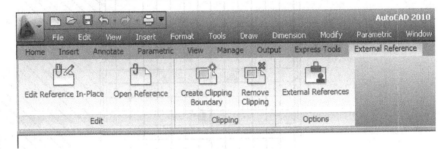

Figure 17.15 – External References via Ribbon - 2

Level 2 Drawing Project (7 of 10) – Architectural Floor Plan

Here we will add some mechanical work in the form of diffusers and return air ducts, all embedded in a reflected ceiling plan. This is not necessarily how an architect would choose to do it, but the point here is CAD practice, not adherence to strict design protocol. Moreover, a residential home wouldn't necessarily have such an air circulating system, which is more typical of an office space. Consider this practice of all types of architectural design rolled into one drawing. We will need additional layers such as M-Diffuser, M-Return-Air, A-RCP and others as you see fit. Note that all interior work except for walls will be frozen for this exercise and we will be working strictly with the structure of the building only. Use the following dimensions for the mechanical features:

- RCP – 2' × 2' grid
- Diffuser and return air ducts – both 2' × 2' as seen on the next page.

The best way to create the RCP is to draw a pline around the border of each room and begin the plan either from the center, or more commonly from one edge and offset out at 2' intervals. You can also use the NET hatch pattern. When this is done, place the diffusers and return air ducts as shown.

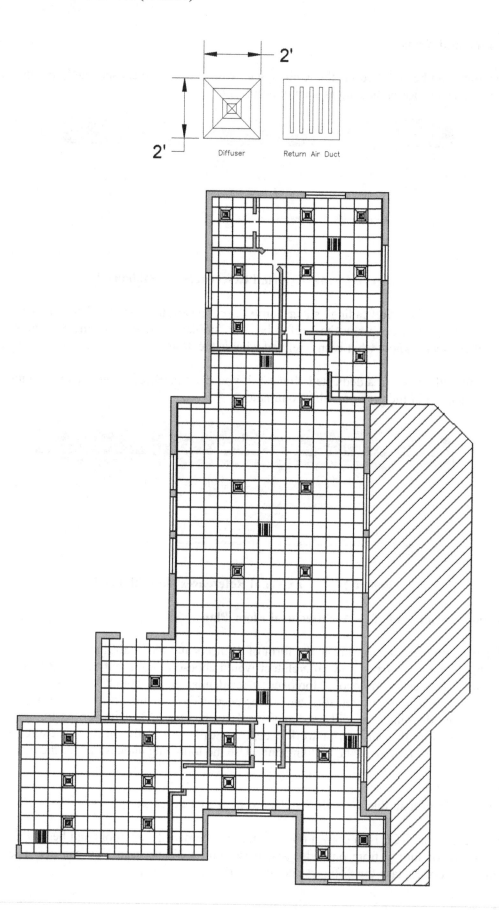

Diffuser   Return Air Duct

Chapter 17

Summary

You should understand and know how to use the following concepts and/or commands before moving on to Chapter 18.

- **Xref command**
  - Inserting
  - Detaching
  - Unloading
  - Reloading
  - Binding

- **Layers in Xref**

- **Nesting Xrefs**

Chapter 17

Review Questions

Answer the following based on what you learned in Chapter 17.

1) What are the benefits of Xref?

2) How do you attach an Xref?

3) How do you detach, reload and bind an Xref?

4) What do layers in an Xref'd drawing look like?

5) What are some nesting options for Xref?

Chapter 17

Exercises

Exercise #1 – Draw the following floor plan in order to practice the Xref command
(Difficulty level: Easy/Moderate, Time to completion: 45-60 minutes)

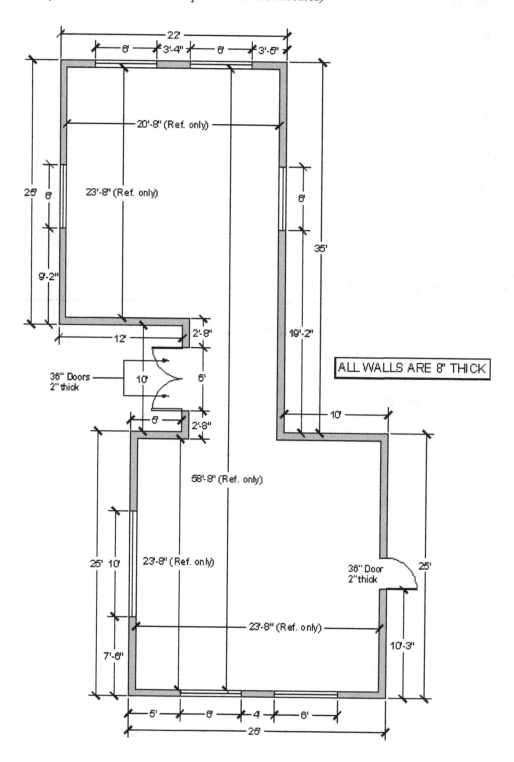

LEVEL 2

CHAPTER  18

Paper Space

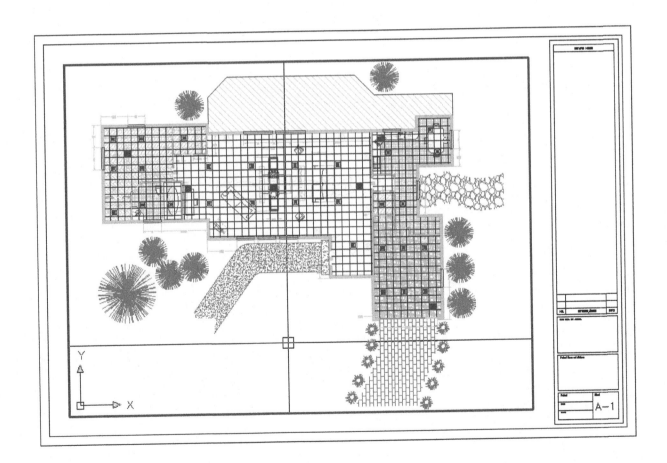

Chapter 18
Learning Objectives

In this chapter we will introduce and thoroughly cover the concept of Paper Space. We will specifically introduce:

- The primary reasons for using Paper Space.
- Entering Paper Space and setting up title blocks (Layouts).
- Viewports and their properties.
- Viewport scale assignments.
- Controlling what is visible in viewports.
- Setting up text and dim styles based on viewport scale.
- The new Annotation feature that simplifies setup of text and dims.

By the end of the chapter you will be well versed in applying this critical concept to your design work.

Estimated time for completion of chapter: 3 hours

Sec 18.1 - Introduction to Paper Space

Paper Space is an extensive and critically important topic in AutoCAD. Learning it is a rite of passage for all AutoCAD designers, and proper set-up and use of Paper Space is a mark of an expert user. The topic is not necessarily difficult, but can be somewhat tedious, requiring care and attention from the designer who really has to "dot the i's and cross the t's" to assure everything is done right. It is also a topic that needs to be understood well in theory first and foremost. The actual "button pushing" process to set things up is much easier when you understand what Paper Space is and why it is used. The result of learning all this however will be a professional drawing set that is accurate, easy to read, understand, and print. This is a necessity and a goal well worth striving for. We will explore Paper Space from all angles, one step at a time. Read and understand each section before moving on to the next. Paper Space knowledge is cumulative.

> **What is Paper Space?**

When you create a design of a building with pencil and paper, this design needs to be drawn to a certain scale, as opposed to full-size, for the simple reason that it would not fit on a piece of paper if you did not do this. This is the reason why architects carry with them an architectural scale, a portion of which is shown below.

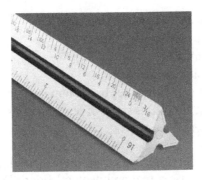

Figure 18.1 – Architectural scale

Then they can simply pick the appropriate scale for the size of their intended design (1/8" = 1'-0" is common), and now due to the fact that 8 feet are represented by 1 inch, the building floor plan can fit nicely on a large sheet of paper.

In AutoCAD however, you do not draw "to scale." Because you now have an infinite size canvas upon which to create your design, everything is drawn one-to-one or "life-sized," meaning that if you were to jump into your design through the computer monitor, you would find yourself in proportion to (and would be able to freely walk around inside) your building.

This is the #1 "Golden Rule" of AutoCAD, and indeed in all cases of computer aided design, and marks a significant departure in philosophy and methods from the previous hand-drafting approach. In fact, attempting to draw "to scale" results in severe complications and potential for error when creating, and later changing, the design, and is simply not done in practice.

Drawing one-to-one, however, creates an immediate problem. How do you output a design that may be dozens or hundreds of feet across, onto a paper that is generally only 3 or 4 feet wide? Furthermore, this design has to be framed by a title block or title page that itself has to be drawn one-to-one, so as to not violate the Golden Rule. The solution is elegant, simple in principle, and has stood the test of time.

The idea is as follows. Draw the design in one area of AutoCAD called Model Space, and then draw the title block in another unconnected and unrelated area called Paper Space. Then connect the two together via a clever concept called a Viewport, which is simply a "window" looking in *from* Paper Space *onto* Model Space. This viewport will then be assigned an appropriate scale and if everything is drawn (and then printed) one-to-one from Paper Space, then the design will come out to scale and the problem is resolved.

This in essence is Paper Space: nothing more than a technique to separate the two worlds and then reconcile them together in a meaningful way to get around the tricky problem of a finite sized piece of paper holding a much larger design.

Be sure you understand the previous few paragraphs completely. The reasons behind everything outlined from here on will be easier to see once you have a good idea of *why* we are doing what we are doing.

Sec 18.2 - Paper Space Concepts

The devil, as they say, is in the details, and the study of Paper Space involves a number of separate concepts, each of which is an important part of the complete picture and needs to be looked at closely.

- *Layouts* – Entering Paper Space and setting up title blocks.
- *Viewports* – Viewports and their properties.
- *Scaling* – Viewport scale assignments.
- *Layers* – Controlling what is visible in viewports.
- *Text and Dims* – Setting up text and dim styles based on viewport scale.
- *Annotation* – A new feature that simplifies setup of text and dims.
- *Summary* – Putting it all together.

These topics are listed roughly in the order a designer would go through them as he or she sets up and does the project. As an example drawing, we will use a real architectural floor plan that you have been working on since the start of Level 2. It happens to fit nicely on a D-sized (36" × 24") sheet of paper at 1/8" scale to illustrate the concepts. If you didn't get a chance to draw it, or want to use one of your own designs, the building is roughly 100 ft by 65 foot at its widest points. You will get the best results in the following discussion if you substitute in a design of similar proportions. If you do have the drawing completed, then turn it counter-clockwise 90° and thaw all the layers. It will be messy, but we will fix that later. Let's begin with the first topic, layouts.

> **Layouts**

This first step involves switching over to Paper Space and familiarizing yourself with the layout tabs. They are found at the bottom of your drawing area, resemble Microsoft Excel tabs, and say Model, followed by Layout 1 and Layout 2 as seen in Figure 18.2.

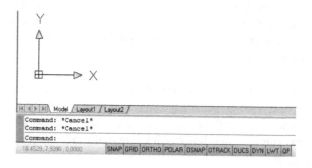

Figure 18.2 – Layout Tabs

Go ahead and switch over into Paper Space by clicking on Layout 1. You will immediately notice a different environment, as shown in Figure 18.3. For starters, the UCS icon is a triangle, Layout 1 tab is highlighted, you background color is white, and finally you have a variety of background features (rectangles, shadows and others) We will somewhat modify this environment in a minute, but in the meantime click on the Model tab and jump back into Model Space. You now know how to enter and exit Paper Space.

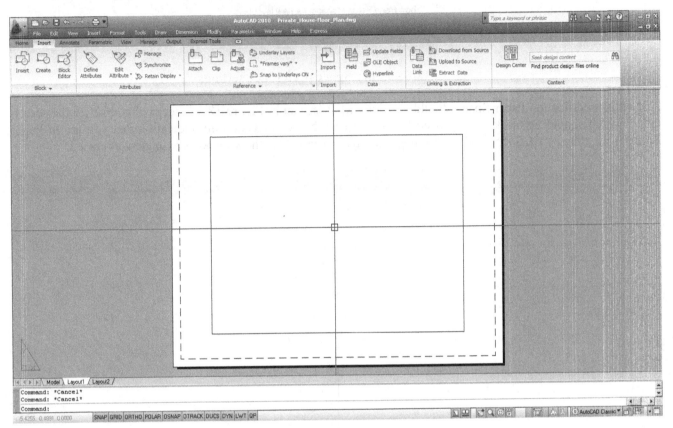

Figure 18.3 – Paper Space

There really is no need for Paper Space to look any different from Model Space. There is nothing special about Paper Space whatsoever. Remember, it is just a technique in AutoCAD to create two separate worlds. You can actually draw in Paper Space as if it's your regular drawing area with no adverse effects, but of course that would defeat the whole purpose of what we are doing.

Based on the above, it is highly recommended that the Paper Space shadow and other features are turned off. Indeed most experienced users prefer their Paper Space to look as close to Model Space as possible. It is ultimately up to you if you want to do this, but all screen shots in the remainder of this chapter will feature the plain Paper Space environment. To do this you need to right-click to Options…, and then select the Display tab and take a look at the Layout elements (lower left). Go ahead and uncheck all but the very first choice as seen in Figure 18.4 and also set the Sheet/layout color to black in Colors… (though in the textbook it will remain as white and the drawing will remain monochrome to conserve ink and for clarity). Finally, click OK.

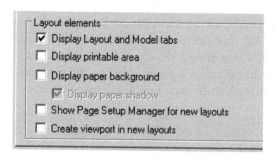

Figure 18.4 – Paper Space preference settings

After this modification Paper Space should look like what is seen in Figure 18.5 (except of course for the screen color, which should be black, as preferred by most users). Notice that the only indication that you are in Paper Space is the triangle UCS icon and the highlighted Layout 1 tab. Also the viewport rectangle was erased.

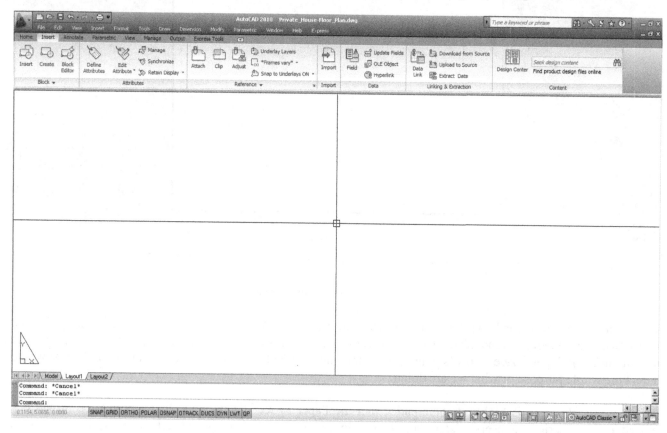

Figure 18.5 – Paper Space, modified

So what are these Layout tabs in real life? They are literally sheets of paper when the job is completed and ready for printing. The design (done in Model Space) will be "presented" in Paper Space with each Layout being one sheet of the total drawing set; i.e., the Architectural, Mechanical, Electrical, Details and so on.

If you are thinking that we need to learn how to make title blocks at this point and perhaps create more Layouts, then you are correct. We will create a title block from a built-in AutoCAD template. This may or may not be the way you will be doing it at your job. If you are in an architecture office, there may already be a pre-drawn title block that you can just retrieve and paste in, but for our learning purposes we will create a new one.

Select the Layout 1 tab (left-click) and then right-click to bring up the menu shown below.

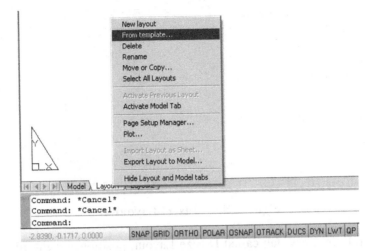

Figure 18.6 – Layout tab menu

We will be working with this menu for the next few pages. Let's first of all try creating a title block from a template by clicking the **From template...** choice. You will get the following dialog box (Figure 18.7). Pick the *Tutorial-iArch.dwt* template and click Open.

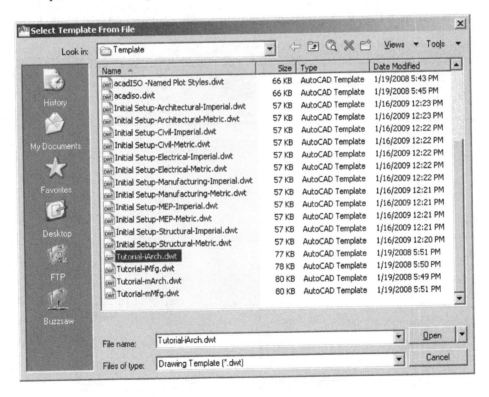

Figure 18.7 – Templates

Another small dialog box, Insert Layout(s), will appear as shown in Figure 18.8. It is simply informing you that the layout is a D-sized one. Click OK.

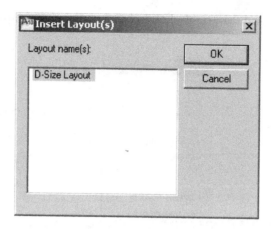

Figure 18.8 – Insert Layout(s)

Notice what happened: you have a new tab called D-Size Layout, positioned right after the other three layouts. Click on it and you will see a decent title block (much improved in AutoCAD 2010 versus older releases). It is a reasonable representation, minus the fancy logos, of what a typical architectural office may use. Inside the title block is a blue rectangle through which you may or may not see your building; this is the viewport, and we will cover this topic next. *For now, go ahead and delete it.* We will make our own soon. *Also, explode the title block as we may want to modify it.*

What you have now is shown in Figure 18.9 (except that your screen background will likely be colored black).

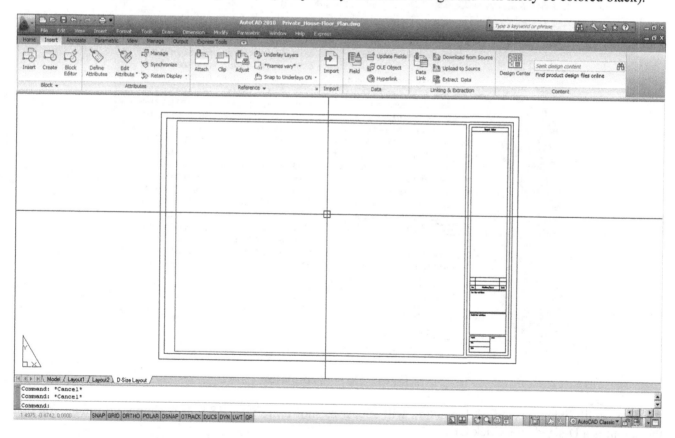

Figure 18.9 – Basic "Tutorial-iArch.dwt" title block

The next step is to modify the layout tabs by renaming them to something more descriptive and removing the unnecessary ones. Right-click on the new "D-Size Layout" tab to get the menu to appear again. Select the **Rename** choice and the tab's name will become editable; type in "Architectural Layout". Let's now get rid of the Layout 1 and Layout 2 tabs, as they are empty and not needed. Begin by clicking on Layout 1, then right-click to get the menu up again. Select **Delete** and press OK when the warning box pops up. Repeat the procedure for Layout 2. After these steps we have only the following two tabs remaining: Model and Architectural Layout as seen in Figure 18.10.

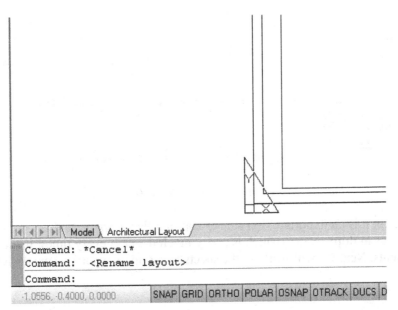

Figure 18.10 – Model and Architectural Layout tabs

Take a moment to look over the title block in the Architectural Layout tab. As a reminder once again, delete the original viewport (the blue rectangle) as we will make our own from scratch. Then explode the title block if you haven't done so yet. Modify anything you may want to look different (for example delete the USER, REVDATE and the FNAME text in the lower left corner and added sheet A1 in the lower right corner of the title box), but this is optional.

Now we need to create more of these title sheets, for say the Mechanical and Electrical layouts. Go back to the menu (right-click on the Architectural Layout tab) and select the **Move or Copy...** choice. The following dialog box (Figure 18.11) will appear.

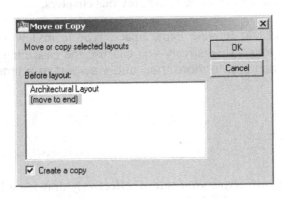

Figure 18.11 – Move or Copy

Be sure to do two things. First of all, check off the "Create a copy" box at the bottom left and also click on to highlight the "(move to end)" choice, which will simply place the new title sheet next in line after the existing Architectural Layout, and may be important later when it come to the printing order. The order can be seen in Figure 18.12. Finally click OK and observe what happens with the tabs. A new one appears, saying Architectural Layout (2), and it is a perfect copy of the previous existing tab. Go ahead and rename it (as outlined earlier) to Mechanical Layout. Repeat the whole procedure to create a third tab called Electrical Layout, and finally a fourth tab called Landscaping Layout. Here is the final result (Figure 18.12).

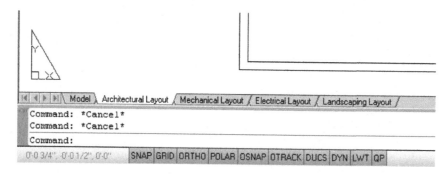

Figure 18.12 – **All tabs**

Review this entire section; it is an important first step to learning Paper Space, and is a common procedure needed to set up a project that has multiple pages (which is pretty much the case every time). It should be clear as to what each sheet (tab) represents. Next we will work on the sheets to set up viewports so the design can be viewed at the proper scale.

> **Viewports**

Viewports, as mentioned in the Paper Space introduction, are simply windows from the Paper Space world looking into the Model Space world. It is as if we ripped a hole in the fabric of Paper Space and observed what was behind there. We will assign scales to these viewports later on and in this manner reconcile the Paper and Model Spaces. For now we need to first learn how to create and modify the viewports and go over some of their properties. Your new layouts already have a premade viewport (the blue rectangle), but as mentioned earlier, you should have now erased it and we will learn to create new ones from scratch.

The following procedure assumes that in Model Space you have the house drawing that you've been working or a design provided by yourself or your instructor. If you are using the house design, make sure it is rotated 90° counter-clockwise. That is the proper plan view orientation, and it was only turned on its edge to better fit the printed paper so you could see the design up close in the previous chapters.

Click on the Architectural Layout tab, entering Paper Space, type in **mview** and press Enter. This is the key command for creating new viewports. Alternatively you can use the cascading menus **View→Viewports→1 Viewport**. Note here that you can create more than one viewport at a time, but we will only do one for now.

- AutoCAD will say:
```
Specify corner of viewport or
[ON/OFF/Fit/Shadeplot/Lock/Object/Polygonal/Restore/LAyer/2/3/4] <Fit>
```

Then simply click and drag the mouse across your screen, effectively drawing a large rectangle from one side of the sheet to the other (avoiding the actual title block in the lower right corner). AutoCAD will say: `Specify opposite corner:` as you are doing this and: `Regenerating model` as you complete the viewport.

You should now be able to see the design inside the "rectangle" you created as shown in Figure 18.13.

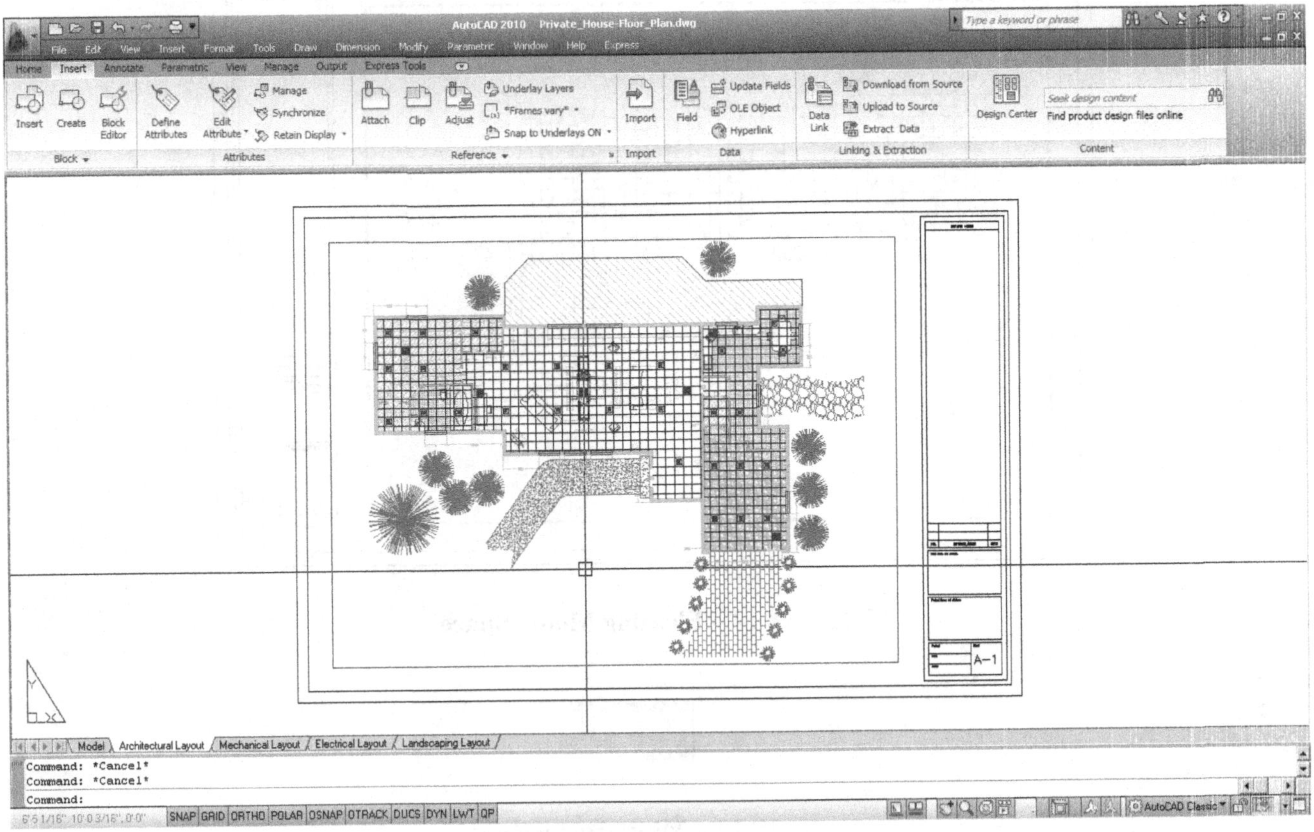

Figure 18.13 – New viewport

You can create more viewports in the other layouts if you'd like for practice, but we will only focus on this one in the Architectural Layout.

Floating Model Space

Double click inside the viewport. Its border will become thicker and you will notice your mouse is "inside" the viewport as shown in Figure 18.14. The advantage of having 100% width crosshairs is evident here as the viewport "clips" the ends of the crosshairs and it becomes clear that you are in this mode. Pan and zoom while inside the viewport and you will notice the design shift around while the Paper Space title sheet stays still. This floating model space mode is sort of a "no man's land" between Model and Paper spaces and is critical in scaling the design as we shall soon see. For now double click anywhere outside of the viewport to get back to Paper Space. Be careful not to double-click the actual viewport itself. That will take you to the "vpmax" mode (maximum viewport size), which can be used for editing, but will not be needed here. If you do accidentally double-click it, then press the button shown in Figure 18.15 to get back (bottom of the screen).

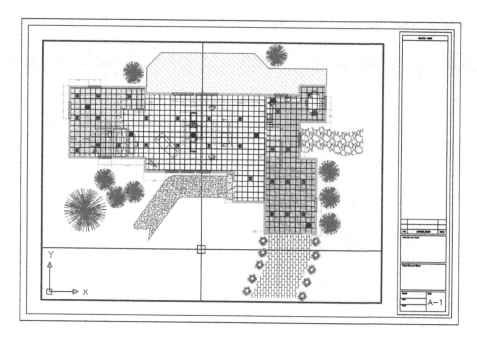

Figure 18.14 – Floating Model Space

Figure 18.15 – Minimize viewport

Practice jumping in and out of the viewport several times; this will be very important for the rest of the chapter. Note also the big picture discussed so far. You are in Paper Space and created the viewport to look onto the design in Model Space. When you are inside the viewport, you are technically in Model Space, but still in Paper Space as well. You will need this for scaling purposes only. Design work on either the floor plan or the title sheet/block will be done in their respective spaces.

Next we need to learn how to modify the viewport(s). Note that they are NOT static displays but rather dynamic tools that you can and should mold and shape to your needs as outlined in the next section.

Modifying the viewport

First, make sure you are in Paper Space, and then click once on the viewport (not inside of it, but the lines of the actual viewport rectangle). The viewport will become dashed and grips will appear. Using them you can resize the viewport to any size desirable by clicking on any grip, activating it (it will turn red) and adjusting the viewport's size and shape. In general you can:

- Scale the viewport(s) up and down to any size
- Adjust the viewport(s) to any shape
- Move the viewport(s) anywhere you'd like
- Erase the viewport(s)
- Copy the viewport(s) as many times as needed
- Make new ones as needed

Take a few minutes to go through ALL of these features. Make your viewport smaller, then copy it several times. Move all three around the screen and then erase one. Finally create it again with mview. In the end try to have what is shown in Figure 18.16, and in general make sure you understand viewport manipulation, the mview command and getting in and out of Floating Model Space.

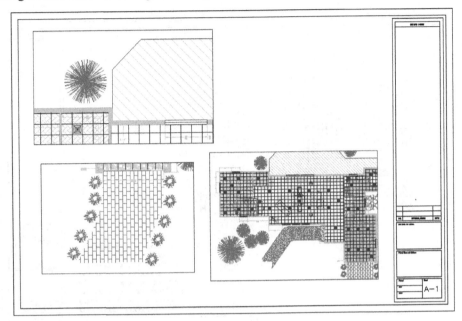

Figure 18.16 – Three random viewports

Polygonal viewport

The shape of the viewport doesn't have to be rectangular or square; it can be any shape necessary by using the Polygonal option. To practice this, click over to the Mechanical Layout and create a new viewport using the mview command, however, before clicking to draw the actual viewport select "P" for Polygonal in the lengthy menu.

- AutoCAD will say: Specify start point: and after the first click Specify next
                    point or [Arc/Length/Undo]:

Click your way around in some random fashion (do not overlap the clicks) and press "C" for Close to close up the viewport. This is what I got after creating a starburst sort of pattern. You can see the design inside the oddly shaped viewport in Figure 18.17.

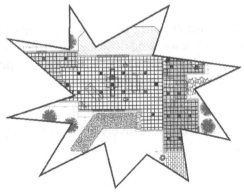

Figure 18.17 – Polygonal Viewport 1

Of course this isn't really what this tool is for. Rather it is to make viewports that weave their way around possible revision blocks or notes on the drawing, thereby freeing you from being forced to use a rectangle and wasting space. An example is shown in Figure 18.18, and this is often the case in real life designs, making Polygonal viewports very important for creating an efficient layout.

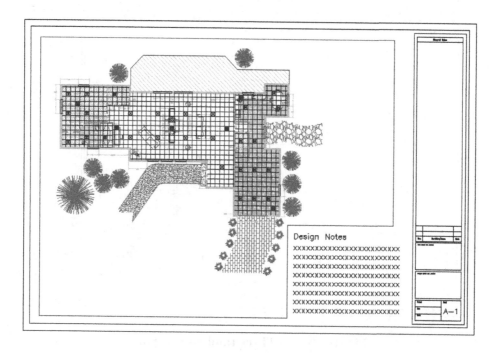

Figure 18.18 – Polygonal Viewport 2

There are a few other important items to cover with viewports such as locking them and making them invisible, but we will introduce these features a bit later as the need arises. What we need to look at next is the "core" of Paper Space: scaling of the viewports and their contents.

➤ **Scaling**

Here we will finally put everything together, because really what we have been doing so far was just the "set up". We have our design in Model Space and our title block/sheet in Paper Space, connected together by one or more viewports. Now we need to give these viewports a scale, which was our goal all along.

To make things simple, let's just have one viewport in our Architectural Layout, so delete any others you may have, including polygonal ones, and leave one large viewport similar to Figure 18.13.

So what is scaling? Well, you may have already guessed it has something to do with zooming. After all, as the design is zoomed in and out (while in Floating Model Space), the scale continuously changes and is always *something*. So scaling the design properly is just a matter of knowing how much to zoom it in the viewport. The big question of course is *how much*? Another way to put it is what scale factor to apply.

There are two ways to assign scale to this or any viewport. The first one is a technique used for just about every AutoCAD release up until the last few. It's a bit more involved but will really get you to understand what's going on. Then we'll cover the other, easier method.

Method 1
Double click inside the viewport (into Floating Model Space), type in **zoom** and press Enter. You will see this familiar menu:

```
Specify corner of window, enter a scale factor (nX or nXP), or
[All/Center/Dynamic/Extents/Previous/Scale/Window/Object] <real time>:
```

It is here you used to pick "E" for extents and see the entire drawing fill up your screen. However did you notice the **enter a scale factor (nX or nXP)** choices? This is what we are interested in right now, specifically nXP.

What nXP means is that you need to enter a ratio relative to Paper Space (which is the 1:1 "anchor" so to speak), and then the design will scale to that ratio. So what is it? Well for 1/8"=1'-0" the "n" is 1/96 (followed by the xP). So type in **1/96xp** and press Enter. The design will zoom to precisely 1/8"=1'-0" scale as shown in Figure 18.19.

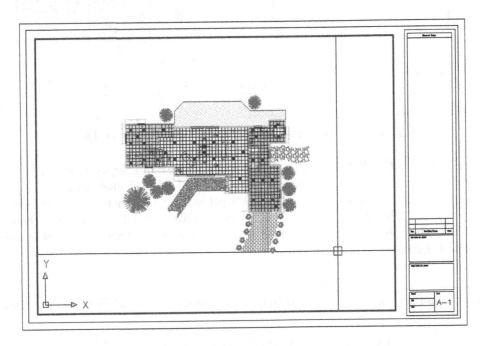

Figure 18.19 – Design zoomed to 1/8"=1'-0" scale (1/96xp)

Now this seems to be too small. No problem, just type in zoom and press Enter again. Then type in **1/48xp** and press Enter again. The design will zoom to 1/4"=1'-0" as shown in Figure 18.20, which seems to be a better size for this design and title sheet.

Some of the landscaping you may notice is outside of the border, but we can fix that later by trimming up the length of the driveway and walkway; they don't need to extend as much as they do. These are the kind of design issues you may be faced with when doing the final output setup. If the items that don't fit are critical to the design, then trimming them out would not be an option and a new scale would have been chosen. Here it's not a big deal and we will just shorten the driveway later on.

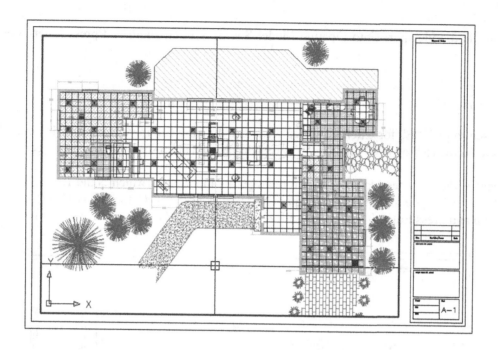

Figure 18.20 – Design zoomed to 1/4"=1'-0" scale (1/48xp)

So where did these 1/96xp and 1/48xp come from? They are just ratios developed from the fact that there are 12 inches in 1 foot and the scale is 1/8 or 1/4. Therefore for 1/8 it is 1/96 (1/8 inch scale × 1/12 inches in a foot = 1/96 relative to the 1:1 Paper Space). For 1/4 it is 1/48 (1/4inch scale × 1/12 inches in a foot = 1/48 relative to the 1:1 Paper Space) and so on. What would 1/2"=1'-0" be? That would be 1/2 × 1/12 = 1/24xp. There are other scales such as 3/8", 5/8" and others; they can all be derived in a similar manner. There is of course a faster and easier way to set up viewport scales.

Method 2
Double-click back out to Paper Space and bring up the Viewports toolbar, which is shown in Figure 18.21.

Figure 18.21 – Viewports toolbar

Now click the viewport *once* (not inside, but the lines of the actual viewport rectangle). When the viewport becomes dashed the scale associated with that viewport will be shown in the white field on the right side of the Viewport toolbar. Click the down arrow and you will see a wide variety of choices. Pick the scale you want and the viewport will take on that scale as shown in Figure 18.22.

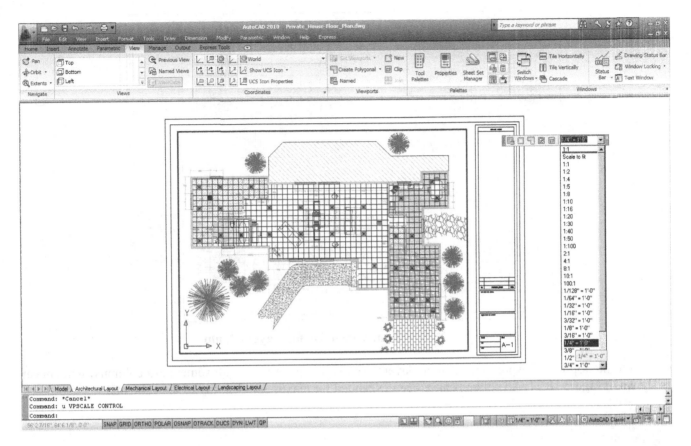

Figure 18.22 – Viewport scaling Method 2

> **Layers**

We have done a lot already in setting up the drawing in Paper Space. We now turn our attention to several important details that are either new or change significantly when dealing with Paper Space.

One of those is Layers. While the fundamental Layer concepts don't change, a new ability is added to the Layers dialog box. It turns out that when viewports are present, you are able to freeze and thaw layers in those viewports independently of each other and the main drawing. This is very significant, as you want to be able to display different items as you go from viewport to viewport.

Perhaps the most important example of that is when you jump between the Architectural, Mechanical and Electrical Layouts. Each tab is a new viewport, but yet all views of the design originate from one actual copy of that design. In each layout you only need to see what is relevant to that layout and it would be a major problem if you could only turn layers on and off in the main Model Space drawing. Then each tab would display the exact same thing.

Fortunately this is not the case, and in each viewport you can turn layers on and off as needed to get the design points across. Therefore the Architectural layout will only have Architectural layers and so on. It is the same idea with multiple viewports on one layout (such as the case with detail sheets). Each viewport is truly a world unto itself. To see an example of this lets create four viewports in the Electrical Layout as shown in Figure 18.23. The scales assigned are random.

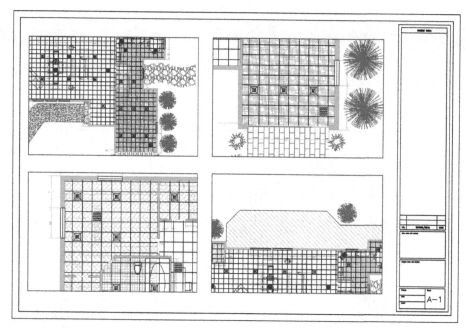

Figure 18.23 – Four viewports for Layers demo

Click into the top left viewport and bring up layers. The layer dialog box (with some extra columns) will appear as shown in Figure 18.24. All columns have been expanded for maximum visibility of the contents.

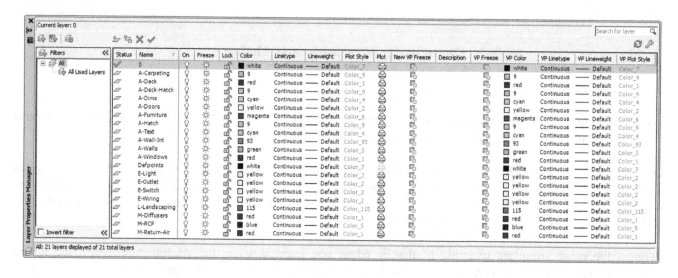

Figure 18.24 – Layers dialog box with View Port columns

Look for the VP Freeze column; this is the only one we will be interested in. In that column you can freeze all the un-needed layers for that layout without affecting either the main layout or the other viewports. Go ahead and freeze all the Furniture and Mechanical layers, as well as A-Dims, A-Carpeting and L-Landscaping layers and press OK. The result is shown in Figure 18.25 (with some color changes for clarity). Note how the neighboring viewports are not affected by what you did in the top left one, and you can clearly see the electrical layout.

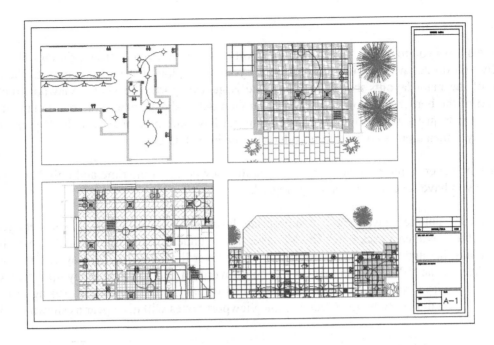

Figure 18.25 – VP Frozen Layers (top left viewport)

Lock VP

We'll now cover two other notable features of viewports. The first is the ability to lock the viewport, which simply means you cannot change the viewport's scale and all zooming attempts will fail (the entire Paper Space drawing will zoom instead). This can be useful to prevent accidental zooming while inside the viewport, and should be set after the design is laid out and set up. To do this bring up the Properties box (refer to previous chapters if you forgot how). Then click the viewport *once*, and select (under Misc.) the *Display locked* option, setting it to "Yes" as shown in Figure 18.26.

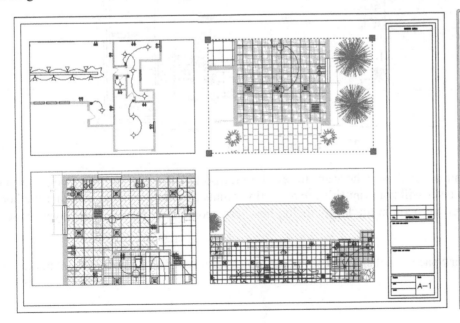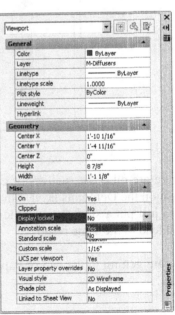

Figure 18.26 – Locking the Viewport

Freeze VP

This next feature allows you to either "freeze" the viewport or simply prevent it from printing. The reason for this is that generally you do not want a big fat rectangle on your final plots that will be going to the client. This is especially true of the main layouts that feature just one large viewport. The visible viewport rectangle can be mistaken for part of the border and is unnecessary. Even with detail sheets, where there are many details on each of them, the trend is to just separate them adequately but not have any rectangles visible. If you "didn't do it by hand in the old days" then there's no need to introduce it now in AutoCAD.

Having said this, however, some designers prefer to see the viewports all the time and only have them disappear upon printing. So you have several options in this regard.

First of all, we need to put the viewports on an appropriate layer if they're not already. Designers generally create a "_VP" layer for them. Notice a trick here. The VP layer has an underscore in front of it. This will force the layer (or *any* layer that has that underscore) to rise to the top of the layer list, above the "0" layer. This makes it easy to find title block/sheet or the VP layers (as long as you don't overdo it and stick too many up there). Go ahead and change the color of the layer to Grey and freeze it. Then finally change all the viewports (using Properties) to that frozen _VP layer. If you did everything correct then the viewport boxes will disappear as shown in Figure 18.27.

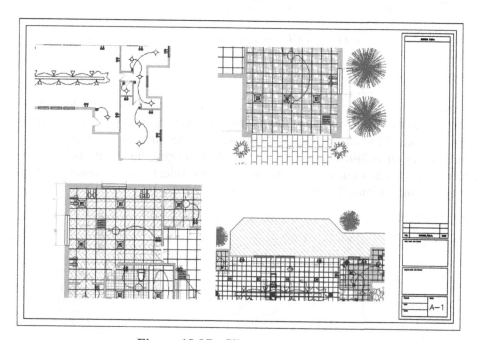

Figure 18.27 – Viewports frozen

There are other alternatives of course for those who want to see the viewports but not print them. You can put them on the Defpoints layer. This layer will not print (it's the definition points from dimensions). Another method is to just put a slash through the printer icon in the layers dialog box corresponding to the viewport layer, which will lead to the same result.

We are getting to the end of Paper Space and there are two major topics left: Text and Dims and the Annotation tools.

> **Text and Dims**

This topic is probably the most confusing one in Paper Space, and really requires the AutoCAD designer to not only understand everything thoroughly, but also work carefully and deliberately. In practice, much of this is only done once and saved as part of a template, but for educational purposes we will assume you are being tasked with the Text and Dims setup from scratch and must know all of the steps.

Our goal here is quite straightforward: we want all of the text and dimensions to be the same size when printed, regardless of what scale viewport they appear in. This uniformity is a basic requirement of a professional drawing. Architects and engineers will specify something to the effect of: "I want the text height to be ¼" on all drawings, both in layouts and in all details," and it will be up to the AutoCAD designer to ensure that this happens. With hand drafting this was not an issue; the text and dims were drawn all at once using the same sized template.

With AutoCAD, due to our use of Model Space and Paper Space, things aren't so simple. The basic fact remains that all text and dimensions have a certain fixed measurable size when drawn in Model Space and will appear as different sizes when viewed through the various scaled viewports just as the design itself will. Some designers try to simplify everything by dimensioning the design in Paper Space to achieve uniformity (AutoCAD will let you do this), but this is NOT correct. The dimensions and all text must remain WITH the design in Model Space. The reasons for this are many, not the least of which is the fact that any movement of the design will misalign everything.

So the text and dims need to stay where they are supposed to be, and we need to understand how to properly size text and dimensions based on the anticipated scale of the viewport they will appear in. This all really just boils down to a game of "match up." The basic theory is as follows. You already know that viewport scale is a function of zooming. Therefore, when you have uniform text in Model Space and view the same set of text through ¼"=1'-0" versus ½"=1'-0" scaled viewport in Paper Space, the text in the ½" viewport will appear two times as large, as the viewport is a *larger scale*. To counter this effect, you need to make sure that the text that appears in that ½" viewport is drawn to be half as big so it appears uniform with the ¼" text (just as an example).

The trick is to be consistent and careful, so that the wrong size text doesn't appear with the wrong viewport scale. So how is this done in practice? How do you know what size text and dims are appropriate for a particular scale? We have tables for this, such as the one shown in Figure 18.28.

To use this chart you have to know what scale viewports you are using. With the main layouts such as the Architectural, Mechanical, etc., done earlier, it is very easy. These layouts typically feature only one viewport of a known scale (1/8" = 1'-0" is commonly used), therefore if the architect desires all text output to be 3/16" in height (also a common size) then by reading the chart from left to right until we meet the 3/16" column we arrive at a height of 18 inches, which is to be used for all text in that drawing. For the 1/4" = 1'-0" scale that same text size will now be 9 inches; half as big, since the new scale, 1/4" = 1'-0" is twice as large. Hopefully, this should all make sense to you.

With details, which may come in all scales listed, the procedure is not any different; you just have to be very careful in knowing ahead of time what scale viewport you will be using, to use the appropriate text height. This will all come with experience, and after much practice you will be able to quickly assess the necessary scale of a given detail to ensure that all aspects of it are visible and clear to whoever is reading the plans. This will be the only factor in determining the appropriate text size.

VIEWPORT SCALE	DRAWING TEXT HEIGHT			SCALE FACTOR	LT SCALE
	1/8	3/16	1/4		
Architectural					
1 1/2" = 1'-0"	1	1.5	2	8	3
1" = 1'-0"	1.5	2.25	3	12	4.5
3/4" = 1'-0"	2	3	4	16	6
1/2" = 1'-0"	3	4.5	6	24	9
3/8" = 1'-0"	4	6	8	32	12
1/4" = 1'-0"	6	9	12	48	18
1/8" = 1'-0"	12	18	24	96	36
1/16" = 1'-0"	24	36	48	192	72
1/32" = 1'-0"	48	72	96	384	144
Engineering					
1" = 10'	15	22.5	30	120	45
1" = 20'	30	45	60	240	90
1" = 30'	45	67.5	90	360	135
1" = 40'	60	90	120	480	180
1" = 50'	75	112.5	150	600	225
1" = 60'	90	135	180	720	270

Figure 18.28 – Text and Scale Standardization Chart

The chart not only shows the text height but also the dimension sizes under the "SCALE FACTOR" column. It is here that you retrieve the appropriate size (96 in the 1/8" = 1'-0" example) and enter that value in the "Use overall scale of:" text field under the Fit tab of the DDIM dialog box. Also shown is the appropriate ltscales for linetypes (36 in the case of 1/8" = 1'-0"). You will need this to ensure you can see a hidden line for example under this scale.

Most of the information contained in the chart of Figure 18.28 needs to be entered into AutoCAD in the form of text styles and dim styles. This is the job of the CAD manager upon initially setting up an architectural or engineering office's CAD standards. Yes, it is as tedious as it sounds; the chart is abbreviated and doesn't contain every scale used, so the work is substantial. Fortunately it only needs to be done once and a template is formed for all users to follow from here on out. Enforcing this usually turns out to be the hardest part of all, as the "system" is only as good as the people who follow it. You must be careful to use the proper style (usually called "Arial_.25" for example), and be sure to set that style as current while working on the relevant design or detail. The payoff is later in the process when you view the design in Paper Space. Uniformity and professionalism is the final goal.

Practice:

Step 1
Go to the **style** dialog box and create three new text styles - all Arial font. They will be set up for a final text height of ¼" and will be viewed through three viewports of the following scales:

- 1/2" = 1'-0" (therefore, size: 6")
- 1/4" = 1'-0" (therefore, size: 12")
- 1/8" = 1'-0" (therefore, size: 24")

Step 2
Go to the **ddim** dialog box and create three new dim styles. All must use the appropriate text from the above procedure (therefore the 1/2" = 1'-0" dimension style will use the Arial_.5 style of 6" height, etc). Then set the fit as shown on the chart and reproduced below.

- 1/2" = 1'-0" (therefore, Use overall scale of: 24")
- 1/4" = 1'-0" (therefore, Use overall scale of: 48")
- 1/8" = 1'-0" (therefore, Use overall scale of: 96")

Step 3
One by one, in Model Space, set each text and dimension style as current and create one set of text and one set of random dimensions next to each other as shown in Figure 18.29.

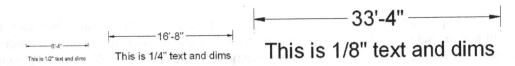

Figure 18.29 – 1/2", 1/4", and 1/8" text and dimensions

This may take you a few minutes to set up and execute. Be sure to work carefully, checking and rechecking all your settings. As expected, the three scales go up in size 2x for each jump from 1/2" to 1/4" to 1/8". We now want to see them uniform in size in Paper Space.

Go into Paper Space and create three viewports next to each other. Use the tools learned previously to set the first one to the 1/2" = 1'-0", the second to 1/4" = 1'-0" and the third to 1/8" = 1'-0". Pan around until the appropriate text/dims combination is in the respective viewport (you may have to go to Model Space and move them apart slightly). The final result is shown in Figure 18.30.

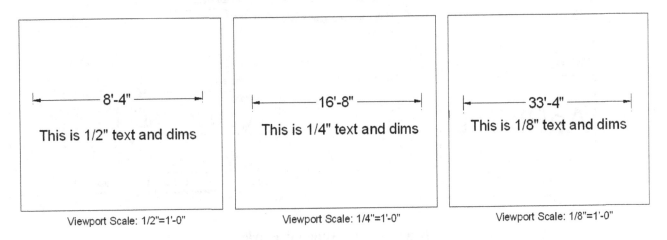

Figure 18.30 – Paper Space text and dims uniformity

Notice the complete uniformity of the text and dims in Paper Space (as opposed to how they looked in Model Space of Figure 18.30. This is the essence of what we are doing here. If you can set up these 3 viewports and all the text/dims correctly you can do any drawing. Be sure to review and learn everything thus far presented. Ask your instructor for assistance if needed.

> **Annotation**

This final Paper Space feature we will briefly discuss is relatively new and first appeared in AutoCAD 2008. It's called annotation, and its purpose is to present an alternate approach to the methods of the previous section "Text and Dims." The new approach is different in that it allows you to set multiple scales for the same text so the proper scales are visible in the proper viewports with no further input from the user. Several items have the ability to be annotated – namely styles, dimensions, blocks and hatches. We will explore text as outlined in the following procedure, but it is similar with dimensions.

Open a brand-new file, type in **style** and press Enter. The familiar style dialog box will come up. Create a new style, calling it **Arial_Anno**, font **Arial**. Check off the Annotative box under Size and leave the Paper Text Height as **0**. The new text will have a triangular symbol next to it as seen in Figure 18.31. Be sure to set it as current.

Now switch to Model Space and create a line of text. The first time you do this a dialog box such as shown in Figure 18.32 will pop up. Set the scale at which you anticipate this text to show up in (such as inside a ¼"=1'-0" viewport as done here). Go ahead and set the previously mentioned scale and press OK. Finish typing the text.

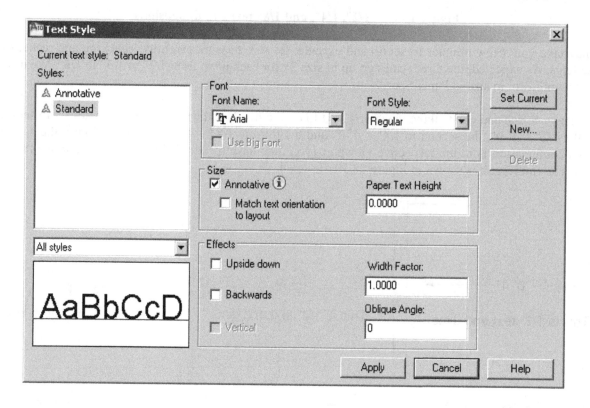

Figure 18.31 – Annotative style

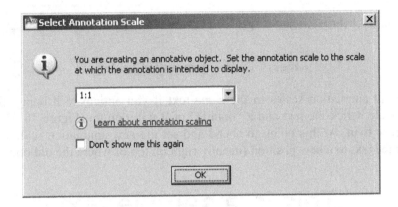

Figure 18.32 – Select Annotation Scale

Now switch to Paper Space and create a viewport and set it to the 1/4" scale. Expand or contract it to fit nicely around the text as shown in Figure 18.33.

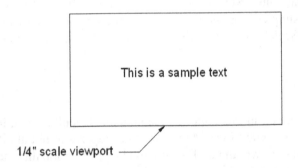

This is a sample text

1/4" scale viewport ——————

Figure 18.33 – Annotation text in sample viewport

Now copy that viewport to the right and change its scale to 1/8". The text will briefly shrink to a smaller size then balloon back up to the same size you set (1/4") as shown in Figure 18.34.

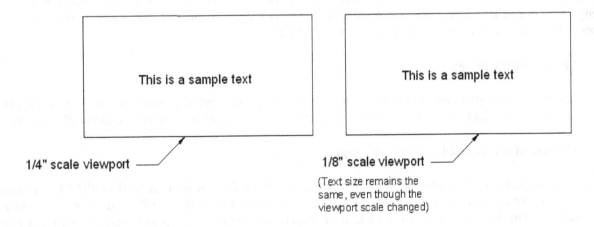

This is a sample text

This is a sample text

1/4" scale viewport ——————

1/8" scale viewport ——————

(Text size remains the same, even though the viewport scale changed)

Figure 18.34 – Annotation text in sample viewports

This is the essence of Annotation. The annotation scale is adjusted at the bottom right of the taskbar as shown in Figure 18.36, so when new text is being created you need to set its anticipated viewport scale right away.

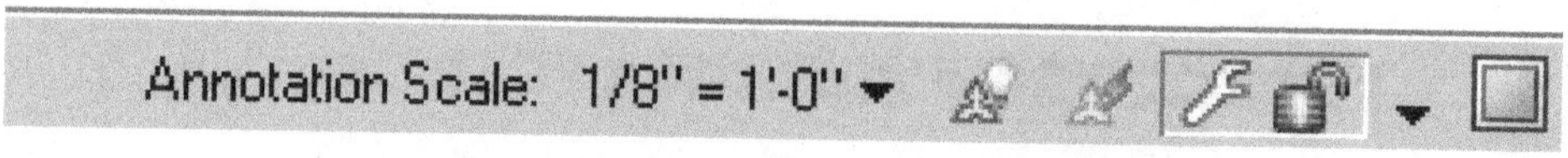

Figure 18.35 – Annotation Scale

You can also set different annotation scales to the *same* text if you anticipate it being visible in two or more viewports of different scale. Click the text and a "double" of it will pop up (Figure 18.37). This double is not a true copy but a temporary twin. At this point go ahead and set the new annotation scale from the pop up list in Figure 18.36 and move the text to a new position (usually right on top of where the old one was).

Figure 18.36 – Annotative text copy

Everything done here with text can be done with dimensions and the other items listed previously. Experiment with this tool to see exactly what AutoCAD's designers had in mind. It is not necessarily easier than the previous method but may be somewhat simpler – opinions vary. Note the Annotation Visibility option (scale symbol with light) next to the Annotation Scale list. This will show all objects in all scales; the symbol next to it is for automatic scale addition.

> **Summary**

We've gone over a lot of information in this Paper Space chapter. Hopefully you have absorbed all of it and will be able to use this valuable and important tool. Many use it improperly or not at all, and being an expert in Paper Space will set you apart from the crowd and add a true level of professionalism to your designs. This summary will serve as a capstone to the chapter, and present the big picture once again, now that the details have been mastered.

Don't think of Paper Space as a rigid or fixed system where buttons are pushed and results pop out. While there is an element of this present, the proper way of looking at it is as a flexible and loose collection of tools that allow you to shape and mold the presentation of your design in a way that is efficient, clear, professional and of course pleasing to the eye. There is no ONE way to do things. Rather there is a constant tug of war between (sometimes conflicting) requirements. For example, you have to consider:

- **The size of the design**

    This will be the first and foremost concern. Your design takes priority over everything else, so focus on only that in Model Space, creating your architectural layout or engineering device as per client need.

- **The size of the available plotter and paper**

    After the design is completed, output is considered. What do you have to work with? Most companies have a plotter capable of 36" x 24" plots. Some have the larger 48" × 36" plotters. Know your paper options. The larger plotters will take E sized paper, the smaller only up to D size. You cannot print on what you don't have unless you outsource to a print shop.

- **The scale needed to clearly see all facets of the design**

   This is the next important step. Take a look at how your design will look on the D or E sized sheets. Can you see everything? Are any parts of the design not visible clearly? Generally a 100 ft × 75 ft building (such as the one used as an example in this chapter) will fit nicely on a D sized sheet at ¼" = 1'-0" scale. If it is bigger, then you can either switch to an E sized sheet (if possible) or switch to another scale. Another scale however will make the small details of the design hard to see, so an option is to stay at the previous scale and "split" the drawing into halves or quarters and use match lines for alignment.

As you can see there are many variables in setting up Paper Space properly, and with experience all this will be in your head at the start of a project and you will know instinctively what paper, scale and viewports to go with based on the size and expected overall shape of the design.

Level 2 Drawing Project (8 of 10) – Architectural Floor Plan

Here you will finally put everything together and produce the four plans you've been working on, combined with you new Paper Space knowledge. You will need to go to each tab (i.e.; Architectural Layout, Electrical Layout, etc.) and create an mview window. Then freeze all layers that aren't necessary for this view. Finally, fill in any information you would like in the titleblock. We will return to the titleblock in the next chapter and add in Attributes. Your output for the Architectural Layout should look similar to the following image. The rest of the layouts are similar, with their respective layers showing.

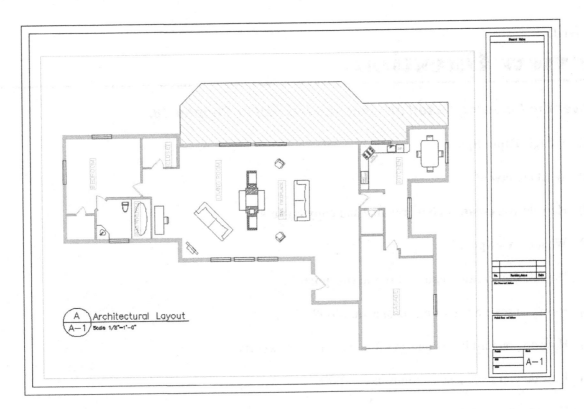

Chapter 18
Summary

You should understand and know how to use the following concepts and/or commands before moving on to Chapter 19.

- **Paper Space**

- **Layouts**

- **Viewports**

- **Scaling a design inside of a viewport**

- **Working with layers in individual viewports**

- **Scaling text and dims in viewports**

- **The annotative concept**

Chapter 18
Review Questions

Answer the following based on what you learned in Chapter 18.

1) What is **Paper Space**?

2) What are **layouts**?

3) How do you **create, delete, rename** and **copy layouts**?

4) What are **viewports**?

5) How do you **create, scale** and **shape viewports**?

6) How you scale a design inside of a viewport?

7) How do you turn layers on and off in individual viewports?

8) How do you scale text and dims in viewports?

9) Describe the **annotative** concept.

Chapter 18
Exercises

Exercise #1 – Set up a total of five title blocks in Paper Space using the Tutorial-iMfg.dwt template file seen below:

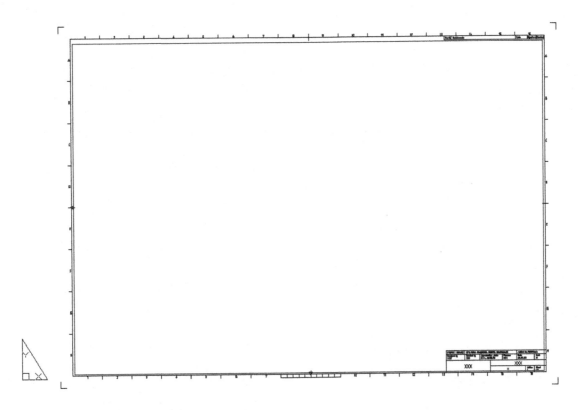

Name the tabs:

- Front View
- Side View
- Section A-A
- Section B-B
- Section C-C

Delete any extra tabs.

(Difficulty level: Easy, Time to completion: 10 minutes)

LEVEL 2

CHAPTER  19

Attributes

Mfg: Sears
Color: Black
Catalog #: 454545

Mfg: IKEA
Color: Black
Catalog #: 12345

Mfg: Sears
Color: Black
Catalog #: 454545

Mfg: IKEA
Color: Black
Catalog #: 12345

Chapter 19

Learning Objectives

In this chapter we will introduce and thoroughly cover the concept of Attributes. We will specifically discuss:

- The purpose of Attributes
- Defining Attributes
- Editing Attributes and Properties
- Extracting Data from Attributes
- Invisible Attributes

By the end of the chapter you will be well versed in applying this critical concept to your design work.

Estimated time for completion of chapter: 1-2 hours

Sec 19.1 - Introduction to Attributes

Attributes is an important advanced topic. An attribute is defined in AutoCAD as *information inside of a block*. The key words here are *information* and *block*. You can have a block without information – you've done this many times before - it's called a regular block or wblock, but not an attribute. You can also have information in the form of text sitting in your drawing. That is just text or mtext, but also not an attribute. However, put the two concepts together and you have what we are now about to explore; essentially an "intelligent" block, one that has useful embedded information.

What is the reason for creating attributes in the first place? Why would you want information inside of blocks, and what can you do with this information? The information itself can be anything of value to the designer or client. For an office chair in a corporate floor plan, one may want to know what make, model, color, and catalog number it may be. Then, if there are a variety of chair types for this office, the designer can keep track of all of them. Another example is a list of connections going into an electrical panel. It would be nice to know what and how many. All this information can be entered into the attribute and displayed on the drawing if one wishes.

However, this is only half the story and displaying information is not the main reason for creating attributes. As a matter of fact, this information is often completely hidden so as not to clutter up the drawing. So what's the main reason then? The key is that this information can be *extracted* out of the attributes and sent to a spreadsheet (among other destinations). Now you have a truly useful system in place, where extensive information is hidden unobtrusively in the drawing, yet can be accessed if needed at the touch of a few buttons, and put into an Excel file or other very useful form for cost analysis, or just plain old keeping track and organizational purposes.

You may have already heard of these tools in use, or envisioned a need for them in architecture with such items as doors and windows schedules. The software keeps track of each door and window in the drawing and links the information to a database that is updated as you add or remove the objects from your design. While basic AutoCAD doesn't do this automatically, other add-on software does. Underneath it all is the concept of embedded attributes, similar to what we are about to discuss.

You should have a good understanding of the preceding few paragraphs. As always, the button pushing is far easier once the concept is crystal clear. We will now go ahead and do the following:

- Create a simple design
- Create a basic attribute definition
- Make a block out of the simple design AND the attribute definition
- Extract information out of our new attribute

The information will be extracted via a multi-step process to two locations, an Excel spreadsheet and into a table. In either case the information will be in an organized tabular form and while not linked bi-directionally, will be an exact reflection of drawing conditions. Let's give it a try.

Sec 19.2 - Creating the Design

We don't need anything complicated to learn the basics here. Go ahead and draw a simple chair made up of several rectangles and rounded corners as shown in Figure 19.1.

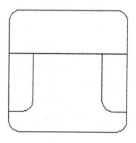

Figure 19.1 – Basic chair design

We would like to embed the following information in it: *manufacturer*, *color* and *catalog number* and put these three items just above the chair. Let's first add some text, a header of sorts, indicating what the three categories are, as seen in Figure 19.2. Use regular text of any appropriate size, font and color (blue in this case).

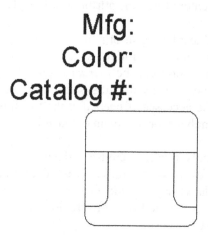

Figure 19.2 – Basic chair design with text

This is not yet the attribute definition, merely the name of the categories so we know what we are looking at. You don't have to do this, but it helps clarify things initially as the attributes will just be plain "Xs". You now have a simple design and the start of some useful information. We're now ready to make an attribute definition.

Sec 19.3 - Creating the Attribute Definitions

Here we will define the new attributes that will contain the actual information. The idea for now is to simply assign them all a value of "x" and change the x to meaningful data later in the process; so what we are doing is creating *generic definitions*, which will allow us to create multiple sets of data from one template.

Keyboard: Type in **attdef** and press Enter.
Cascading menus: **Draw→Block→Define Attribute**
Toolbar icon: none
Ribbon: **Insert tab→** Define Attributes

Start up the attribute definition via any of the above methods. The following dialog box will pop up as shown in Figure 19.3.

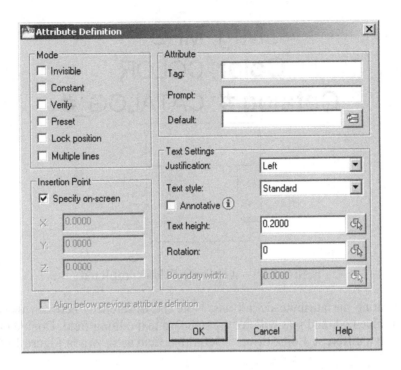

Figure 19.3 – Attribute Definition

We are only interested in the upper right of the dialog box for now. Under the Attribute category let's focus on "Tag:", "Prompt:" and "Default:".

For:

- Tag – type in **Mfg** for Manufacturer (this is the actual category)
- Prompt – type in **Please enter manufacturer** (this is just a reminder for the user)
- Default – type in **X** (this is the generic placeholder)

…as seen in Figure 19.4.

Figure 19.4 – Attribute fields

The rest of the information in the dialog box is to be left alone for now, though we will discuss a feature under Mode (upper left) later in the chapter. Insertion Point and Text Settings can be left as is for now (though you should try to match the attribute size the size of the blue header text).

Press OK and your attribute definition will appear attached to your crosshairs. Go ahead and place it next to the "Mfg" text. Note that it will be in capital letters by default. Go ahead and go through the same procedure for the remaining two categories, changing the Tag value and slightly modifying the Prompt. The default will remain as X. Note also that you cannot have blank spaces in the tag, hence the CATALOG_# underscore. The final results are shown in Figure 19.5.

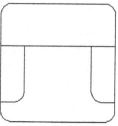

Mfg: MFG
Color: COLOR
Catalog #: CATALOG_#

Figure 19.5 – Attribute fields completed

We are almost done creating the attribute definitions. Note the differences between the text and the attribute. Double click the regular blue text, and you will get the standard text editing field. Double click the black attribute definition and you will get a different, Edit Attribute Definition field as shown in Figure 19.6.

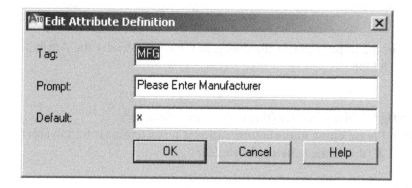

Figure 19.6 –Attribute editing fields

At this point it may not be obvious why we have two identical sets of categories next to each other or why we entered the X. After this next step what we're doing should become much clearer.

Sec 19.4 - Creating the Attribute Block

This next step should be familiar; create a regular block named Sample Chair via any method you prefer. Give it a name, select the objects (both the chair *and* all the text!), and pick an insertion point. Finally click OK, and the dialog box shown in Figure 19.7 will show up.

This Edit Attribute dialog box allows you to enter specific values for those placeholders (marked as X), but we will NOT do this at this point. Simply press OK and your new attribute is officially born (Figure 19.8). Notice all the previous categories are now just X and ready for customization. This is the process for creating a generic attribute. In the example of the office chairs, you would create several copies (the different types of chairs) and customize the information for each. Then copy and distribute the types as needed on the floor plan as we will do next.

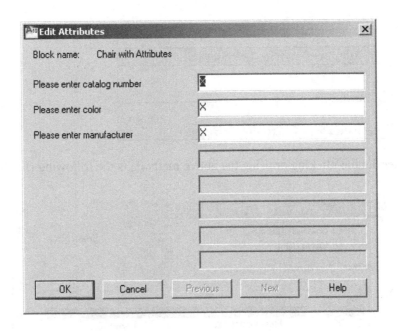

Figure 19.7 – Edit Attributes dialog box

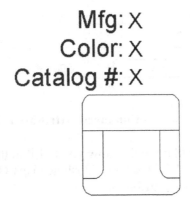

Figure 19.8 – Completed attribute

Sec 19.5 - Attribute Properties and Editing

Now that you have a block with attributes, you need to learn how to edit its fields in order to add or delete information from the attribute as well as change its color, font and just about any other property. Some of these techniques overlap each other and duplicate effort, so we will need to clearly state what is used for what.

First of all, if you only want to change the *content* of the attribute (in other words, what it says) you will need to get back to the dialog box shown in Figure 19.7. It is basic and simple, but won't change fonts and colors. Also, the only way to get to it is by typing. The command is called **attedit**, and will take care of the content. Indeed this is the way you should "populate," or fill in, the fields once we start copying the chair. Try it right now and edit the chair to say "Sears" for the manufacturer, "Brown" for color, and "S-1234" for catalog #.

Now what if you wanted to change an attribute's properties as discussed above? Well, the method here changes slightly as you will need to deal with the Enhanced Attribute Editor. Although a full command matrix is shown next, the easiest way to get to it is just to double-click the attribute.

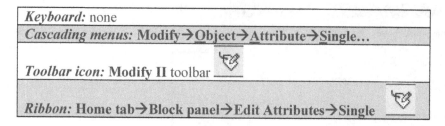

Keyboard: none	
Cascading menus: Modify→Object→Attribute→Single...	
Toolbar icon: Modify II toolbar	
Ribbon: Home tab→Block panel→Edit Attributes→Single	

What you get after either double-clicking or using the above methods is the following (Figure 19.9).

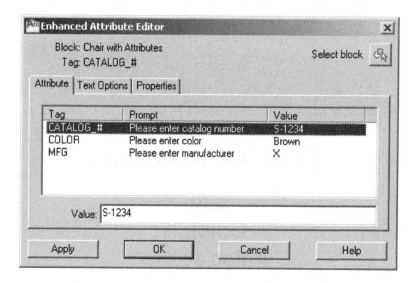

Figure 19.9 – Enhanced Attribute Editor

Here you can do quite a lot. This editor will not only allow you to fill in the fields, but also a variety of properties (colors, layers, fonts, etc.). Click through the three tabs, Attribute, Text Options and Properties to explore all the options. Here are some additional pointers on attributes.

Exploding Attributes
Attributes can be easily exploded, but this action will destroy the blocks and turn them into what you see in Figure 19.6, where you have just generic placeholders unattached to a physical design. You can change a few fields and reassemble the block, which is a perfectly valid technique, especially if the attribute is giving you a hard time. Be careful though; exploding complex attributes can suddenly fill your screen with an enormous amount of data, so keep one finger on that undo button!

Inserting Attributes
As you learned in this chapter, attributes are blocks; therefore you can insert them as needed as an alternative to copying them. When you insert a block that contains an attribute you will be prompted to fill in the X placeholders at the command line. Indeed this is how titleblocks are often filled out when a generic template block is inserted into a specific job file.

Hopefully, everything makes sense now that you see the big picture of what was done. Review the chapter up to now if not. Our next step is to learn how to extract the useful information. For this you need to duplicate the attribute to simulate having a number of them in a real drawing. Then go ahead and change some of the values. The final result is shown in Figure 19.10.

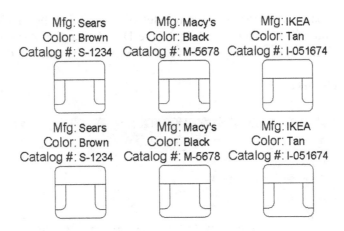

Figure 19.10 – Varied attributes

Sec 19.6 - Attribute Extraction

Extracting the information is a multistep process, necessary because of the number of options offered along the way. Our target extraction will be both an Excel spreadsheet and a table embedded in AutoCAD. Remember of course that this is a basic generic extraction and more complex spreadsheets and tables can be created to hold this data, which will be left to the student to explore.

Step (Page) 1 - Begin

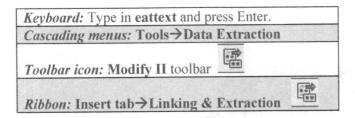

Keyboard: Type in **eattext** and press Enter.	
Cascading menus: **Tools→Data Extraction**	
Toolbar icon: **Modify II** toolbar	
Ribbon: **Insert tab→Linking & Extraction**	

Start up the data extraction process via any of the above methods. The Data Extraction – Begin (Page 1 of 8) dialog box shown in Figure 19.11 will appear.

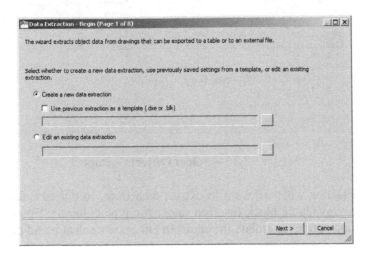

Figure 19.11 – Data Extraction – Page 1

You're going to create a new data extraction so leave the default top choice, but uncheck the "Use previous…" box. Then Press Next> and you will be taken to a File→Save Data Extraction As dialog box. Save your file as Sample.dxe anywhere you will find it easily on your drive. You will then be taken to the next page.

Step (Page) 2 – Define Data Source

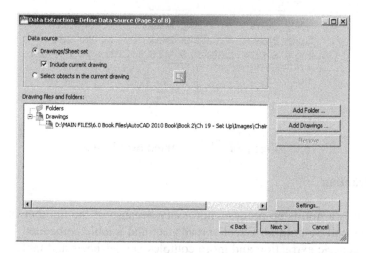

Figure 19.12 – Define Data Source – Page 2

Here we are interested in the data source, or location of the file to be looked at. Leave the default choice with the "Include current drawing" checked off. Press Next>.

Step (Page) 3 – Select Objects

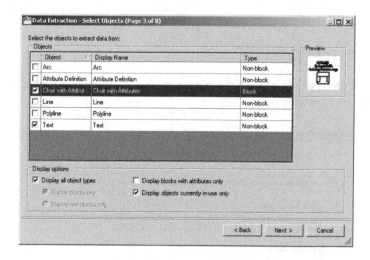

Figure 19.13 – Select Objects – Page 3

Here we are interested in selecting what we want to extract data from. In our case there is only one choice, the sample chair, as it is the only available block (as seen under the type column). You can also click on "Display blocks only" under the Display options to isolate the chair. In either case select it and press Next>.

Step (Page) 4 – Select Properties

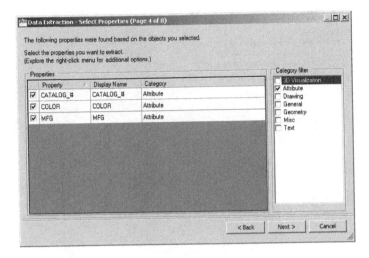

Figure 19.14 – Select Properties – Page 4

Here you will initially see many property choices. Use the category filter on the right to uncheck all the choices except Attribute. You will then easily see the three properties on the left that we created earlier. Check off all three and press Next>.

Step (Page) 5 – Refine Data

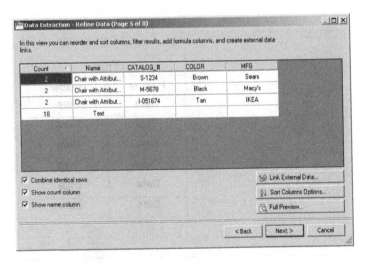

Figure 19.15 – Refine Data – Page 5

This step is for you to refine the data if needed. It contains some interesting features such as Link External Data… (to tie in this output to another Excel file) as well as some tools for adjusting and tweaking the columns. Much of this and more can be done in Excel itself after extraction is completed, but can be useful to adjust the look and content of the table extractions as tables are less flexible. If you don't want to change anything, press Next> for the final few steps.

Step (Page) 6 – Choose Output

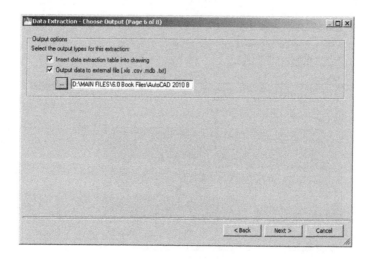

Figure 19.16 – Choose Output – Page 6

Here we are interested in our destination output. Although the Excel spreadsheet is the one that gets all the attention, you can also extract to a table that will be inserted into the drawing itself. Tables were first introduced in Chapter 15, so review that section if necessary. Check off both boxes if you haven't done so yet and finally select the location of the Excel file by using the three dot browse button. When ready, press Next>.

Step (Page) 7 – Table Style

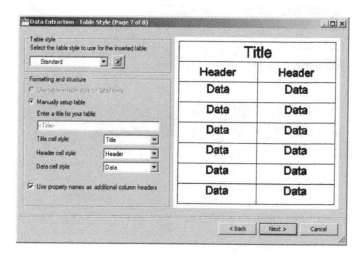

Figure 19.17 – Table Style – Page 7

We will not do anything to the table, leaving it as is, but you should review the options available and press Next> when done.

Step (Page) 8 – Finish

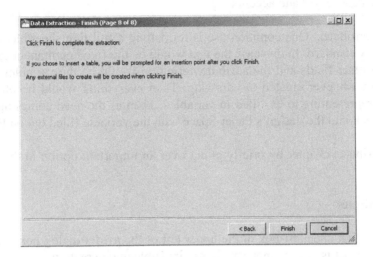

Figure 19.18 – Finish – Page 8

On the last and final page there isn't much to do except read what AutoCAD is telling you and press Finish. You will be asked for an insertion point and notice the table attached to the mouse. You may have to zoom in to see it better. Finally when you can see it, click to place the table. A table will be inserted into your drawing as shown in Figure 19.19.

Count	Name	CATALOG_#	COLOR	MFG
2	Chair with Attributes	I-051674	Tan	IKEA
2	Chair with Attributes	M-5678	Black	Macy's
2	Chair with Attributes	S-1234	Brown	Sears
18	Text			

Figure 19.19 – Table insertion (AutoCAD)

It isn't fancy but it gets the information across. The main output, however, is the Excel file. Look for it wherever you may have dropped it off and double-click on it to start up Excel. Then after some cleanup (cell borders, column width, fonts, colors, etc.) select the cells and Copy/Paste it into AutoCAD, as you've done in Chapter 16.

Count	Name	CATALOG #	COLOR	MFG
2	Chair with Attributes	I-051674	Tan	IKEA
2	Chair with Attributes	M-5678	Black	Macy's
2	Chair with Attributes	S-1234	Brown	Sears
18	Text			

Figure 19.20 – Table insertion (Excel)

This in essence is Attribute Data Extraction. You should hopefully recognize the value in what you did and the potential it holds for easy data access and accounting.

There are other uses for attributes. One common use is formatting a full title sheet/block with generic attributes and saving it as a company standard. In this case the text would be the fixed permanent categories such as "date", "sheet", "drawn by" and other fields and the attribute definition would be the variables such as the actual date, sheet # and the initials of whoever created the drawing. Then everything would be blocked out and saved for future use, with the X's representing to be filled-in variables. Then as the need comes up for a new title block it would be inserted as a block into the design's Paper Space with the variable filled out on the spot.

We will conclude the attributes chapter by briefly going over an important option in the **attdef** command dialog box.

Sec 19.7 - Invisible Attributes

This option appears as the first checkoff box in the upper left of the **attdef** dialog box, and is the most important of all the ones in that column. It is used to make attributes invisible upon creation.

Now why would you do this? As mentioned at the beginning of this chapter, this is a relatively common way of creating attributes for the simple reason that you don't need to see the text and the attribute information. If you know what you created, then there's no need to clutter up the drawing with that list over and over again. This is especially true if there is a great deal of information (electrical termination panels come to mind) and revealing all would create a dense blanket of text on the drawing, obscuring everything else.

To use the invisible option simply start the attribute creation process with **attdef,** check off the invisible option, then proceed as outlined before. The attribute will of course remain visible until the block is created and disappear afterwards. To see what you have there is no problem: simply double-click the block or type in **attedit** as outlined previously.

As you can see, being invisible for an attribute is no problem at all as you can easily access it for editing if needed, and it's unobtrusively out of the way until then; best of both worlds. The only downside of course is that you don't know that the attribute is there if you were not the one who set it up. It's always a good habit to double-click blocks just to see what pops up on an unfamiliar drawing, and indeed this is a standard investigative procedure in these cases.

Figure 19.21 shows the Invisible option checked off in the **attdef** dialog box.

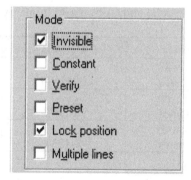

Figure 19.21 – Invisible option

Level 2 Drawing Project (9 of 10) – Architectural Floor Plan

We will continue the main project here by applying attributes to create an intelligent titleblock. This was mentioned in the main text of this chapter, but the working example employed was the chair, as that was valuable to demonstrate extraction.

Using attributes to create a company titleblock, however, is no less important an application, and many design firms create a generic set of titleblock "templates" and insert them into the Paper Space layouts of new jobs. The multitude of information can then be filled out quickly and easily. Note, however, that this info is generally *not* extracted; rather it's a convenient way to simply fill out the needed information quickly. The titleblock also remains a block, not individual lines of information.

We will take the basic existing titleblock you've created in the previous chapter and focus on creating the attributes with at least the following information:

- Name of drawing
- Number of drawing
- Drawn by
- Checked by
- Approved by
- Date
- Sheet number

Here is what the basic Architectural Layout titleblock would look like after initial setup using **attdef**.

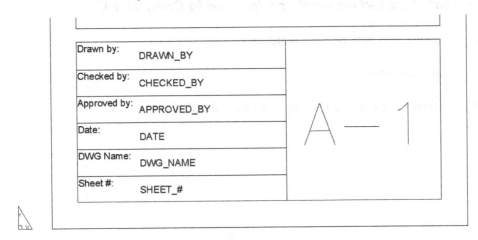

Next, you would of course make a block out of, and the blue attributes will turn to "X's". Then you can either fill them out right away or insert the titleblock into Paper Space (after deleting the old one), and fill in the data as you go.

Chapter 19
Summary

You should understand and know how to use the following concepts and/or commands before moving on to Chapter 20.

- **Attribute Definitions**

- **Attribute Blocks**

- **Editing Fields and Properties**

- **Attribute Extraction**

- **Invisible Attributes**

Chapter 19
Review Questions

Answer the following based on what you learned in Chapter 19.

1) What is an **attribute** and how do you create one?

2) How do you **edit** attributes?

3) How do you **extract data** from attributes? Into what form?

Chapter 19
Exercises

Exercise #1
   a) Draw the following doors:

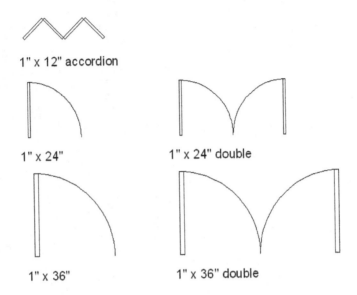

   b) Next add text attributes to each as shown by the sample below:

   1" × 12" accordion:

   Type: Accordion Double
   Size: 1×12
   Color: Natural Wood
   Mfg: XYZ Door Company
   Stock #: 654-A-1x12

   Create realistic attribute data for the remaining four doors. Vary the colors and the stock numbers.

   c) Make a block out of each one and fill in the attribute information. Copy the entire set once for a total of 10 doors.

   d) Extract the data to Excel

(Difficulty level: Intermediate, Time to completion: 30 minutes)

LEVEL 2

CHAPTER 20

Advanced Output and Pen Settings

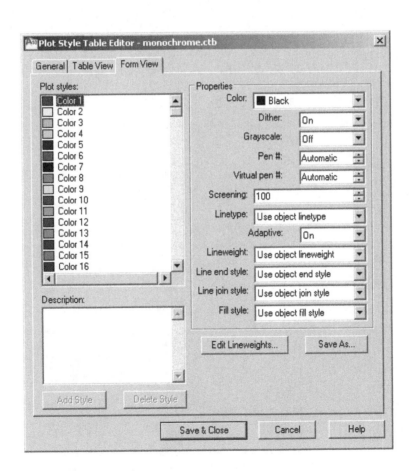

Chapter 20

Learning Objectives

In this chapter we will introduce and thoroughly cover the concept of Advanced Output, which includes the concept of the "ctb" pen settings file. We will specifically discuss:

- CTB Files and their purpose
- Editing Lineweights
- Screening
- The LWT option

By the end of the chapter you will be well versed in applying this critical concept to your design work.

Estimated time for completion of chapter: 1 hour

Sec 20.1 - Introduction to Advanced Output and Pen Settings

This final chapter of Level 2 is fittingly about the final task you have before you as a CAD manager or advanced user – creating advanced output. Although general printing and plotting has already been explored in Chapter 10 of Level 1, a crucial piece of information was omitted. This information concerns pen settings and other advanced output techniques, essential knowledge for professional looking output. The theory (and reasons why) we are doing this is outlined next, and flows directly from the time honored techniques of hand drafting.

Most designs are rather complex when everything has been added to the drawing. Lines cross other lines and text is everywhere. The eye needs to be able to quickly and easily focus on what is essential – for example the outlines of the walls of a floor plan, or the outline of the mechanical part among the many other elements in the design.

This is analogous to a mix of a rock song. Not every instrument can be the loudest, something had to give, and a balance needs to be found so the nuances and color of the music can stand out. Usually when the singer sings, other instruments have to be turned down a bit, but when a guitar solos, its volume gets turned up and so on. Otherwise it would be a loud, obnoxious mess with all sounds competing for attention.

Such is the situation with a pencil and paper drawing. All lines cannot be the same thickness and darkness. Such a monochrome mess would be confusing to look at. Draftsmen have known this for as long as the profession has been around, and as a result have numerous pencils at their disposal ranging from lightest (hard lead) to darkest (soft lead) and everything in between. They use the darkest for the primary design and the lighter leads for all else.

AutoCAD has a similar philosophy. Lines must have differences in thickness for some to stand out more than others. But how to do this as we no longer have pencil leads? The answer is use color. Colors will then be assigned line thicknesses, and assuming of course a consistent and intelligent pattern of use on the part of the designer, differences in line thickness can easily be incorporated into the design by assigning certain colors to the primary geometry. As you can see, colors serve a dual role. They allow you to not only tell apart geometry on the screen, but also on paper. The challenge then is to set up a table that assigns colors to certain line thickness, and, most importantly, stick to using these colors properly and consistently!

Sec 20.2 - Setting Standards

The very first thing you need to determine before anything gets touched in AutoCAD is what colors will be used to represent the most important geometry. You've actually done this back in Chapter 3 when layers were introduced. This of course is something done once by the CAD manager and will usually be applied to all drawings. All we are really doing is establishing a "pecking order," so to speak, for colors. This order will correspond to the most important (hence thickest) to the least important (the thinnest) pen setting. There are no hard rules here as far as *what* colors to use; it is more or less arbitrary, though convention favors some colors over others.

Let's say (based on what the author has seen over the years and what was shown in Chapter 3), that green will be the wall layer color in our architecture plan and will be assigned the thickest line weight. The windows layer is typically red and the door layer yellow. Both of those objects are almost as important as the walls, but can be set a bit thinner. The furniture and appliance layer will be magenta and thinner still. Text and dim layers will be cyan and thinner yet and finally in our hypothetical example is hatch. Those dense patterns really need to be light so as not to overwhelm the drawing, so we will assign them a grey color and make it the thinnest line weight of all.

Though this is just an arbitrary example, it's also exactly the thought process you need to go through to have a rough idea of how things will look. Once this is generally documented, you can move on to the actual settings.

Sec 20.3 - The CTB File

The ctb file is the file that needs to be modified to set the pen settings. It is where you assign pen thickness to a color, and where we will be spending our time for the remainder of the chapter. We will open up the ctb file, give it a name, modify settings and save. Then as you set up plots, you will assign the ctb file to those plots, which is a permanent, "do only once" action.

There are two ways to set up the ctb file; either by accessing it in the Page Setup Manager or as a separate action, which is what we will do. Select **Tools→Wizards→Add Plot Style Table...** from the drop-down menu. The following dialog box will appear.

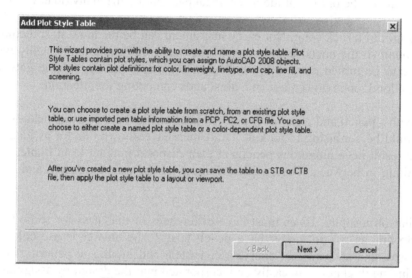

Figure 20.1 – Add Plot Style Table

Press Next> and you will be taken to the next page of this dialog box (Figure 20.2) where we you will select the first "Start from scratch" option. This will begin a new line setting procedure from the default base values.

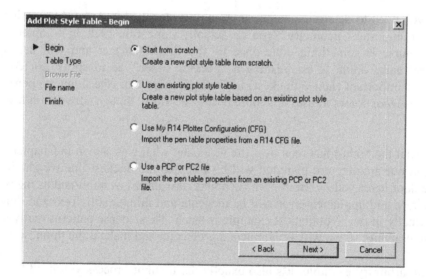

Figure 20.2 – Add Plot Style Table - Begin

Press Next> and you will be taken to the next page of this dialog box (Figure 20.3) and select the Color Dependent Plot Style Table choice. This is the only choice we will be looking at; the named plot styles will not be covered and are rarely used.

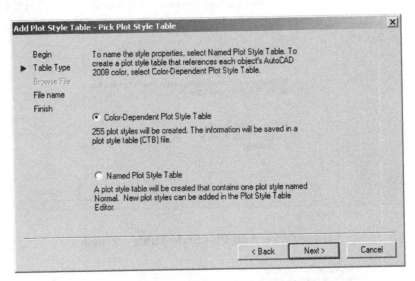

Figure 20.3 – Add Plot Style Table – Pick Plot Style Table

Press Next > and you will be taken to the next page of this dialog box (Figure 20.4). Give the new ctb file a descriptive name like StandardCTB, ColorCTB, or anything else that will be recognized.

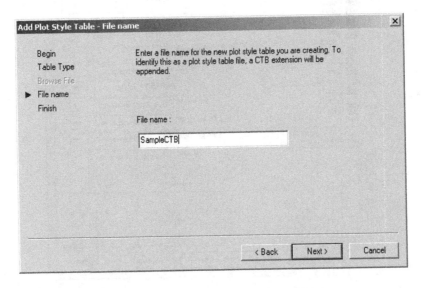

Figure 20.4 – Add Plot Style Table – File name

Press Next> for the final step (Figure 20.5). Here we are interested in the "Plot Style Table Editor…" to actually set the lineweights. Press that button and you will be taken to our final destination, the actual editor shown in Figure 20.6.

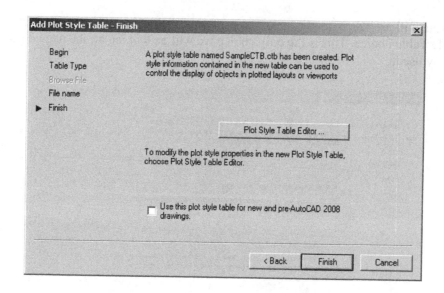

Figure 20.5 – Add Plot Style Table – Finish

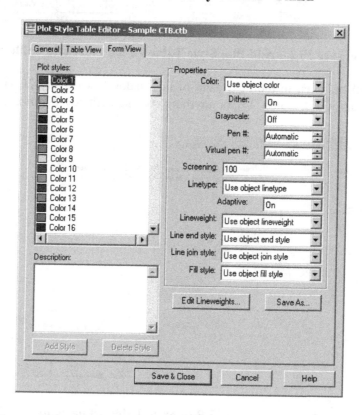

Figure 20.6 – Plot Style Table Editor for the Sample CTB file

It is here that we will set the colors, weights and a few other tricks. Stay in the default Form View (referring to the tabs on top) as we will not need the information in any other format. Familiarize yourself with the editor. On the left is the palette of the available 255 colors and on the right is a variety of effects and settings you can impart on whatever color or colors you select from the palette. All the tools are there; you just need to proceed carefully.

Step 1
The first thing you need to do is to set all the colors in the color palette to *black*. As the default settings are now, each color will be interpreted literally, which means that if you are sending your design to a color plotter, all the linework will come out in color, exactly as seen on the screen! This is a very desirable outcome in certain situations but not in most (we will talk about this more later on). For now we need to monochrome everything so we can work on line thicknesses.

- Select all the colors (1-255) in the palette by clicking your mouse in the empty white area just to the right of the colors, holding down the button and sweeping top to bottom. Holding down the shift key and clicking the first and last entry will also accomplish the same thing.

- Then as all the colors are highlighted blue, select the very first drop-down choice in Properties called "Color:" and select Black (Figure 20.7). All colors will be assigned black ink.

- Clear the palette choices by clicking randomly anywhere in the white empty area of the left side of the palette, and test out your changes by clicking random colors. They should all be black.

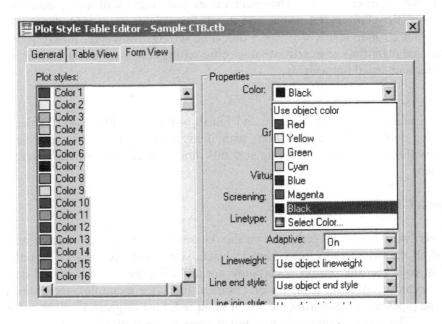

Figure 20.7 – Selecting black for all colors

Step 2
Next we need to set the actual thicknesses. This will be done in the "Lineweight:" drop-down menu, about half way down the list on the right. Currently it says "Use object lineweight" which simply means nothing is set specifically and all lineweights are the same (the default value is .2000 mm).

Here we need to discuss a major point you must understand in working with this feature. The actual line sizes are of secondary importance to the relative *differences* between the sizes. This means that you don't have to worry too much about how big .2mm or .15mm is – though that is the range of sizes that happen to look just right on paper. What is more important is that there is a large enough difference between the choices (.2mm vs. .15mm as opposed to .2mm vs. .18mm) so the eye can notice this difference. You can just as easily use .25mm and .3mm, but as you get up in the higher numbers all lines will end up being overly thick so it's best to stick to the range of .25mm on the upper end and .05mm on the lower.

So let's put this all to use and set some values:

- Click on and select color #3 (Green) – set Lineweight to .2500 mm
- Click on and select color #1 (Red) – set Lineweight to .2000 mm
- Click on and select color #2 (Yellow) – set Lineweight to .2000 mm
- Click on and select color #4 (Cyan) – set Lineweight to .1500 mm
- Click on and select color #6 (Magenta) – set Lineweight to .1500 mm
- Click on and select color #9 (Grey) – set Lineweight to .1000 mm

Now to put it all together, recall the Level 1 apartment drawing. The A-Walls layer was colored Green and will appear darkest (.2500mm) as required. The A-Doors and A-Windows layers were Yellow and Red respectively and will be slightly lighter (.2000mm), followed by the A-Text and the A-Appliances (or A-Furniture) layers, colored Cyan and Magenta respectively, lighter still (.1500mm). Last were the Grey colored hatch patterns coming in at (.1000mm).

So will these particular lineweight choices look good? The drawing will certainly look more professional and features will stand apart from each other. The exact values and colors will all be determined by the architect, engineer or designer as they inspect the results and tweak lineweights as needed. This is an extensive trial and error procedure and not all individuals will select the same settings; however, in any professional office, these techniques must be used regardless of exactly what specific settings are set. Nothing screams amateur more than a monochrome drawing with equal lineweights.

Step 3
To complete the procedure, press Save & Close and Finish in the "Add Plot Style Table – Finish" dialog box. Then open a drawing such as the Level 1 floor plan if available. Type in Plot and set the drawing up for plotting/printing as outlined in Chapter 10, being sure this time to select Sample.ctb instead of monochrome .ctb in the appropriate field.

Sec 20.4 - Additional CTB File Features

We will briefly cover two other useful features you should know of. The first one was already alluded to earlier. When you selected all the colors and turned them to black, you could have left a few out. These colors would have then plotted in color. So this action is certainly not an "all color" or "all black" deal. You can mix and match, and what is typically done in a piping or electrical plan (for example), the entire floor or site plan is monochrome, but the pipes or cables/wiring are left in vivid red, blue or green (yellow and cyan do not show up well on white paper). The intent of the design is much clearer. It's a great effect, though it can obviously be derailed by not having a color capable printer or plotter.

The second feature has to do with the screening effect. What this does is lighten any color of your choosing so less ink gets on the paper during printing. It is literally "faded out" and is a very useful trick as we will see with an example. The color most often chosen to be screened is color #9 (Grey) as it is a natural for this effect and already looks like it's been faded, although any color can be used.

To screen a color, select it in the editor as outlined just previously, and go to the Screening field (as seen in Figure 20.8). Then use the arrows or type in a value. Typically the screening effect is most useful around 30% - 40%. Then proceed as before with saving and plotting.

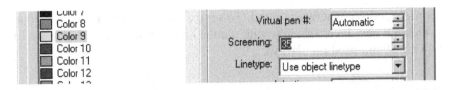

Figure 20.8 – 35% Screening of Color #9

So what is the effect like? Here are two solid fills – same grey color, one with no screening (a) meaning it's 100% and one with the 35% screening (b) as viewed through the plot preview window.

Figure 20.9(a) and (b) – Plot preview without and with 30% screening

It's a great trick to use. Here are some ideas where you can do this:

- In a *key plan*, where the area of work is grayed out and screened, yet the underlying architecture is still visible.
- In an *area of work on a demolition plan*, where the area to be demolished for a redesign is grayed out, with all the features still visible.
- As a rendering tool, to create *contrasting shading*.
- As a rendering tool to give a sense of *depth or solidness* to exterior elevations

…and many more applications…

Sec 20.5 - The LWT Option

There is an additional Lineweight (LWT) option available to you, and it is one that is often noticed first because of its visibility (found all the way on the right among your bottom push-in menu choices) as shown in Figure 20.10.

Figure 20.10 – LWT (Lineweight Option)

This is NOT the same thing as what we covered in this chapter so far. The ctb file and setting the line weights in that manner will get you only *plotted* results. This means that on-screen the lines will all remain the same size, regardless of settings, the results will be viewed on paper.

In contrast the LWT option allows you to set line thickness that will actually be visible on-screen. It's a rather interesting effect. Once set, the lines will remain at that thickness regardless of how much you zoom in or out. Opinions on this effect vary between AutoCAD users. Some think it's very useful and accurately reflects what will be printed, while others (author included) see it as a distraction and are perfectly fine remembering what colors are set to what line thickness via the ctb method.

It's up to you to experiment and see where you stand. To set the line thickness using this option you need to go to the layer dialog box, create a new layer (usually with a color) and set the value under the Lineweight header as seen in Figure 20.11 (with .50mm used for clarity).

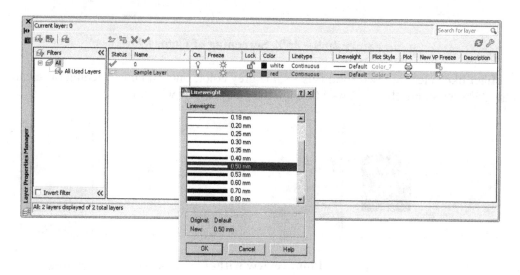

Figure 20.11 – Lineweight set to .50mm

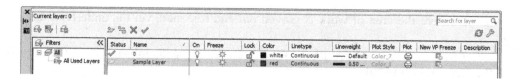

Then, after setting this layer as current, draw something. You will notice the line thin as if nothing happened. Now press in the LWT button and the line will suddenly turn thicker as seen in Figure 20.12. This is exactly how it will print. Note however that ctb settings will supersede the LWT settings, and you would leave the ctb settings as default if you were inclined to use this LWT feature instead.

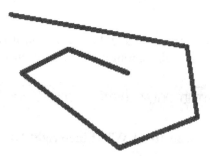

Figure 20.12 – Line drawn with the .50mm LWT setting

Level 2 Drawing Project (10 of 10) – Architectural Floor Plan

The final step of the Level 2 project will be the easiest. Your task is to do what you learned in this chapter and set up the lineweights for the floor plan. Assume the standard we have been following, and make green the darkest line, followed by yellow and red, and then all others. Print out and compare the drawing versus just using monochrome.ctb.

Chapter 20

Summary

You should understand and know how to use the following concepts and/or commands before completing this textbook.

- **CTB File**

- **Editing Lineweights**

- **Screening**

- **LWT**

Chapter 20

Review Questions

Answer the following based on what you learned in Chapter 20.

1) What is the CTB file?

2) Describe the process of setting up a CTB file.

3) What is screening?

4) What is LWT?

Chapter 20

Exercises

Exercise #1

a) Create a brand new "ctb" file. Call it Ch_20_Sample_File.
b) Monochrome all colors.
c) Assign the following thickness to the listed colors:
Color 1 – Red - .2500
Color 2 – Yellow - .2000
Color 3 – Green - .1500
Color 4 – Cyan - .3000
Color 5 – Blue – Use object lineweight
Color 6 – Magenta - .0900
Color 7 – White - .1000
Color 9 – Grey – Use object lineweight

d) Screen color 9 to 30%

(Difficulty level: Easy, Time to completion: 10 minutes)

Spotlight On: Engineering

Much as in architecture, AutoCAD finds widespread use in the various engineering professions. Though engineers have a variety of additional design and analysis software tools at their disposal (see Appendix B), there is still a need for extensive 2D drafting and some basic 3D modeling that AutoCAD is well suited for. For this "Spotlight On:" we will take a look at the various engineering specialties and how they use AutoCAD.

Engineering is defined as "the application of science and mathematics by which the properties of matter and the sources of energy in nature are made useful to people." There are dozens of engineering specialties, but the main four are generally recognized as:

- **Civil engineering** – Perhaps the oldest of the engineering specialties, civil engineers specialize in design and construction of bridges, roads, buildings, dams, canals and other infrastructure. Some principles of civil engineering were already well established during the building of the pyramids in ancient Egypt and aqueducts in Rome. Civil engineering has a number of subspecialties including:

  - Environmental engineering
  - Geotechnical engineering
  - Structural engineering
  - Transportation engineering
  - Municipal or Urban engineering
  - Construction
  - Water resources…and many others

- **Mechanical engineering** – Also one of the oldest, and today one of the broadest, engineering specialties. Mechanical engineers focus on design of mechanical systems and devices, large and small. From lawn mowers to locomotives, this discipline has designed nearly every device or vehicle you see around you. Mechanical engineering has numerous subspecialties, some of which have become major branches of engineering in their own right including:

  - Aerospace engineering
  - Naval/submarine engineering
  - Propulsion engineering
  - Robotics
  - Rail design…and many others

- **Electrical engineering** – A relatively new engineering field, it appeared in the 19th century when the secrets of the electron were being discovered. Today electrical engineers design everything from iPods and computers to massive power stations. Some of the subspecialties of electrical engineering include:

  - Electronics and microelectronics
  - Power generation and transmission
  - Controls and signals…and many others

- **Chemical engineering** – Another relatively new field that broadly deals with the process of converting raw materials or chemicals into more useful or valuable forms. Chemical engineers design, construct and operate chemical plants that use processes to create the materials we see around us. They are also involved in development of new materials and compounds used in engineering design such as composites, fibers, adhesives and much more. Chemical engineering has many subspecialties such as:

  - Biochemical and biomedical
  - Minerals and Metallurgy
  - Petroleum engineering
  - Plastics engineering
  - Textile engineering…and many others

In the Unites States, engineers typically have to attend a 4-year ABET accredited engineering school for their entry level degree, a Bachelor of Science. While there, all engineers go through a somewhat similar program in their first two years, regardless of future specialization. Classes taken include extensive math, physics and some chemistry courses, followed by statics, dynamics, mechanics, thermodynamics, fluid dynamics and material science.

In their final two years, engineers specialize by taking courses relevant to their chosen field. For aerospace engineering this would include aerodynamics, structures, controls, propulsion and orbital mechanics. For mechanical engineering that would be machine design, vibrations and heat transfer. For electrical engineering it would be circuit analysis, signals, controls, electronics, and much more. For civil engineering, classes would include advanced structures, concrete, soil, advanced materials and hydro. Finally, for chemical engineering you would take an extensive array of chemistry and processes classes.

Upon graduation, engineers can immediately enter the workforce or go on to graduate school. Though not required, some engineers, especially in civil and mechanical engineering, choose to pursue a Professional Engineer (P.E.) license. This first involves passing a Fundamentals of Engineering (F.E.) exam, followed by several years of work experience, and finally the PE exam. Average starting pay for most engineers runs in the $45,000 - $65,000 range, with the highest belonging to petroleum/chemical engineers and some electrical/computer science specialties as well as those with master's degrees.

So how do engineers use AutoCAD and what can you expect? AutoCAD finds the most widespread use among electrical and, to a lesser extent, mechanical engineers. The reasons become evident when you look at what electrical engineers deal with: mostly 2D schematics, AutoCAD's forte. Mechanical engineers have mostly moved to 3D solid modeling software such as CATIA, NX, Pro/Engineer, SolidWorks, Inventor and others, but there is still some work done on AutoCAD.

Civil engineers also have some use for AutoCAD, especially in site plan work and building design. That profession however has many dedicated 3D packages specially geared to classic civil engineering projects (bridges, road, etc.) such as MicroStation, GEOPAK and InRoads from Bentley. If AutoCAD is used, it is often paired with additional software like Civil 3D, Land Desktop or various GIS add-ons.

Aerospace, naval and chemical engineering probably has the least use for AutoCAD as it takes more high-end design software to model and test aircraft, and naval vessels. Chemical engineering projects, which are often plants, also require more sophisticated packages to track processes and complex layouts. On the next few pages we will take a look at some examples of AutoCAD use in the engineering profession.

AutoCAD layering is far simpler in mechanical and electrical engineering than in architecture. There is no AIA standard to follow, only an internal company standard, so there is room to improvise. As always the layer names should be clear as to what they contain.

In electrical engineering, you will not likely need advanced concepts such as Paper Space and Xref, but often you will need to build attributes that will hold information on connectors and wires. You will also need to have a library of standard electrical symbols handy so nothing needs to be drawn from scratch every time. The key in electrical engineering drafting is accuracy (lines straight and connected) as the actual CAD work is quite simple. A related field, network engineering, is similar. Here you may make use of plines or splines that will indicate cables going into routers and switches. Shown below is an example of an electrical schematic. Notice the extensive use of symbols for capacitors, diodes, transformers and switches.

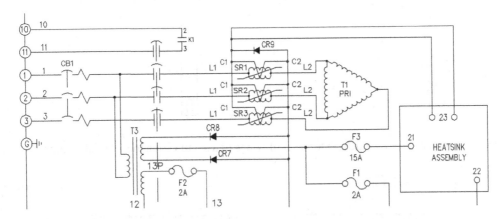

Shown below is an example of some racked equipment in a network/electrical engineering drawing. There is nothing complex in this image - just careful drafting. You've seen this earlier when "wipeout" was discussed.

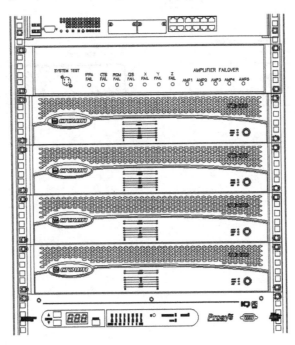

In mechanical engineering, AutoCAD is often used in situations where actual manufacture is not needed, and the company only assembles a product. Then, using AutoCAD, which has no CNC or true solid modeling ability, makes sense as the high-end products are a bit of overkill to simply show a visualization of a design. Often companies use AutoCAD when they really need to upgrade to solid modeling, and it is just a matter of time, effort and money allocated. Shown below are some basic AutoCAD mechanical parts that can be done in 2D. Nothing here is complex either, only accuracy is needed. You've seen some of these parts as exercises in previous chapters.

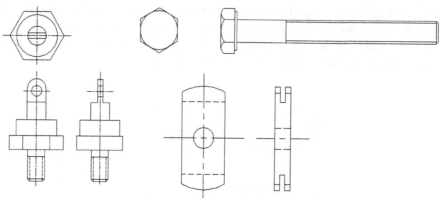

Here is a rare aerospace application of AutoCAD for a set of wings of a small Unmanned Aerial Vehicle (UAV).

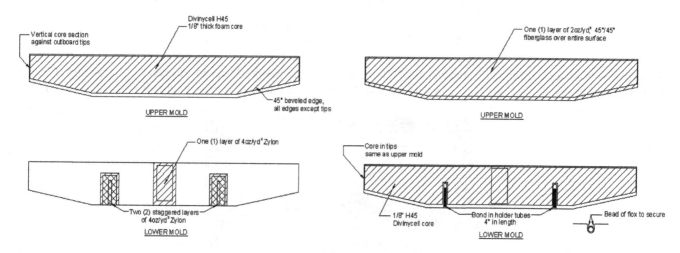

Image courtesy of DynaWerks LLC.
www.DynaWerks.com

Appendix A
Additional information on AutoCAD

Appendix B
Other CAD software and Design/Analysis tools and concepts

Appendix C
File Extensions

Appendix D
Custom Linetypes and Hatch Patterns

Appendix E
Principles of CAD Management

Appendix F
AutoLISP Basics and Advanced Customization Tools

Appendix G
PC Hardware, Printers/Plotters and Networks

Appendix H
AutoCAD Certification Exams

Appendix I
AutoCAD Humor, Oddities, Quirks and Easter Eggs

Appendix A

Additional information on AutoCAD

> ### Who makes AutoCAD?

As you may already know, AutoCAD is designed and marketed by Autodesk®, Inc. (NASDAQ: ADSK) a software giant headquartered in San Rafael, CA. Its website (a good resource center) is **www.autodesk.com**. The company was founded in 1982 by programmer John Walker and his associates, and went public in 1985 (see history). Walker was one of AutoCAD's early developers, building on original source code purchased from another programmer, Michael Riddle. Autodesk today is headed by President and CEO Carl Bass, has over 7,300 employees (down after some layoffs in early 2009) and posted revenue of $2.315 billion in FY 2009. Although the company now has more than 90 software applications, AutoCAD remains its flagship product and biggest source of revenue.

The installed base of AutoCAD (the number of software copies purchased) worldwide (as of 2007) is 3.9 million of AutoCAD and 3 million of AutoCAD LT. Due to piracy, the actual installed base is probably significantly higher. Also, the number of seats can vary depending on how you count educational versions and vertical software purchases. Either way though, these numbers indicate that AutoCAD is indisputably the world's #1 drafting software with a market share often estimated to be around 80%.

> ### What is AutoCAD LT?

Contrary to popular belief, LT does not stand for "lite" (or as we joke in class, "AutoCAD with less calories"). Instead it means "Lap Top" and was originally meant for slower and weaker early laptop computers - a sort of portable AutoCAD if you will. Today laptops are just as powerful as desktops and LT exists to compete in the lower-end CAD market with similarly priced software. It is essentially stripped-down AutoCAD. What is missing is detailed exhaustively on the Autodesk website, but it essentially boils down to no 3D ability, no advanced customization and no network licensing. In addition to what is listed above, some other minor missing features are scattered throughout the software, but the differences are NOT that great and aside from lack of 3D, it takes a trained eye to spot LT. It really looks and acts like AutoCAD, and with good reason. It was created by taking regular AutoCAD and "commenting out" (and thereby deactivating) portions of the C++ code that Autodesk wanted to reserve for its pricier main product.

Having said this, however, AutoCAD LT is an excellent product, and is recommended for companies that do not have ambitious CAD requirements. The simple truth (and something Autodesk doesn't promote) is that some users will never need the full power of AutoCAD and will do just fine with LT, while paying a fraction of the cost of full AutoCAD. Electrical engineers especially need nothing more than LT for basic linework and diagrams, as do architects and mechanical engineers who deal with schematics. The size of the files you anticipate working on, by the way, is of no consequence. LT lacks certain features, not the ability to work with multi-story skyscraper layouts if needed. It truly is a hidden secret in CAD. Autodesk must have been taken by surprise with the strong LT sales since 2002, and tried to dramatically raise prices on the software from $700 to $1,200 in 2008 and 2009, but as of the 2010 release had a change of heart (or a dose of economic reality), and prices are back down to $699.

> ### How is AutoCAD purchased and how much does it cost?

A number of years ago, you could not have purchased AutoCAD at a retail store or from Autodesk itself. You had to contact an authorized dealer (or reseller) and the sale was then usually arranged by phone, paid for with a credit card, and AutoCAD was shipped via registered mail. You can still do this, and some customers continue to, however since Release 2000 Autodesk's website was configured for direct purchase. Look under "Products", then "AutoCAD" and then "How to Buy" in the main section of the site. AutoCAD is still not available at retail outlets, although AutoCAD LT always has been, and continues to be, sold in some retail computer stores.

How much AutoCAD costs is a trickier question. If purchasing from a dealer, prices may vary widely but are generally the lower "street prices." If purchasing from the Autodesk site, however, the prices are fixed. Currently AutoCAD 2010, the latest release, is shipping (or downloading if you prefer) for $3,995. An upgrade from 2009 is $476, and higher from older versions (an incentive, and not a well-liked one, to force customers to upgrade now for fear of paying more later on). AutoCAD LT, as mentioned before, is now $699 retail. In general, use the above prices only as guidelines, they may be lower with dealers, and of course this is all subject to change.

There are also variations in cost if you are buying only AutoCAD versus one with vertical add-ons, as well as how many copies (for large volume purchases). With a dealer the purchase may come bundled with telephone or (sometimes) with an on-site support contract for any anticipated problems or training needs. Sometimes these are sold separately. Autodesk in general, like other software firms, has moved toward a subscription model where you pay a yearly fee to own AutoCAD, but upgrades are free. The subscription price for AutoCAD 2010 is $4,445, shipped or downloaded. This makes more sense to larger firms that anticipate upgrading often. If you are tasked with purchasing AutoCAD for your company, explore all your options and shop around. You also don't need to upgrade as often as Autodesk suggests!

> ### Are there significant differences between AutoCAD releases?

This is a common question at the start of training sessions. Many schools are Authorized Training Centers, and as such have the latest version of AutoCAD installed. That may or may not be the case with the students, and their respective companies. No issue seems to cause as much confusion and anxiety as this one. To put your fears to rest, if you are learning AutoCAD 2010 (this book) and have 2009, 2008, 2007 or earlier versions at home or at work, do not worry. The differences, while real, are for the most part relatively minor, and cosmetic (such as the introduction of the Ribbon in 2009). While AutoCAD has certainly been modernized, it has NOT fundamentally changed since it moved to Windows and away from DOS over 16 years ago and finally stabilized with R14 in 1997. If you were to open up that release, you would feel right at home with 90% of the features. Purists and AutoCAD fanatics may scoff, but although there have been many great enhancements, the fact remains that the vast majority of what is truly important has already been added to AutoCAD and new features are incremental improvements. Most of what you learn in one version will be applicable to several versions backwards and forwards, especially at a beginner's level.

> **A brief history of Autodesk® and AutoCAD**

The father of Autodesk and AutoCAD is John Walker. A programmer by training, he formed Marinchip Systems in 1977, and by 1982, seeing an opportunity with the emergence of the personal computer, decided to create a new "software only" company with business partner Dan Drake. The plan was to market a variety of software including a simple drafting application they acquired earlier. Based on code written in 1981 by Mike Riddle and called MicroCAD (later on changed briefly to Interact), the software would sell for under $1000 and would run on an IBM PC. It was just what they were looking for to add to their other products.

Curiously, the still unnamed company did not see anything special in the future of only MicroCAD, and focused its energy on developing and improving other products as well. Among them were Optical (a spreadsheet), Window (a screen editor) and another application called Autodesk or "Automated Desktop," which was an office automation and filing system meant to compete with a popular, at the time, application called Visidex. This Automated Desktop was seen as the most promising product and when it came time to finally incorporate and select a name, Walker and company (after rejecting such gems as "Coders of the Lost Spark") settled on Autodesk Inc. as the official and legal name.

Written originally in SPL, MicroCAD was ported by Walker onto an IBM machine and recoded in C. As it was practically a new product now, it was renamed AutoCAD which stood for Automated Computer Aided Design. AutoCAD-80 made its formal debut on November 1982 at the Las Vegas COMDEX show. As it was new and had little competition at the time, the new application drew much attention. The development team knew a potential hit was on their hands and a major corner was turned in AutoCAD's future. AutoCAD began to be shipped to customers, slowly at first, then increasing dramatically. A wish list of features given to the development team by customers was actively incorporated into each succeeding version – they were not yet called releases.

In June 1985 Autodesk went public and was valued at $70 million, in 1989 it was $500 million and in 1991 $1.4 billion. With each release, milestones large and small, such as introduction of 3D in Version 2.1 and porting to Windows with Release 12 and 13, were achieved. In a story not unlike Microsoft, Autodesk, through persistence, innovation, firm belief in the future of personal computers and a little bit of luck, came to dominate the CAD industry.

Shown below is perhaps the most famous AutoCAD drawing of all – a nozzle drawn in 1984 by customer Don Strimbu on a very early AutoCAD version. This was the most complex drawing on AutoCAD up to that point and was used by the development team as a timing test for machines and was featured on the covers of many publications including Scientific American in September 1984.

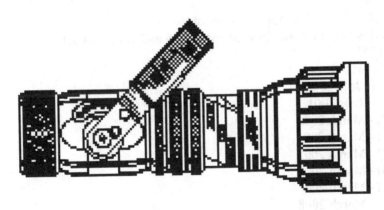

➤ **AutoCAD releases**

Autodesk first released AutoCAD in December of 1982, with Version 1.0 (Release 1) on DOS, and has continued a trend of releasing a new version sporadically every 1 to 3 years until R2004, when they switched to a more regular once per year schedule. A major change occurred with R12 and R13 when AutoCAD was ported to Windows and support for UNIX, DOS and Macintoshes was dropped. R12 and R13 were actually transitional versions (R13 is better left forgotten!), with both DOS and Windows platforms supported. It wasn't until R14 that AutoCAD regained its footing as sales dramatically increased. Shown below is AutoCAD's timeline followed by a release history.

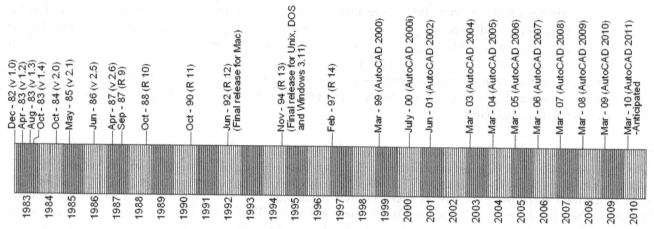

AutoCAD History Timeline

Version 1.0 (Release 1) - December 1982
Version 1.2 (Release 2) - April 1983
Version 1.3 (Release 3) - August 1983
Version 1.4 (Release 4) - October 1983
Version 2.0 (Release 5) - October 1984
Version 2.1 (Release 6) - May 1985
Version 2.5 (Release 7) - June 1986
Version 2.6 (Release 8) - April 1987
Release 9 - September 1987
Release 10 - October 1988
Release 11 - October 1990
Release 12 - June 1992 (last release for Apple Macintosh)
Release 13 - November 1994 (last release for Unix, MS-DOS and Windows 3.11)
Release 14 - February 1997
AutoCAD 2000 (R15.0) - March 1999
AutoCAD 2000i (R15.1) - July 2000
AutoCAD 2002 (R15.6) - June 2001
AutoCAD 2004 (R16.0) - March 2003
AutoCAD 2005 (R16.1) - March 2004
AutoCAD 2006 (R16.2) - March 2005
AutoCAD 2007 (R17.0) - March 2006
AutoCAD 2008 (R17.1) - March 2007
AutoCAD 2009 (R17.2) - March 2008
AutoCAD 2010 (R17.3) – March 2009
AutoCAD 2011 – Anticipated March 2010

➢ **Major Autodesk® products**

Autodesk markets 90+ software products, some major and some obscure. Only one application sold today is the company's original product – that is AutoCAD. The rest have been developed in-house from scratch over the years, purchased outright from the original developers, or as is often the case, acquired in takeovers and mergers. Many products are designed to "hitch a ride" on the back of AutoCAD, and extend its power to serve a niche market with additional functionality and menus not found in the base product. Those are often home-grown applications and are referred to as "verticals." In other cases Autodesk markets radically different software such as one used for high-end animations and cartoons. Those have been acquired to extend the company's reach into new markets.

The following lists Autodesk's flagship products that you should know about, as you may come across them in the industry. Also listed are the popular add-on products that are purchased with AutoCAD.

Flagship products:

- **AutoCAD** – Autodesk's best selling and biggest product for Computer Aided Design.
- **AutoCAD LT** – A limited functionality version of AutoCAD for an entry-level market.
- **3ds Max** – A powerful high end rendering and animation software product.
- **Autodesk AliasStudio** – High end surface modeling, sketching and illustration product.
- **Inventor** – A mid-range parametric solid modeling and design package for engineering.
- **Revit** – High end parametric software for architectural and building design.
- **Maya** – Very high end animation, effects and rendering software.

Add-ons to AutoCAD (Verticals):

- **AutoCAD Architecture** – Additional functionality for architecture design applications.
- **AutoCAD Civil** – Additional functionality for civil engineering applications.
- **AutoCAD Electrical** – Additional functionality for electrical engineering applications.
- **AutoCAD Land Desktop** – Additional functionality for land/civil/environmental design.
- **AutoCAD Mechanical** – Additional functionality for mechanical engineering.
- **AutoCAD P&ID** – Additional functionality for creating Piping & Instrumentation Diagrams.

These are just some of the major software applications in Autodesk's inventory. A significant number of products, including all the less known or obscure ones, have been omitted. A full and comprehensive list of all of them can be found on the Autodesk website. You're encouraged to look them over to stay ahead of the curve and know what other design professionals may be using.

> ➤ **AutoCAD related websites**

Below is a listing of important and valuable websites for AutoCAD and general CAD-related information. On-line content has grown from occasionally useful to indispensable, and an experienced AutoCAD user needs to have at least the following at their fingertips when the need arises. Note that link fidelity is not guaranteed, and except for the few major ones at the start of the list, the sites may and do change!

Autodesk Inc.
www.autodesk.com
Mentioned previously, it is home-base for the makers of AutoCAD. Here you will find locations of AutoCAD dealers and training center as well as endless information on AutoCAD and other software products and services. Well worth visiting occasionally.

Autodesk User Group International
www.augi.com
The Holy Grail for all things related to AutoCAD. This site is the officially sanctioned user community for Autodesk and its products. AUGI puts out numerous publications and runs conferences and training seminars all over the United States and worldwide. Its website is packed full of tips, tricks, articles and forums where you can ask the community your hardest AutoCAD questions. Claiming over 100,000 members, AUGI has the distinction of being the only organization that can officially submit "Wish Lists" directly to Autodesk, meaning that if you are a member and have an idea for a new feature or how to improve AutoCAD, they just may listen to you.

Cadalyst Magazine
www.cadalyst.com
A great magazine dedicated to all things CAD including solid modeling and rendering, but with a large amount of coverage dedicated specifically to AutoCAD. A must read for AutoCAD managers, the magazine features informative articles on all aspects of the job, including programming, industry trends and advanced topics. Subscription is free and can be applied for on-line; highly recommended. In 2009 the magazine went "all digital" pending further notice, so check the website for the latest.

CAD Block Exchange Network
http://cben.net
Many sites post helpful articles; this site posts actual AutoCAD files. Blocks, blocks and more blocks! If you need it and really don't want to draw it, check here first. Great resource for people, trees and vehicle blocks for civil and landscape engineering, in plan, elevation and 3D, though just about everything else can be found there too.

AutoCAD blocks resource site
www.autocadblock.com
With lots of free, and for purchase, blocks in every category, this is a useful and growing site, if a bit too commercial. However, if you're looking to spend some money on entire libraries of blocks, this is the place to go.

Lynn Allen's blog
http://lynn.blogs.com
You may have never heard of Lynn Allen if you're just starting out with AutoCAD, but hang around long enough, you probably will. She is what the AutoCAD community refers to as an Autodesk Technical Evangelist, or an expert on all things AutoCAD. Though there are certainly many experts around, what sets Lynn apart is the fact that she actually works for Autodesk and gets to learn AutoCAD directly from the developers and her blog is the place to go to when a new release is about to hit the market. She will likely have already analyzed it and written a book by then. Lynn is also a prolific speaker appearing at just about every AutoCAD conference and has been a columnist for Cadalyst for many years as well. She writes and speaks well and employs humor in her approach to thorny technical issues.

AutoCAD Lessons on the Web
www.cadtutor.net
A well organized and extensive site for learning AutoCAD which also includes some tutorials on 3d max, VIZ, Photoshop and web design.

The Business of CAD – Part I
www.upfrontezine.com
The website describes itself thus: "Launched in 1995, the upFront.eZine e-newsletter is your prime independent source of weekly business news and opinion for the computer-aided design (CAD) industry". The site is a no-nonsense, cut through the hype source for CAD (mostly AutoCAD but others too) information, latest releases and analysis of Autodesk business decisions; a great resource for a CAD manager.

The Business of CAD – Part II
http://worldcadaccess.typepad.com
A spin-off blog from the above site, it also addresses news in the media about all things CAD and how it will affect you; recommended for the serious CAD manager or user.

John Walker's site
www.fourmilab.ch
This site, while not essential AutoCAD reading, is still interesting, fun and often quite informative. As co-creator of AutoCAD, former programmer John Walker has put together THE definitive early history of how it all began. Fascinating reading from a fascinating man who is now "retired" in Switzerland and pursuing Artificial Intelligence research. His site is chock full of readings on numerous subjects as you would expect from a restless and inquisitive mind; truly someone who changed computing history in a significant way.

As with many topics in this brief appendix, what is listed here only scratches the surface. Remember that you are not the only one out there using AutoCAD so feel free to reach out and see what else is out there if you have questions or would just like to learn more. At a minimum it is recommended to join AUGI and get a subscription to Cadalyst.

Appendix B

Other CAD software and Design/Analysis tools and concepts

While it is true AutoCAD is the market leader for drafting software, it certainly doesn't operate in a vacuum and is not the only game in town. The list below summarizes other major players in the drafting field. The reason you should know about them is because you may run into them in use, and may have to exchange files if collaborating. Saying "I've never heard of that" is usually not a good way to start a technical conversation in your area of expertise, so take note of the following products.

MicroStation
www.Bentley.com

A major (and really the only serious) competitor to AutoCAD, it is produced and marketed by Bentley Systems. For historical reasons the software is widely used in government agencies and in the civil engineering community. It comes in full and PowerDraft variations (akin to AutoCAD and LT). The current version is MicroStation V8 XM. Its file extension is DGN. AutoCAD does not open these files directly, though that may soon change. MicroStation, on the other hand, does open AutoCAD's DWG files. There are of course many techniques and even third party software to quickly and easily convert one file format to another.

MicroStation is more icon-driven, but otherwise similar to AutoCAD in many ways. Its 3D capabilities are considered excellent, but its interface can be a bit cumbersome to use for a novice. Its main disadvantage is that it is not an industry standard; it is otherwise excellent software, and its ability to handle Xrefs (multiple external file attachments) is unparalleled. Its users are a sometimes fanatical bunch, promoting its virtues against AutoCAD at every opportunity. An objective view of both applications yields an opinion that both have their pros and cons, and anyone that disparages one over the other usually doesn't know what they are talking about.

ArchiCAD
www.graphisoft.com

A popular application among architects, geared toward the Macintosh, but also available for Windows. Developed by a Hungarian firm called Graphisoft at around the same time as AutoCAD, ArchiCAD took a different approach, allowing for a parametric relationship between objects, as well as easier transitions to 3D, as all objects are created inherently with depth. The software has a loyal installed user base of about 100,000 architects worldwide and can interoperate fully with AutoCAD files. The current release is ArchiCAD 11.

TurboCAD
www.turbocad.com

A mid-range 2D/3D drafting software, first developed and marketed in South Africa in 1985, and brought to the U.S. the following year. It was marketed as an entry level basic drafting tool (once sold for as low as $49), but with constant upgrading and development TurboCAD has grown over the years to a $1200 - $1500 software product for the architectural and mechanical engineering community. It operates on both Windows and Macintosh. The current release is TurboCAD 15.

Some other notable CAD applications include **RealCAD**, **VectorWorks** (formerly MiniCad), **IntelliCAD**, **Form Z** and **Rhino**, though the last two are often used only for modeling and visualization. This list is not exhaustive by any means, as there are literally hundreds of other CAD programs available.

Many of my students are interested in other aspects of design, and questions often come up as to how AutoCAD fits into the "big picture" of a much larger engineering/design world. To address some of these questions, the following information, culled from a chapter written in 2007 as part of an undergraduate Machine Design Theory workbook, is included. AutoCAD students, especially engineers, may find this informative. The subject matter is extensive and this brief overview only scratches the surface. Hopefully, if the reader was not aware of these concepts and the engineering design/analysis software described, this will serve as a good introduction and a jumping off point to further inquiry.

- **CAD** stands for *Computer Aided Design*. It is a broad generic term for using a computer to design something. It can also be used in reference to a family of software that allows you to design, engineer and test a product on the computer prior to manufacturing it. CAD software can be used for primarily 2D drafting (such as with AutoCAD), or full 3D parametric solid modeling. The latter is the software that aerospace, mechanical, automotive and naval engineers commonly use. Examples include CATIA, NX, Pro-Engineer, IronCAD and SolidWorks among others. Included under the broad category of CAD is testing and analysis software.

- **FEA** stands for *Finite Element Analysis* and is used for structural and stress analysis, testing thermal properties and much more as related to frames, beams, columns and other structural components. Examples of FEA software include NASTRAN, ALGOR, ANSYS and a few others.

- **CFD** stands for *Computational Fluid Dynamics* and is software used to model fluid flow around and through objects to predict aerodynamic and hydrodynamic performance. Examples of CFD software include Fluent, Flow 3-D and others.

- **CAM** stands for *Computer Aided Manufacturing*. The basic idea is: OK we designed and optimized our product, now what? How do we get it from the computer to the machine shop to build a prototype? The solution is to use a CAM software package that takes your design (or parts that make up your design) and converts all geometry to information useable by the manufacturing machines so the drills, lathes and routing tools can get to work and create the part out of a chunk of metal. Examples of dedicated CAM software include Mastercam, FastCAM and many others.

- **CAE** stands for *Computer Aided Engineering*. Strictly defined, CAE is the "use of information technology for supporting engineers in tasks such as analysis, simulation, design, manufacture, planning, diagnosis and repair." We will use this term to define an ability to do this all from one platform, as explained below.

As you read the CAD/CAM paragraphs you may be wondering why CAD companies have not provided a complete solution to design, test and manufacture a product using one software package. They have, to varying degrees of success. For a number of years an engineering team designing an aircraft, for example, had to typically create the design (and subsystems) in a 3D program like CATIA, then import the airframe for FEA testing into NASTRAN, followed by airflow testing in CFD software (and a wind tunnel), and finally send each component through CAM software like Mastercam to manufacture it (a greatly simplified version of events of course).

Obviously data could be lost or corrupted in the export/import process, and a lot of effort was expended in making sure the design transitioned smoothly from one software application to another. While the final end-product was technically a result of CAE, software companies had something better in mind: "one stop shopping." As a result, CAE today means one source for most of your design/manufacturing needs. CATIA for example has "workbenches" for FEA analysis and CAM. Code generated by CATIA will be imported into the CNC machines that will (typically after corrections and editing), produce virtually anything you can design (with some limitations of course).

As you may imagine, not everything is rosy with this picture. As the specialized companies got better and better at their respective specialties (CAM, FEA or CFD), it became harder and harder for the CAD companies to catch up and offer equally good solutions embedded in their respective products. Mastercam is still the undisputed leader in CAM worldwide, and NASTRAN is still the king of FEA. CATIA's version of these has gotten better, but still a notch behind. The situation is the same with Pro/E, NX and other CAD software.

These high-end 3D packages are also referred to as **Product Lifecycle Management** (PLM) software, a term you will hear a lot of in the engineering community. As described above, these software tools aim to follow a product from design and conception all the way through testing and manufacturing all in one integrated package.

> **CNC/G-Code** stands for **Computer Numerical Code**. This is the language (referred to as G-Code) that is created by the CAM software and read by the CNC machines. G-Code existed for decades. It was originally written by skilled machinists based on the geometry of the component to be manufactured (and sound manufacturing practices). This code can still be written by hand. It is not hard to understand. Most lines start with "G", hence its name, and they all describe some function that the CNC machine does as related to the process. After some practice the lines will start to make sense and you will be able to roughly visualize what is being manufactured, just by reading them.

Here are some more specific details on the CAD software mentioned in the preceding section. You may run into many of these throughout your engineering career. It's well worth being familiar with these products. The information is up to date as of press time.

CATIA
www.3ds.com

Computer-Aided Three dimensional Interactive Application (CATIA) is the world's leading CAD/CAM/CAE software. Originally developed in France by Dassault for their Mirage fighter jet project (though originally based on Lockheed's CADAM software), it is now a well-known staple at Boeing and many automotive, aerospace and naval design companies. It has even been famously adopted by Architect Frank Geary for certain projects. CATIA has grown into a full PLM solution, and is considered the industry standard, though complex and with a not always user friendly reputation. The current version is CATIA V6 R2010.

NX
www.ugs.com

Another major CAD/CAM/CAE PLM software solution from a company formerly called UGS. This package is a blend of the company's Unigraphics software and SDRC's popular I-DEAS software that UGS purchased and took over. Siemens AG now owns UGS and the NX software. Used by GM, Rolls Royce, Pratt & Whitney, and many others. The current version is NX 6.

Pro/Engineer
www.ptc.com

The third major high-end CAD/CAM/CAE PLM software package, Pro/E is historically significant. It originally caused a major change in the CAD industry when first released (in 1988) by introducing the concept of parametric modeling. Rather than models being constructed like a mound of clay with pieces being added or removed to make changes, the user constructs the model as a list of features, which are stored by the program and can be used to change the model by modifying, reordering, or removing them. CATIA and NX also now work this way. The current version is Pro/E Wildfire 5.0.

SolidWorks

www.solidworks.com

A mid-range CAD/CAM/CAE product from Dassault, the makers of CATIA. Marketed as a lower-cost competitor to major CAD packages, the software still has extensive capabilities and has caught on quickly in smaller design companies. The current version is SolidWorks 2008.

Inventor

www.autodesk.com

Another mid-range product, from the makers of AutoCAD, for solid modeling and design. Considered a low-to-mid-range product when first introduced, Inventor has matured to become a major competitor to SolidWorks and even Pro/Engineer. Current version is Inventor 2010.

IronCAD

www.ironcad.com

Another vendor in the mid-range market. The software has gained ground in recent years due to a user friendly reputation and a more intuitive approach to 3D design. The current version is IronCAD 10.

Solid Edge

www.solidedge.com

Also a mid-range product from UGS (now owned by Siemens AG), the same company that makes the high-end NX package, it is comparable to SolidWorks in cost and functionality. Current release is Solid Edge V20.

Listed next are some of the major software products in FEA and CFD, as mentioned earlier.

NASTRAN

www.NEiNastran.com

Originally developed for NASA as open source code for the aerospace community to perform structural analysis, *NASa sTRuctural Analysis* or **NASTRAN** was acquired by several corporations and marketed in numerous versions. NEiNastran is just one (commonly used) flavor of it. NASTRAN in its pure form is a solver for finite element analysis (FEA), and cannot create its own models or meshes. Developers, however, have added pre- and post-processors to allow for this. The current release for NEiNastran is V9, though other companies have different designations for their products.

ANSYS

www.ansys.com

A major FEA product for structural, thermal, CFD, acoustic and electromagnetic simulations.

ALGOR

www.algor.com

Another major FEA product for structural, thermal, CFD, acoustic and electromagnetic simulations.

Fluent

www.Fluent.com

The industry standard for CFD software, holding about 40% of the market. It imports geometry, creates meshes (using GAMBIT software) and boundary conditions, and solves for a variety of fluid flows, using the Navier-Stokes equations as theoretical underpinnings. The company was recently acquired by ANSYS. The current release is Fluent 6.3.

Appendix C

File Extensions

File extensions are the bane of the computer user's existence. They number in the many hundreds (AutoCAD alone claims nearly 200) and even if you pare the list down to a useful few, it's still quite a collection. Here we will focus on those extensions that are important to an AutoCAD user, as chances are you will encounter most, if not all of them while learning the software top to bottom. They will be listed alphabetically (with explanations and/or descriptions), but grouped in categories so as to bring some order to the chaos and prioritize what you really need to know and what you can just glance over. After AutoCAD's list there is a further listing of extensions from other popular and often-used software, which you may find useful if you are not already familiar with most of them.

AutoCAD Primary Extensions

- **.BAK (BAcKup file)**

BAK is an AutoCAD backup file that is created and updated every time you save a drawing, provided that setting is turned on in the Options dialog box. The BAK file is the same name as the main DWG file, generally sits right next to it, and can be easily renamed to a valid DWG if the original DWG is lost.

- **.DWF (Design Web Format file)**

DWG is a format for viewing drawings on-line or with a DWF viewer. The idea here is to share files with others that don't have AutoCAD in a way that enables them to look but not modify. If this sounds like Adobe PDF, you are correct; DWF is a competing format.

- **.DWG (DraWinG file format)**

DWG is AutoCAD's famous file extension for all drawing files. This format can be read by many of Autodesk's other software products as well as by some competitors. The format itself changes somewhat from release to release, but these are internal programming changes and invisible to most users.

- **.DWT (DraWing Template)**

DWT is AutoCAD's template format, and is essentially a DWG file. The idea is to not repeat setup steps from project to project and use this template. It is equivalent to just saving a completed project as a new job and erasing the contents.

- **.DXF (Drawing eXchange Format)**

DXF is a universal file format developed to allow for smooth data exchange between competing CAD packages. If an AutoCAD drawing is saved to a DXF it can then be read by most other design software (in theory anyway). This of course isn't always the case, and some CAD packages open DWG files anyway without this step, but the DXF remains an important tool for collaboration. DXF is similar in principle and intent to IGES and STEP for those readers who may have worked with solid modeling software and understand what those acronyms mean.

AutoCAD Secondary Extensions

- **.ac$** - A temporary file generated if AutoCAD crashes. Can be deleted if AutoCAD isn't running.
- **.ctb** – A color table file created when pen settings and thickness are set (Chapter 20 topic).
- **.cui** – Customizable User Interface file for changes to menus and toolbars (Chapter 14 topic).
- **.err** – An error file that is generated upon an AutoCAD crash. You can delete it.
- **.las** – Layer States file that can be imported/exported (Chapter 12 topic).
- **.lin** – A linetype definition file that can be modified and expanded (see Appendix D).
- **.lsp** – A LISP file. This is a relatively simple language used to modify, customize and automate AutoCAD to effectively expand its abilities (see Appendix F).
- **.pat** – A hatch definition file that can be modified and expanded (see Appendix D).
- **.pgp** – A program parameters file that contains the customizable command shortcuts (Chapter 14 topic).
- **.scr** – A script file used for automation (mentioned in Chapter 12).
- **.sv$** - Another temporary file for the auto save feature. Will appear if auto save is on, and will remain in the event of a crash. It can be then renamed to DWG or deleted.

Misc. Software Extensions

These extensions are for software and formats likely to be encountered daily or at least occasionally by AutoCAD users and are always good to know. Some extensions can be found in AutoCAD as Export/Import choices.

- **.ai** – Adobe Illustrator files.
- **.bmp** – Bitmap, a type of image file format that, like JPG, uses pixels.
- **.doc** – MS Word files (also WordPerfect and WordStar).
- **.dgn** – MicroStation file extension. This competitor to AutoCAD can open native DWG files. Appendix C of Level 1 discusses this software.
- **.eps** – Encapsulated PostScript file.
- **.jpg** – JPEG (Joint Photographic Experts Group) is a commonly used method (algorithm) of compression for photographic images. It is also a file format. JPGs are discussed in Chapter 16.
- **.pdf** – The famous Adobe Acrobat file extension. You can print any AutoCAD drawing to a PDF assuming Acrobat is installed.
- **.ppt** – MS PowerPoint file extension
- **.psd** – The famous Photoshop file extension.
- **.vsd** – MS Visio drawing file extension. Visio is a simple and popular "drag and drop style" CAD package for electrical engineering schematics and flowcharts among other applications. It accepts AutoCAD files.
- **.xls** – MS Excell file extension

This list is by no means complete, but covers many of the applications discussed in Chapter 16.

Appendix D

Custom Linetypes and Hatch Patterns

This appendix will outline the basic procedures for creating custom linetypes and hatch patterns if you need something that isn't found in AutoCAD itself or with any third party vendor. All linetypes and hatch patterns were originally "coded" by hand and there is nothing inherently special or magical about making them; they just take some time and patience to create. Though it's admittedly rare for a designer to need something that isn't already available, creating these entities is a good skill to acquire. We will just present the basics here and you can certainly do additional research and come up with some fancy designs and patterns.

Linetype Definitions (Basic)

All linetypes in AutoCAD (about 38 standard ones to be precise) reside in the acad.lin or the acadiso.lin file. These are Linetype Definition Files that AutoCAD accesses when you tell it to load linetypes into a drawing. In AutoCAD 2008 these files can be found in the Program Files\AutoCAD 2008\backup folder. Anything you do to these files including adding to them immediately shows up next time you use linetypes. Our goal here will be to open up the acad.lin file, analyze how linetypes are defined and then make our own.

Locate and open (usually with Notepad) the acad.lin file and take a look at it closely. Here is a reproduction of the first 6 linetype definitions Border (Standard, 2 and X2) and Center (Standard, 2 and X2).

```
*BORDER,Border __ __ . __ __ . __ __ . __ __ . __ __ .
A,.5,-.25,.5,-.25,0,-.25
*BORDER2,Border (.5x) __ . __ . __ . __ . __ . __ . __ . __ .
A,.25,-.125,.25,-.125,0,-.125
*BORDERX2,Border (2x) ____ ____ . ____ ____ . ___
A,1.0,-.5,1.0,-.5,0,-.5


*CENTER,Center ____ _ ____ _ ____ _ ____ _ ____ _ ____
A,1.25,-.25,.25,-.25
*CENTER2,Center (.5x) __ _ __ _ __ _ __ _ __ _ __ _ __
A,.75,-.125,.125,-.125
*CENTERX2,Center (2x) _____ __ _____ __ _____
A,2.5,-.5,.5,-.5
```

Let's take just one linetype and look at it closer. Here's the standard size center line:

```
*CENTER,Center ____ _ ____ _ ____ _ ____ _ ____ _ ____
A,1.25,-.25,.25,-.25
```

The two lines that you are seeing are the header line and the pattern line; both are needed to properly define a linetype. The header line consists of an asterisk, followed by the name of the linetype (CENTER), then a comma and finally the description of the linetype (Center). In this case they are one and the same.

The pattern line is where the linetype is actually described and the cryptic looking "numerical code" (A,1.25,-.25,.25,-.25) is the method used to describe it, and needs to be discussed for it to make sense.

The "A" is the alignment field specification and is of little concern (AutoCAD only accepts this type so you will always see an A). Next are the linetype specification numbers, and they actually describe the linetype using dashes, dots and spaces. Let's go over them as really this is the key concept.

The basic elements are:

- Dash – (Pen down, positive length specified. Ex: .5)
- Dot – (Pen down, length of 0)
- Space – (Pen up, negative length specified. Ex: -.25)

Now you just have to mix and match and arrange these in the order you want them to be in to define a linetype that you envisioned.

What would this look like?
(A,.75,-.25,0, -.25, .75)

Well you know that dashes are positive, so everywhere you see a .75 will be a dash of that length. Spaces are negative, so everywhere you see a -.25 will be a space of that length, and finally a 0 is a dot. This is the result:

___ . ___

This is the essentials of creating basic linetypes. AutoCAD, however, also allows for two other types to be created and they are part of the Complex Linetype family: String Complex and Shape Complex.

Linetypes (String Complex and Shape Complex)

The problem with the previously mentioned linetypes is obvious; you only have dashes, dots and spaces to work with. While this is actually enough for many applications, you may occasionally (especially in civil engineering and architecture) need to draw boundary or utility lines that features text (Fence line, Gas line, etc.) or you may want to put small symbols between the dashes. It's here that strings and shapes come in.

String Complex linetypes is how you insert text into the line. Here's a gas line example:

```
*GAS_LINE,Gas line ----GAS----GAS----GAS----GAS----GAS----GAS--
A,.5,-.2,["GAS",STANDARD,S=.1,R=0.0,X=-0.1,Y=-.05],-.25
```

The header line is as discussed before, and the pattern line still starts with an A. After the A we see a dash of length .5 followed by a space of length .2; so far it's familiar. Next is information inside a set of brackets, defined as follows:

["String", Text Style, Text Height (S), Rotation (R), X-Offset (X), Y-Offset (Y)].

Finally there is a .25 space and the string repeats again.

The meaning of the information inside the brackets is as follows:

- String – This is the actual text, in quotation marks (GAS in our example).
- Text Style – The predefined text style (STANDARD in our example).
- Text Height – The actual text height (.1 in our example).
- Rotation – The rotation of the text relative to the line (0.0 in our example).
- X-Offset – The distance between the end of the dash and the beginning of the text, as well as the end of the text and the beginning of the dash (a space of .1 in our example).
- Y-Offset – The distance between the bottom of the text and the line itself. This function centers the text on the line.

With these tools you can create any text string linetype.

Shape Complex linetypes are basically similar, aside for one difference; instead of a text string we have a shape definition inserted. Shape definitions are a separate topic in their own right and are not covered in detail in this textbook. Briefly however, shape files are descriptions of geometric shapes (such as rectangles or circles but can be more complex) and are written using a text editor and compiled using the compile command. They are saved under the *.SHP extension. AutoCAD will access these when it reads the linetype definitions. If you are familiar with shapes, then you can proceed in defining Shape Complex linetypes as with String Complex ones.

Hatch Pattern Definitions (Basic)

Knowing the basics of linetype definitions will leave you on familiar ground when learning hatch pattern definitions. Though they are more complex, the basics and styles of description are similar.

Hatch pattern definitions are found in the acad.pat and acadiso.pat files, and they reside in the exact same folder as the "lin" files. Open up the acad.pat file and take a look at a few definitions. Some of them can get long and involved due to the complexity of the pattern (Gravel for example), but here is simpler one to use as a first example; Honeycomb:

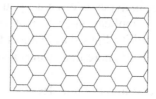

The pattern definition file that describes it looks like this:

*HONEY, Honeycomb pattern
0, 0,0, .1875,.108253175, .125,-.25
120, 0,0, .1875,.108253175, .125,-.25
60, 0,0, .1875,.108253175, -.25,.125

The first line (header line) is as discussed previously with linetypes. The next three lines are the Hatch Descriptors, and they have somewhat different definitions as shown below:

Angle, X-Origin, Y-Origin, D1, D2, Dash Length

where (in this particular example),

- Angle – Angle of the pattern's first hatch line from the x-axis (0° in our example).
- X-Origin – X coordinate of hatch line (0,0 in our example).
- Y-Origin – Y coordinate of hatch line (.1875 in our example).
- D1 – The displacement of second line (.1083 in our example).
- D2 – The distance between adjoining lines (.125 in our example).
- Length – Length of dashes and/or spaces

The pattern is repeated two more times to form the overall shape and repeated many more times in the overall hatch pattern.

Appendix E

Principles of CAD Management

So you've been hired on as a CAD manager, or promoted to this role, or more likely forced into it due to your substantial AutoCAD skills (or was it because no one else wanted the job?). Either way, you are now the top dog, head honcho...no, wait! You are just an architect or engineer and someone had to take the bull by the horns (many bulls actually, that run around and draft whatever way they please just to get the job out). Then again, maybe you're none of the above and instead simply a student finishing off Level 2. The bottom line is that you could use some ideas and a short discussion on this elusive science called CAD management. Well, you've come to the right appendix.

All jokes aside, this is an important topic. You must take everything you learned in Levels 1 and 2 and apply it, while keeping others in line. You must be the keeper and the enforcer of standards and the technical expert that can resolve any issue. A true CAD manager that does nothing else isn't all that common. You may have to multi-task and do your own work while doing the management, especially in smaller companies.

Here are a few rules, ideas and thoughts gathered together from the author's 12+ years experience as a CAD consultant, educator and CAD manager. They outline just some of the challenges and issues (and provide some possible answers) to those in charge of the software that is used to design a lot of what is around us.

Part 1 - Know the 7 Golden Rules of AutoCAD

1) **Always draw one-to-one (1:1).**
2) **Never draw the same exact item twice.**
3) **Avoid drawing standard items.**
4) **Use layers correctly.**
5) **Do not explode anything.**
6) **Use accuracy: Ortho and OSNAPs.**
7) **Save often.**

Notice that most of those rules were discussed in some way, shape or form in the first few chapters of Level 1. It's the basics people screw up, not some fancy new feature. Let's briefly go over these one by one; they bear repeating.

1) *Always draw one-to-one (1:1)* – This is the most important one. Always draw items to real life sizes. Endless problems arise if users decide to scale items as they draw them. Leave scales and scaling to the pencil-and-paper folks. A drawing must pass a simple test. Measure anything that is known to be a certain value (a door opening, for example, that is 3 ft. wide) by using the list command. AutoCAD should tell you its value as 36 inches or 3 ft, and nothing else. You must enforce this as a CAD manager.

2) *Never draw the same exact item twice* – In hand drafting just about everything needed to be drawn one item at a time. Some inexperienced CAD users do the same in AutoCAD. That is not correct! AutoCAD and computer drafting in general requires a different mindset, philosophy and approach. If you have already drawn something once, then copy it, don't create it again. While this one may be obvious, another overlooked tool is the block or wblock command. If you anticipate using the new item again (even if there is a small chance), then make a block out of it! As a CAD manager you need to sometimes remind users of this. Remember: "Time=$$".

3) *Avoid drawing standard items* – This, like the rule before, aims to prevent duplicate effort. Before starting a project, as a CAD manager, you need to see to it that any standard parts (meaning ones that are not unique but are constantly reused in the industry) are available for use and are NOT drawn and redrawn. Chances are good that you either already have these parts, or can get the "cads", as they say, from the part supplier. Another resource is the web. Sites such as www.cben.net and many others have predrawn blocks available for use. Buy them, borrow them or download them, but do not redraw items that someone already has done before. Again: "Time=$$".

4) *Use layers correctly* – Layers are there to make life easier and they are your friend! While overdoing it is always a danger (500+ layers is pushing reason), you need to use layers and place items on the appropriate ones. As a CAD manager you need to develop a layering standard (more on that later) and enforce it. Users in a hurry don't always appreciate what having all items organized in layers does ("it all comes out the same on paper!" they may say), so you must stand firm and enforce the standard. And do not use layer zero! It cannot be renamed with a descriptive name, so why use it?

5) *Do not explode anything* – How do you destroy an AutoCAD drawing and render weeks of work and thousands of dollars in man-hours useless? Why, explode everything, of course. Short of deleting the files, this is the best way to ruin them. While this is generally a far-fetched scenario, remind users to not ruin a drawing in small ways, like once in a while exploding an mtext, dimension or hatch pattern. They are liable to do so if these items don't behave as they are supposed to and users explode them to manually force them into submission. Not a good idea; lack of skill is not an excuse. Remind them to ask you for help if a dimension just won't look right. There is always a way to fix something. Leave the exploding for the occasional block or polygon.

6) *Use accuracy: Ortho and OSNAPs* – Drawing without accuracy isn't drafting, it's doodling. Granted most users learn the basics, such as how to draw straight lines and connect them, you may still see the occasional project manager (who only gets to draft on average once a month) fix a designer's linework and forget to turn on OSNAPs. When zoomed out, the lines look OK, and will print just fine (reason enough to throw accuracy out the window for the occasional user), but a closer inspection reveals they are not joined together, causing a world of problems for a variety of functions (hatch, area, distance and other commands). Besides it is just plain sloppy. How accurate is AutoCAD? Enough to zoom from a scale model of the earth to a grapefruit (more on that later), so be sure to use all the available accuracy!

7) *Save often* – Let's state the obvious. There is no crash-proof software. AutoCAD does and will eventually lock up, get the equivalent of a stroke and crash. Remind users to save often, and do daily and weekly file backups. But you already knew this, right?

Part 2 – Know the Capabilities and Limitations of AutoCAD

As a CAD manager you have to know what you have to work with. All Computer Aided Design software including AutoCAD has an operating range for which the software is designed and optimized. Some tasks it excels at, and you will need to leverage that advantage, and some tasks are outside the bounds of expectation. AutoCAD can't be all things to all people (though you would be hard pressed to tell based on marketing). What follows is a short discussion on the capabilities and limitations and how you need to best take advantage of what you have.

Capabilities:

Often as a CAD manager you will be asked to provide an estimate on how long to complete the drafting portion of a job or just recreate something in CAD. Sometimes these requests come from individuals (senior management) that have had extensive experience in hand drafting, and they will be expecting an answer in the form of a significantly smaller time frame. AutoCAD has an aura of magic about it, as if a few buttons just need to be pushed and out pops the design. Where IS that "finish project" icon??

The surprising truth is that a professional hand drafter can give an average AutoCAD designer a run for the money if the request is a simple (small) floor plan. After all, he's not burdened with setting up layers, styles, dims and other CAD prep work. But this is not where AutoCAD's power lies. Once you begin to repeat patterns and objects, AutoCAD takes a commanding lead. A layout that took one hour to draft by hand or via computer can be copied in seconds on the computer but will add another hour to the hand drafter's task.

Don't be surprised by the seeming obviousness of this. The author has met many users and students that don't always see where exactly CAD excels; they think it's always faster on a computer. It may not be, and may lead to erroneous job estimates. But in all cases with patterns, copying and use of stored libraries, CAD is much quicker.

So when estimating a job, look for:

- Instances where the offset command can be used.
- Instances where there are identical objects.
- Minimal cases of unique, uneven and non-repetitive lines work.
- Minimal cases of abstract curves.
- Abundance of standard items.

Knowing what to look for will enable you to make more accurate estimates. Remember, the more repetition, the greater the CAD advantage.

This of course is just one example. Other advantages include accuracy (discussed next) and ease of collaboration. Another interesting advantage noticed by the author is that non artistic people or ones that would never have been able to draft with pencil, paper and T-square often excel at CAD, and skills such as neatness and a steady hand become less critical.

In support of Golden Rule #6, a brief discussion on accuracy follows. This is one often brought up in class. Search the web for a drawing called SOLAR.DWG. Download and open the file. It is a drawing of the solar system created in the early days of AutoCAD by founder John Walker and has become quite legendary over the years. You see, the drawing is to scale. This means that the orbits of all the planets are drawn to full size as well as correct relative to each other. Zoom in to find the Earth. Then zoom in to see the moon around the earth. Zoom in further and find the lunar landing module on the moon. Zoom in again and find a small plaque on one of the legs of the landing module. On it is a statement from mankind to anyone else that may read it. What does it say? And what is the date?

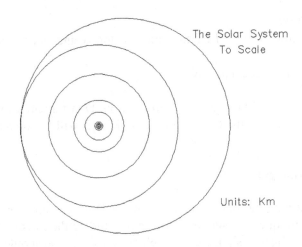

The Solar System
To Scale

Units: Km

Think about what you just witnessed. You have zoomed in from the orbit of Pluto to look at a 5" × 4" plaque. That is quite an amazing magnification. To put things into perspective, this is the equivalent of looking at the earth and then at a grapefruit. Or looking at a grapefruit, and then an atom! Can you design microchips in AutoCAD? Safe to say, yes. Can you connect walls together in a floor plan? Absolutely.

Limitations:

One of the chief limitations of AutoCAD is its 3D capabilities. Not that they are shabby, but sometimes they are pushed to do what they were not created for. The essence is that AutoCAD is not modeling and analysis software, rather, it is visualization software when it comes to 3D. It was not designed to compete with CATIA, NX, Pro/Engineer and SolidWorks. Parametric design, dimension driven design, FEA and CFD analysis is not what AutoCAD is about. These packages are dedicated engineering tools. AutoCAD was intended to be powerful 2D software with tools for presenting a design in 3D when necessary. As such, it is good at pretty pictures, to sell or present an idea.

Another limitation of the software, an obvious one that a CAD manager needs to be aware of is that AutoCAD will not do the thinking for you. It contains little to no error checking, interference checking or design fidelity tools. Garbage in will definitely be garbage out. Be aware as a CAD manager that AutoCAD will let users do almost anything, and the word "anything" can swing in either a positive or a terribly negative direction. And finally understand the learning curve of this software. It WILL take a few months for a novice user to get up to speed. Keep an eye on them until then!

Part 3 – Maintain an Office CAD Standard

A standard is a necessity for a smooth running office, especially for one with more than a few AutoCAD users. A standard ensures uniformity and a professional look to the companies output, minimizes CAD errors and makes it easy for new users to get up to speed. A manual needs to be created, published and given to all users that may come in contact with AutoCAD. This manual should have at a minimum:

- The exact location of all key files and the overall office computer file structure, indicating the naming convention for all jobs and the locations of libraries, templates and standard symbol folders.
- A listing of the office layering convention. This may follow the AIA standard in an Architecture office or just an accepted internal company standard.
- A listing of acceptable fonts, dimension styles and bhatch patterns.
- A listing of print/plot procedures, including *.ctb files.

- A listing of proper X-Ref naming procedures.
- A description of how Paper Space is implemented. List title blocks, logos and necessary title block information.
- A policy on purging files, deleting backup files and auto save.

Keep the manual as short as possible while including all the necessary information. Don't attempt to cover every scenario that could possibly be encountered as the longer the manual the less likely it will be looked at cover to cover.

Part 4 – Be an Effective Teacher and Hiring Manager

Part of being a CAD manager is sometimes explaining the finer points of AutoCAD to new hires or updating the occasional user as well as promotion of better techniques and habits. If you have studied the Level 1 and 2 books cover to cover, you may have noticed a concerted effort to simplify as much as possible. You should do the same with any potential students. While this is not meant to be a teaching guide (the author may publish a full essay on this topic in future editions of this text), some basic points apply. If you are tasked with training new hires, then try to observe the following:

- Cover basic theory first; creating, editing and viewing objects.
- Cover accuracy next, such as ortho and osnap.
- Move on to the fundamental topics in the order of layers, text, hatch, blocks, arrays, dimensions, and print/plot, similar to Level 1.
- If advanced training is needed, cover advanced linework, followed by Xref, attributes and Paper Space in that order. The rest of Level 2 can wait, as some of it is geared mainly to CAD managers anyway.
- Stress to users the secrets to speed: the "pgp" file, right-click customization and intense practice of the basics; remember the 90/10 rule from Level 1.
- Accurately assess the skill level of users and do not assign advanced tasks to those not yet ready, unless the drawings are not critical for a particular project. The author has personally seen the subtle damage to a drawing that can be caused by an inexperienced drafter.

Also, be ready to test potential hires with an AutoCAD test. This test does not have to be long; 45 minutes to an hour should be the absolute max, although truthfully, an experienced user will be able to tell if another person knows AutoCAD within 3 minutes of watching him/her draft. The trick is to maximize every minute and squeeze the most amount of demonstrated skill out of a potential recruit, which means do not have them engage in lengthy drawing of basic geometry, but rather give them broad tasks.

For example the floor plan of Level 1 is a good test for an architectural drafting recruit (the drawing can be simplified further if necessary by eliminating one of the closets or even a room). It should take an experienced user about 15 minutes total to set up a brand-new drawing (fonts, layers, etc.), and draft the basic floor layout (most students can do it in an hour after graduating Level 1 – not too bad). Then another 15 minutes to add text, dims and maybe some furniture, followed by another 5 to 10 minutes of hatch and touchups, followed by a printing. So it's roughly a 45 minute test, and runs the recruit through just about every needed drafting skill. When looking over the drawing, note the layers chosen, accuracy demonstrated, and how much was completed in what time frame.

Note the following. It is NOT critical if a new hire doesn't know Xrefs or Paper Space. These can be explained in a few minutes and learned to perfection "on the job." However, nothing but "hard time" spent drafting will teach someone how to draft quickly and efficiently with minimal errors. This IS a requirement, and don't hire anyone that doesn't have that, as it will take time climbing that learning curve. Exceptions granted of course to individuals hired for other skills, with AutoCAD secondary.

The bottom line (in the author's hiring experience) is that a fast, accurate drafter is far more desirable than a more knowledgeable one with a sloppy and slow manner of drafting. Be very careful of recent AutoCAD drafting class graduates. Extensive knowledge is no substitute for lack of experience, and while many drafters are fine workers and can come in and be productive on an entry level, be sure you understand what you're getting. The author has for a long time encouraged constant practice in class, even at the cost of occasionally not covering an advanced topic or two, knowing full well what these students will face in a saturated market full of experienced AutoCAD professionals.

As a final word, be sure the AutoCAD test is fair but challenging. Leave units in decimal form as the student needs to demonstrate a grasp of nuances and situational awareness and be able to change units to architectural if the foot/inch input doesn't work. Include in the test:

- Basic linework to be drawn
- A handful of layers to set (colors, linetypes, etc.)
- Some text to write (set fonts, sizes, etc.)
- Some hatches to create
- Some arrays to create
- Some blocks to create
- Some dimensions to put in
- Plot of output

Part 5 – Stay Current and Competent

While we are on the subject of technical competency, YOU, as a recent advanced class graduate or CAD manager are not off the hook for continuing to add to your skill set. It is much too easy to settle into a comfortable routine of drafting and/or designing and not learn anything outside the comfort zone (and not just with AutoCAD either). Some thoughts on this topic follow.

First of all, realize that it is virtually impossible for one individual to know everything about AutoCAD (many claim to come close). It is a truly an outrageous piece of software, developed and updated constantly by hundreds of software programmers and designers. This statement should hopefully make you realize that there is always more to learn (not throw your hands up in despair, however!), and you won't run out of new, easier and creative ways to do a task. The main trait shared by all successful AutoCAD experts is that they are genuinely interested in the software and in the concept of Computer Aided Design in general.

To maintain your skills or advance further, look through the following list and determine what applies to you and take the steps needed to acquire that knowledge. Some of this you will learn as part of studying all three books in the series, others you will have to initiate on your own.

- Learn the 3D features. Many users don't know 3D well, and it's a good way to stand out from the crowd. It's fun to learn and is an important side of AutoCAD.

- Learn basic AutoLISP. This relatively simple programming language will allow you to customize AutoCAD and write automation routines among other uses.

- Explore AutoCAD's advanced features that are sometimes overlooked such as Sheet Sets, Dynamic Blocks, eTransmit, Security, Dashboard, CUI and much more. Keep notes of key features and effects (a "cookbook" in software lingo).

- Read up on the latest and greatest from AutoCAD, new tips and tricks, and what the industry is talking about. Cadalyst.com and Augi.com are two good sites for this. You should also get a Cadalyst subscription (it's free).

- Don't work in isolation; talk to others in your field and see how they do things. Also be sure to take an AutoCAD update class if upgrading to a new release.

- Be curious about other design software including AutoCAD add-on programs (verticals) such as AutoCAD Mechanical, Electrical, Civil, MEP and others. Be knowledgeable about as much software as possible. In the world of CAD, Revit and SolidWorks are two software packages to keep an eye on in the future.

- Enjoy what you are doing, if not seek other work duties in your profession. AutoCAD is very miserable to deal with if you hate it. Occasionally you may think it is just one big software virus out to get you; just don't let it be a daily thought.

Appendix F

AutoLISP Basics and Advanced Customization Tools

You may have heard of AutoLISP, VisualLISP, VBA, .NET Framework, Active X and ObjectARX. What exactly are they? These are all various programming languages, environments and tools used to, among other things, customize and automate AutoCAD beyond what can be done by simpler methods like shortcuts, macros and the CUI, all explored earlier in this book.

To properly discuss all of these methods would take up quite a few volumes of text, and indeed much has been written concerning all of them. Our goal here will be to introduce the very basics of AutoLISP and only gloss over the rest so you at least have an idea of what they are. If advanced customization or software development work interests you, you can pursue it much further with other widely available publications.

> ### Overview I - AutoLISP

This is really the very beginning of the discussion. AutoLISP is a built-in programming language that comes with AutoCAD, and is itself a dialect of a family of historic programming languages collectively referred to as LISP (List Processing Language). LISP was invented by an MIT student in 1958 and found wide acceptance in Artificial Intelligence among other research circles. The language and its developer, J. McCarthy have pioneered many expression and concepts now found in modern programming.

AutoLISP is a somewhat scaled back version of LISP and is uniquely geared toward AutoCAD. It allows for interaction between the user and the software, and indeed one of the primary uses of AutoLISP is to automate routines or complex processes. Once the code is written it is saved and recalled as needed. The code is then executed as a script (no compiling needed) and performs its operation.

One of the advantages of AutoLISP over other methods (discussed shortly) is that it requires no special training except some familiarity with syntax, and is thus quite suited for non-programmers. While it is not a compiled, object oriented language like C++ or Java, (therefore not suitable for writing actual software applications), it also doesn't need to be for its intended use.

> ### Overview II – Visual LISP

AutoLISP has over the years morphed into a significantly more enhanced version called Visual LISP. More specifically, Visual LISP was purchased by Autodesk from another developer and became part of AutoCAD since release 2000. It is a major improvement and features a graphical interface and a debugger. Visual LISP also has Active X functionality (more on that later). However, Visual LISP still reads and uses regular AutoLISP, and that is still where a beginner needs to start before moving up. We will briefly touch upon Visual LISP and its interface (the vlide command) later on. The subject is an entire book all in itself.

> ### Overview III – VBA, .NET, Active X and ObjectARX

There is evidence to suggest Autodesk is moving away from using Visual LISP toward other, more advanced tools for customization and programming.

VBA or Visual Basic for Applications is a language based on Visual BASIC, a proprietary Microsoft product. VBA allows you to write routines that can run in a host application (AutoCAD), but not as standalone programs. It has some additional power and functionality over Visual LISP.

.NET Framework is another Microsoft software technology that was intended as a replacement for VBA. It features precoded solutions to common programming problems and a virtual machine to execute the programs.

Active X, also developed by Microsoft, is used to create software components that perform particular functions, many of which are found in Windows application. Active X controls are small program building blocks and are often used on the Internet such as for animations and unfortunately also for some viruses and spyware applications.

ObjectARX is the highest level of customization and programming you can do with AutoCAD, short of working on developing AutoCAD itself. Unlike scripting type languages and methods discussed so far, ObjectARX (ARX stands for AutoCAD Runtime Extension), is an API or an Application Programming Interface. Using C++ programmers can create Windows DLLs (Data Link Libraries) that interact directly with AutoCAD itself. Everything needed to do this is part of the ObjectARX Software Development Kit and is freely distributed by Autodesk. This is not easy to master, as you would expect from professional-level software coding tools. Typical users are third party developers who need a way to write complex software that interacts seamlessly with AutoCAD. ObjectARX is specific to each release of AutoCAD, and even requires the use of the same compiler Autodesk uses for AutoCAD itself.

Hopefully this discussion gave you some insight into what is out there for AutoCAD customization. This is of course only a cursory overview, and you can pursue these topics much further. Any journey to learn advanced AutoCAD customization, however, begins with AutoLISP, and this is where we will go next to introduce some basics of this programming language.

> **AutoLISP Fundamentals**

As mentioned before, despite the proliferation of new tools, AutoLISP is still very much in use and there are thousands of chunks of code still being written and easily found on-line for just about any desired function. The main reason for AutoLISP's popularity remains the fact that AutoCAD designers are not programmers and AutoLISP is accessible enough for almost anyone to jump in and learn. The value it will add to your skill set is substantial. Let's cover some basics and get you writing some code.

AutoLISP code can be implemented in two distinct ways:

- You can type and execute a few lines of code right on the command line of AutoCAD. This is good for a simple routine that is only useful in that drawing.
- You can load saved AutoLISP code and execute it; usually the case with more complex and often used routines.

We will generally be using the second method of writing and saving our routines. At this stage it would be helpful to know how (and where) to save AutoLISP code, and by extension how to load what's written.

All AutoLISP code is saved under the *.lsp extension. Once the code is written, it can theoretically be stored just about anywhere (AutoCAD 2008's Express folder contains a sizable amount of built-in routines). To load the code, go to **Tools→Load Application…** or **Tools→AutoLISP→Load Application...** (or just type in **appload**) and the following dialog box will appear as shown next.

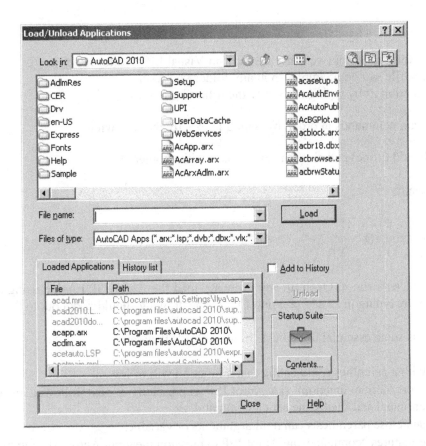

Then you just browse around, find the AutoLISP code you're interested in and press Load. Once loaded, you can close the dialog box and the code is ready to execute with a command input. There is a way to automatically load all existing code upon starting up AutoCAD by adding it to the Startup Suite (lower right corner, the briefcase icon).

Now that you know how to save new code and where to find it, we should try writing a simple example that illustrates some AutoLISP basics. Here are two critical points that need to be understood before you begin:

- AutoLISP is closely tied to AutoCAD commands, so in order to write something useful, you need to know the commands well. If you don't know how to do it in AutoCAD manually, AutoLISP is of no use!
- Approach AutoLISP as an answer to a specific problem. State what it is you want to solve or automate, then write out the steps to do it manually. Only then, using proper syntax, write and test the AutoLISP code.

Following this philosophy, let's state a problem we would like to solve. We would like to modify the Zoom to Extents command by having it zoom to only 90% of the screen, not 100%. That way we can see everything and nothing gets lost at the edges of the screen. How would we do this by hand first of all?

You would of course type in zoom, press Enter and select e for Extents and press Enter again. The drawing will be zoomed all the way. To bring it back a bit you would repeat the zoom command, but type in .90x instead of "e". The drawing will then zoom out a bit to fill only 90% of the screen.

Let's now automate these steps so all you have to do is type in "zz" (a random choice for a command name, you could have picked anything), instead of the two step process described earlier.

We need to:

1. Open up the Notepad editor (we'll use the built-in Visual LISP editor later).
2. Save the blank file as ZoomSample.lsp in any folder you wish.
3. Define the function zz by typing in exactly the following:
   ```
   (defun C:zz ()
   ```
4. List the relevant commands in order by typing in exactly the following:
   ```
   (command "zoom" "e" "zoom" ".90x")
   ```
   ...and finally add (princ) and close the expression with another)

So here is the final result:

```
(defun C:zz ()
(command "zoom" "e" "zoom" ".90x")
(princ))
```

Save the file again and let's run it. Open up any drawing, load the AutoLISP routine using appload or the drop down menu and run it by typing in zz and pressing Enter. The drawing should zoom out to 90% as expected.

This simple routine has some essential ingredients of all AutoLISP code:

- It starts out with a parenthesis, indicating that a function will follow.
- The "defun" or "define function" (an AutoLISP expression) is next followed by C indicating its an AutoCAD command and finally the actual command, zz.
- The set of open/close parenthesis can hold arguments or variables.
- The next line features "command" an AutoLISP expression meaning AutoCAD commands (in quotations, meaning it's a string) will follow.
- The (princ) function is added so the routine doesn't constantly return a "nil" value.
- Finally another matching parenthesis closes out the program.

Here are two simple routines that the author uses often to rotate the UCS crosshairs to align them to some geometry (this general task was covered in Chapter 15).

The first routine does the rotation and is called "perp":

```
(defun C:perp ()
(command "snap" "ro" "" "per" pause "snap" "off")
(princ))
```

The second one brings the crosshairs back to normal and is called "flat":

```
(defun C:flat ()
(command "snap" "ro" "" "0" pause "snap" "off")
(princ))
```

The "perp" and "flat" AutoLISP routines feature two new concepts. The first one is the pause function, which allows for user input. In this case it is for the user to select the line for the crosshairs to be aligned to. Then the function resumes and turns off snap. The other concept is the two quotation marks back to back with nothing between them. They simply force an Enter.

Study these three examples closely. While they are simple, many AutoLISP routines are, as that is all that is needed to get the job done. Here's what we learned so far:

- String: Anything that appears in quotes (often an AutoCAD command).
- Expression: AutoLISP built-in commands such as "defun"
- Functions: Can be user defined (perp, flat) or built-in (pause, princ, command). Functions are sometimes followed by arguments in ().

Variables and Comments

Let's now move on to other concepts. To take a step up in sophistication from what we did so far we need to introduce program variables. Variables can be numbers, letters or just about anything else that can change. The reason they are needed is because they allow for user interaction. Because user input is not known when the code is first written, a variable is used. When that information is acquired, the program proceeds accordingly. Variables are a fundamental concept in all programming.

Variables in AutoLISP can be global (defined once for the entire routine), or local (defined as needed for one time use). Local is usually how variables get defined in AutoLISP. The function that creates variables is setq. To set a variable equal to whatever the user inputs (such as radius of a circle) you would write: setq RadCircle. The next logical step with variables is to pick up on whatever the user inputs. For this we need the getstring function that returns anything the user types in.

Here is how the above concepts would look in practice, with the \n simply meaning that a new line will be started.

(setq RadCircle (getstring "\nCircle Radius: "))

Besides getstring there are many other "get" functions such as getfiled (asks user to select a file using a dialog box), getangle, getdist and others.

As in any programming language, you are allowed to comment your work, so you or someone else can understand your intent when examining the code at a later time. In AutoLISP comments are preceded by a semi colon ";" You can of course add more than one semi colon to indicate sections or headers.

```
;This is a comment
;;So is this
```

Advanced Features and the VLIDE command

We have only scratched the surface of AutoLISP, and if you enjoyed writing and running these simple routines then consider taking a full class on this subject or reading specialized books. There is much more to learn. For example:

- The IF→THEN statements. This is a common feature in all of programming. If certain conditions are met then the program does something; if not it does something else.

- The WHILE function. Another important concept in programming. This function will continue to do something until a condition is met.

and much more…

Finally we come to the vlide command. This brings up the Visual LISP editor as shown next:

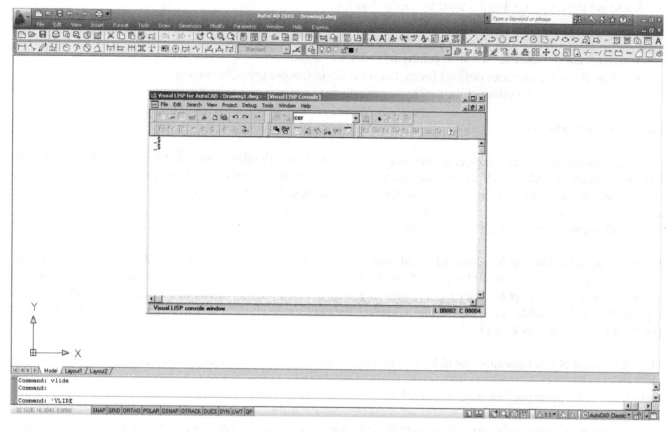

Everything you typed earlier into Notepad can be typed into the Visual LISP editor, which has many additional useful features such as debugging, error tracing and checks for closed quotations and parenthesis (formatting tools). Everything you type is also color coded for easy identification.

In-depth coverage of Visual LISP is beyond the scope of the book, but the student is encouraged to pick up a specialized textbook and try to progress with this invaluable skill.

Appendix G

PC Hardware, Printers/Plotters and Networks

This is another collection of topics that can easily span volumes, and indeed much has been written about all of the above. We will try to focus on hardware from an AutoCAD user's point of view exclusively, which may be a less covered topic (if only slightly).

Hardware, in the author's opinion, goes hand in hand with the software. And while computers, like cars, have become so reliable that a "hands off" approach works fine, remember that CAD in general and AutoCAD specifically, isn't ordinary software. It's a high-end, powerful application that places heavy demands on the associated hardware, and will not run at its peak on a clunky PC. Much like the owner of a Porsche is more likely to be interested in (and look under the hood of) their car versus the owner of a Toyota, so should an AutoCAD designer express interest in what's "under the hood" of their computer. After all, they are manipulating and creating multi-megabyte sophisticated drawings or 3D models, not merely typing up a memo on Word, and it's in their best interest to use a PC that meets or exceeds AutoCAD's demands.

> ### PC Hardware

By its nature, AutoCAD is not as fussy as some other CAD software out there (3D solid modeling applications come to mind), but it does need a high-end PC to operate optimally. Here's a rundown of a recommended set-up with comments.

Processor – The "brain" of the computer, also referred to as a chip, its where all the calculations and operations are performed. Intel and AMD are the major players in the PC market. Which is better is a moot point; while Intel was once the only game in town, these days they constantly one-up each other with performance benchmarks. At this point both manufacturers produce quality products, and the Intel versus AMD comparison is essentially meaningless to the average user. The latest Intel chipset (as of this writing) is the Intel Core 2 Extreme with a 3.33 GHz clock speed. AMD has a similar Opteron series chip. Autodesk recommends a much tamer 2.2 GHz Pentium 4 as a minimum, so you have plenty of room to shop on price between these two extremes.

RAM – Random Access Memory is the next consideration, and the more the merrier. RAM has a direct effect on performance as it is there that software resides while in operation. More available RAM will allow for more functions to execute faster. Autodesk recommends 1 GB, but 2 GB would be better.

Hard Drive – Also known as storage disk, main drive, etc, it needs to be at least 750 MB to install AutoCAD. Fortunately this is rarely a problem as most machines have disks on the order of 50-100 GB.

Video Graphics Card – Perhaps the most overlooked part of the PC, the video or graphics card is similar to a processor, but is specially designed to control the screen images, or graphics. It needs to be beefed up to deliver the kind of graphics AutoCAD is capable of producing (essential in 3D rendering). Autodesk makes no recommendations, but get the best possible under the allowed budget. Quality graphic cards aren't cheap; a high-end card will top out over $2,000. One in the $600-$900 range should be just fine.

OS – Operating Systems that AutoCAD can run on number in the single digits (actually precisely one - Windows). AutoCAD won't run on Unix, Linux or a Mac. Windows XP (with Service Pack 2) is recommended until Vista becomes more stable. Though AutoCAD is officially certified for Vista, students have reported sporadic operational problems. Maybe as of this writing things improved.

Other PC components – These include the motherboard (must be picked according to processor requirements), the sound card/speakers (if needed), the CD/DVD and the power, cooling and wiring systems. These systems are all as per requirement of other hardware pieces, and the DVD capability is needed as AutoCAD now comes on DVD when you buy it.

Monitor – Get the best and largest flat screen monitor that you can afford; your eyes will thank you many times over.

Mouse/Keyboard – Get a laser precision mouse (forget roller ball) and the best soft-touch keyboard you can find. Don't try to cut corners here as you will be in physical contact and using these devices constantly while working on the PC, so get the good stuff.

Whether building your own, or ordering an assembled one, take the PC purchase seriously and try to set realistic budget expectations. CAD workstations are in the same family as gaming stations (in fact the author hand-built what amounts to a gaming machine, even though no games are on it) and you will need to budget $2000 to $3000 (or more) for a serious machine, not including the monitor. Some well known vendors include Dell, but if you are inclined to assemble your own, check out on-line stores such as TigerDirect.com, OutledPC.com and many more.

A typical CAD station may consist of the following pieces (as built by the author in 2006):

- AMD Athlon 6400 processor
- 2 GB RAM
- Asus Motherboard
- ATI FireGL series Video Card
- Creative Labs X-FI Sound Card
- 160 GB Disk Drive
- CD-R/DVD

> **Printers/Plotters**

There are essentially two ways to output something from AutoCAD, a printer or a plotter. Printers are generally Inkjet or LaserJet, black and white or color capable. Prices can range from a $49 desktop Inkjet to a "sky's the limit" corporate high-speed color LaserJet. Which to get depends entirely on budget and desired level of quality for output, as AutoCAD doesn't care. Once AutoCAD is installed it will automatically sense any printer that is attached to the PC (assuming they work and have drivers in the first place). Be aware that most small desktop printers can't handle 11" × 17" paper, but larger LaserJet's can. The 11" × 17" format is very important in AutoCAD, as many jobs are handed in on this sized paper, so be sure your printer can handle it.

Plotters are a familiar sight at architecture or engineering offices. Anything sized above an 11" × 17" sheet of paper needs to be routed to a plotter, all the way up to a 48" × 36" or larger in some cases. Plotters as a rule are more expensive than printers, and buying one should be a carefully researched decision. Many plotters are color capable and accept almost any type of paper, although bond or vellum is the norm. Hewlett-Packard (HP) is the premier manufacturer of plotters (some models on sale are well over $20,000 – though their 500, 800 and 1055 series plotters are much more affordable), with other companies such OCE, Canon and Encad rounding out the field. The typical average cost of a plotter as encountered by the author hovers around $6,000. Do your research carefully when selecting one, and be aware of recurring costs such as ink and paper.

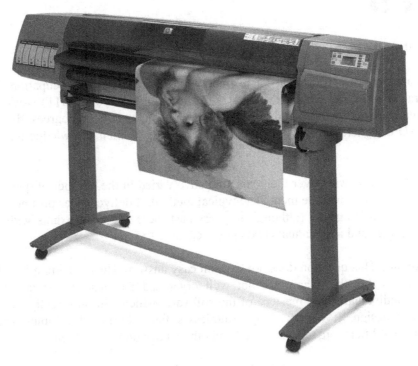

A Typical Plotter

> **Networks**

This topic of course is an entire profession all to itself. A computer network is defined as a group of interconnected computers. Networks can range from local area network (LAN) to city, country or world-wide (WAN or wide area network). Networks allow for collaboration and sharing of information from computer to computer and have their own set of hardware familiar to any network engineer, such as routers, hubs, switches, repeaters and other gear. Networks are operated via a protocol or set of instructions, such as the IEEE 802.3 protocol running on Ethernet technology for a typical LAN. The medium of data transfer between computers is usually a standards cat5e cable, but can also be wireless or fiber optic for larger networks.

As an AutoCAD designer, you will likely encounter a Local Area Network consisting of a server (host) with some other computers (clients) connected to it at your company. All the data is stored on the server and once you log on to the network your computer will be able to access what it needs.

In regard to networks, AutoCAD can be installed on a per seat or network license method. Per seat is cost effective to a certain limit (usually 10 seats). If your company requires more, a site license is purchased which is cheaper, and allows more computers to run the software, which now resides on the server only.

Appendix H

AutoCAD Certification Exams

It seems like there is an exam or certification for just about every profession or occupation. Although AutoCAD and drafting in general is not really part of the greater Information Technology (IT) world, where certification exams are the benchmarks of competence and are almost as valuable as college degrees, there still exists a series of exams for AutoCAD designers to take if they wish to "formalize" their knowledge and hang up a piece of paper in their office.

These exams have changed somewhat over the years, and have varied in the number of questions, difficulty level and the amount of time given to complete them. The typical method of delivery was on-line, with users logging in and taking the exam in a timed fashion (although in years past there had been exams with no time limit). You were able to use the help files and any manuals/texts you were able to find.

Are these exams important? This question does not have an easy answer. The overriding belief among established AutoCAD professionals is that the exams are not well organized and test for trivial knowledge while not accurately assessing an individual's true mastery of the software, which can only really be tested by having the person draw a complex design. It was partially Autodesk's fault for not developing these exams into the powerhouse institutions that Microsoft, Cisco, Novell and others have done with theirs.

Then again, however, there was little demand for the exams in the past. As AutoCAD was exploding in popularity anyone that was reasonably good was hired and allowed to mature on the job. In recent years, as supply of knowledgeable AutoCAD designers finally caught up with demand, employers wanted ready-made experts and needed a way to separate the "wheat from the chaff." This would explain the resurgence in popularity of testing for AutoCAD competence, and the renewed interest from students as to how to study for and pass these exams. So, yes, these exams may be important to get a foot in the door with certain companies, but having passed the exam does not necessarily guarantee a knowledgeable AutoCAD expert, and the author has never used them as the sole reason to hire someone. Let's go over the exams as they exist now for AutoCAD 2010. There are two exams:

1. **AutoCAD Certified Associate Exam**
2. **AutoCAD Certified Professional Exam**

The Certified Associate Exam (CAE), which is taken first, consists of 30 questions with the following breakdown by topic:

- Introduction to AutoCAD – 1 question
- Creating Drawings – 4 questions
- Manipulating Objects – 6 questions
- Drawing Organization and Inquiry Commands – 3 questions
- Altering Objects – 2 questions
- Working with Layouts – 1 question
- Annotating the Drawing – 3 questions
- Dimensioning – 2 questions
- Hatching Objects – 2 questions
- Working with Reusable Content – 3 questions
- Creating Additional Drawing Objects – 2 questions
- Plotting – 1 question

The passing score is 70% and you have a 60 minute time limit.

The Certified Professional Exam (CPE), which is taken second, consists of 20 questions with the following breakdown by topic:

- Manipulating Objects – 5 questions
- Drawing Organization and Inquiry Commands – 1 question
- Altering Objects – 2 questions
- Working with Layouts – 1 question
- Annotating the Drawing – 3 questions
- Dimensioning – 3 questions
- Hatching Objects – 1 question
- Working with Reusable Content – 1 question
- Creating Additional Drawing Objects – 2 questions
- Plotting – 1 question

The passing score is 80% and you have a 90 minute time limit.

The CAE focuses more on multiple choice and matching questions and basic knowledge, while the CPE is more performance based and focuses on drawing (you will need to download, open and work on actual AutoCAD files while taking the exam). You will then be asked to draw something and answer a question based on this drawing.

The intent here is to have an individual take the CAE shortly after completing a training course, and then the CPE a few months later after some practical drawing skills are gained.

There are practice sample tests available, and it is highly recommended to go over these to at least see what type of questions to expect. The cost for taking these exams is $50 per exam, but that is of course subject to change, so check the latest information on the Autodesk website.

Appendix I

AutoCAD Humor, Oddities, Quirks and Easter Eggs

Real AutoCAD Users...

- Have been doing AutoCAD for six or more years.
- Never bother to keep .BAKs.
- Don't do drawings on a dot/matrix printer.
- Have a graphics card that can support two monitors
- Have two SVGA monitors
- Have games on their system.
- Get REAL pissed when someone else uses their machine just to "Look Around."
- Know AutoLISP.
- Know how to edit an AutoLISP file.
- Wonder why Autodesk gives you so much worthless stuff with a software upgrade.
- Wonder why Autodesk charges so much for an upgrade, and then AutoCAD gets pirated.
- Maintain at least three older versions of AutoCAD on their system.
- Never ask for the newest copy of AutoCAD - they already have it.
- Have a free subscription to CADALYST.
- Don't really bother to read CADALYST.
- Could write an article that could go in CADALYST.
- Don't run AutoCAD on a MAC.
- Don't have friends that run AutoCAD on a MAC.
- Despise MACs.
- Are not concerned with losing a drawing.
- Use the PURGE command about 60 times in a drawing
- Use the spacebar, not return.
- Have either drawn a picture of:
  - Their car,
  - Their house,
  - Their dream house,
  - Or their dream house with their dream car parked in the drive.
- Like to waste time, such as like typing ZOOM E twice in a large drawing.
- Can talk on the phone and draw at the same time.
- Drink lots of liquid while drawing.
- Go to the bathroom a lot.
- Get some exercise while at work.
- Are not afraid to eat while drawing.
- Can leave the room while plotting.
- Don't care that the work system is on 24 hrs a day; it's not their machine.
- Don't really like the system they run AutoCAD on; they are sure there is a better one.
- Would be damn lucky if they got the system they have at work, for their home.

- Will always complain that AutoCAD on their system is too slow.
- Work in the dark with AutoCAD screen color - black.
- Get pissed when someone turns on the light.
- Get REAL pissed when people say – "How can you work in the dark like this?"
- Listen to free downloaded mp3s while working.
- Are reading this while they should be working...

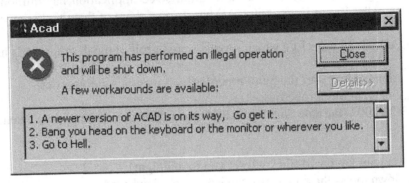

The AutoCAD Monkey joke that never goes away...

A tourist walked into a pet shop and was looking at the animals on display. While he was there, another customer walked in and said to the shopkeeper, "I'll have an AutoCAD monkey please." The shopkeeper nodded, went over to a cage at the side of the shop and took out a monkey. He fitted a collar and leash, handed it to the customer, saying, "That'll be $5000." The customer paid and walked out with his monkey.

Startled, the tourist went over to the shopkeeper and said, "That was a very expensive monkey. Most of them are only a few hundred dollars. Why did that one cost so much?"

The shopkeeper answered, "Ah, that monkey can draw in AutoCAD - very fast, clear layouts, no mistakes, well worth the money."

The tourist looked at a monkey in another cage. "That one's even more expensive! $10,000! What does it do?"

"Oh, that one's a design monkey; it can design systems, layout projects, mark-up drawings, write specifications, some even calculate. All the really useful stuff," said the shopkeeper.

The tourist looked around for a little longer and saw a third monkey in its own cage. The price tag around its neck read $50,000. He gasped to the shopkeeper, "That one costs more than all the others put together! What on earth does it do?" The shopkeeper replied, "Well, I haven't actually seen it do anything, but it says it's an engineer."

Oddities and Quirks...

- AutoCAD has an *oops* command...try it to see what it does.
- AutoCAD had an *end* command. It used to shut down your drawing, even if you meant that as an osnap command and just forgot to type in line and press Enter first. End has been permanently discontinued for obvious reasons.
- AutoCAD has a battman command. It has nothing to do with the caped crusader and everything to do with the Block ATTribute MANager (similar to the Enhanced Attribute Manager, of Chapter 19).

Easter Eggs…

Easter eggs are surprises that one can find in software, hence the analogy to the Easter Egg hunt. These surprises can be anything from a silly message to an entire video game buried in the code. To find and activate Easter Eggs you really just have to be told how, as the complex combination of keys that need to be pressed in correct sequence is just too much for a random guess. Why are they in there? Just for fun is the simple answer. Also because programmers can get away with it; even a medium sized application has millions of lines of code in which to hide the much smaller Easter Eggs. AutoCAD has its own share of them that are well documented.

Here's a known AutoCAD 2008 Easter Egg which reveals a credit roll of the design team's names:

1) On the AutoCAD Command line start the Sun Properties by typing in sunproperties.
2) Click the magnifying glass in Sky Properties.
3) Change the date to 3/23/2007 (the date AutoCAD 2008 was released) via the pop-up calendar. Press Enter to set it.
4) Click on the time.
5) Press the Home key on your keyboard (this will set you to midnight).
6) With the CTRL key down, press the down arrow twice on your keyboard.
7) Now you should see the credits roll with music.

There are other Easter Eggs in virtually all releases of AutoCAD. The more you look, the more you'll find.

Index

Printed in the United States
By Bookmasters